Boundaries between Promotion and Progression during Carcinogenesis

BASIC LIFE SCIENCES

Ernest H. Y. Chu, Series Editor

The University of Michigan Medical School
Ann Arbor, Michigan

Alexander Hollaender, Founding Editor

A Continuation Order Plan is available for this series. A continuation order will bring delivery of each new
volume immediately upon publication. Volumes are billed only upon actual shipment. For further information
please contact the publisher.

Boundaries between Promotion and Progression during Carcinogenesis

Edited by

Oscar Sudilovsky
Case Western Reserve University
Cleveland, Ohio

Henry C. Pitot
McArdle Laboratory for Cancer Research
Madison, Wisconsin

and

Lance A. Liotta
National Institutes of Health
Bethesda, Maryland

Plenum Press • New York and London

Library of Congress Cataloging-in-Publication Data

Boundaries between promotion and progression during carcinogenesis /
  edited by Oscar Sudilovsky, Henry C. Pitot, and Lance A. Liotta.
      p.   cm. -- (Basic life sciences ; v. 57)
    Based on proceedings of a conference held Sept. 28-30, 1988, in
  Cleveland, Ohio.
    Includes bibliographical references and index.
    ISBN 0-306-44031-8
    1. Carcinogenesis--Congresses. 2. Cancer invasiveness-
  -Congresses. 3. Cocarcinogenesis--Congresses.   I. Sudilovsky,
  Oscar. II. Pitot, Henry C., 1930-   . III. Liotta, L. A. (Lance
  A.) IV. Series.
    [DNLM: 1. Carcinogens--congresses. 2. Cell Transformation,
  Neoplastic--genetics--congresses. 3. Precancerous Conditions-
  -genetics--congresses.   W3 BD255 v.57 / QZ 204 B765 1988]
  RC268.5.B68 1991
  DNLM/DLC
  for Library of Congress   91-24270
                                                                  CIP

Based on proceedings of a conference on the Boundaries
between Promotion and Progression during Carcinogenesis,
held September 28–30, 1988, in Cleveland, Ohio

ISBN 0-306-44031-8

© 1991 Plenum Press, New York
A Division of Plenum Publishing Corporation
233 Spring Street, New York, N.Y. 10013

Printed in the United States of America

PREFACE

The purpose of this conference was not to define the two areas that are being bound, which might be a well nigh impossible proposition. Rather, its focus was to concentrate on the mechanistic similarities between promotion and progression. Are the areas involved within the boundaries a continuum? Are these two simultaneous processes? Or are some of the affected cells in the stage of promotion when at the same time others have undergone irreversible changes that position them in the stage of progression? Or are these two stages the same thing, but called by different names?

To explore such concepts we assembled investigators with various backgrounds and asked them to specifically address these and other questions about "The Boundaries", within the context of the session to which they contributed. The conference lasted two and a half days, from Wednesday to Friday. There were at least four speakers per session with morning and afternoon sessions each day, except on Friday when the meeting ended at noon. The first day, each speaker had 25 minutes to present a position, followed by five minutes of discussion. At the end of the session there were 40 or 50 minutes of exchange on all the issues examined. For the remaining days, there were 25 minutes of presentation and 15 minutes of discussion. There was a moderator and a discussion leader who played a very active part in bringing the differences to the fore and in zeroing in on key problems. The role of the discussion leaders was critical because they contributed and elicited discussions, not only in their assigned sessions but in all others for which they felt qualified. Active participation of the audience was encouraged because their queries and comments generated new concepts for further research in the gray area of "The Boundaries." After 3 p.m. every afternoon, the invited participants met in a colloquium to develop reasonable, concise statements of areas that needed further research.

We want to express our gratitude to the companies and organizations listed in the Acknowledgment for their contributions to the funding of this conference and to Ms. Christine Sciulli and Ms. Regina Stoltzfuss, of the Ireland Cancer Center, without whose help the meeting could not have been organized properly. In particular, Ms. Sciulli was outstanding in all aspects related to the publication of this volume. For that, the editors are extremely thankful. A special thanks to Phyllis Best and Ann Pearson for their help in typing this manuscript.

Oscar Sudilovsky
Lance A. Liotta
Henry C. Pitot

ACKNOWLEDGMENTS

The conference acknowledges the following companies and organizations who
have contributed towards our costs:

National Cancer Institute
Bristol-Myers Company
The Council for Tobacco Research
Hoffman LaRoche
Lilly Research Laboratories
Merck Sharp & Dohme Research Laboratories
Mobil Research & Development Corporation
Pfizer Inc.
The Proctor & Gamble Company
Schering Corporation
SmithKline Beckman Corporation
Triton Biosciences Inc.
The Upjohn Company

CONFERENCE ADVISORY COMMITTEE

Ivan Damjanov	Jefferson Medical College
Philip Frost	M.D. Anderson Hospital and Tumor Institute
Gloria Heppner	Michigan Cancer Foundation
Peter Nowell	University of Pennsylvania School of Medicine
Lance A. Liotta	National Cancer Institute
Henry C. Pitot	McArdle Laboratory for Cancer Research
Oscar Sudilovsky	Case Western Reserve University School of Medicine

PARTICIPANTS

George T. Bowden	University of Arizona Health Science Center
Jeff A. Boyd	National Institute of Environmental Health Sciences
Allan Bradley	Baylor College of Medicine
Robert Callahan	National Cancer Institute
Webster K. Cavenee	Ludwig Institute for Cancer Research, Montreal Branch
Ivan Damjanov	Jefferson Medical College
Peter Duesberg	University of California at Berkeley
Emmanuel Farber	University of Toronto
Philip Frost	M.D. Anderson Hospital and Tumor Institute
Russel G. Greig	Smith, Kline and French Labs
Joe W. Grisham	University of North Carolina School of Medicine
Henry Hennings	National Cancer Institute
Meenhard Herlyn	Wistar Institute
Peter Howley	National Cancer Institute
Barbara B. Knowles	Wistar Institute

Michael W. Lieberman   Baylor College of Medicine
Lance A. Liotta   National Cancer Institute
Joseph Locker   University of Pittsburgh School of Medicine
George Michalopoulos   Duke Medical Center
Peter C. Nowell   University of Pennsylvania School of Medicine
G. Barry Pierce   University of Colorado Medical Center
Henry C. Pitot   McArdle Laboratory for Cancer Research
Thomas G. Pretlow   Case Western Reserve University School of Medicine
Joan R. Shapiro   Memorial Sloan-Kettering Cancer Institute
Harry Rubin   University of California at Berkeley
Jose Russo   Michigan Cancer Foundation
Thomas J. Slaga   University of Texas System Cancer Center
Helene S. Smith   Peralta Cancer Research Center
Oscar Sudilovsky   Case Western Reserve University School of Medicine
George H. Yoakum   Albert Einstein School of Medicine

Disclaimer

The papers in this volume are based on presentations and supplements presented at a conference held in Cleveland, OH , on "The Boundaries between promotion and progression during carcinogenesis." The symposium was funded in
part by the National Institutes of Health PHS #CA48762, Bristol-Myers Company, The Council for Tobacco Research, Hoffman LaRoche, Lilly Research
Laboratories, Merck Sharp & Dohme Research Laboratories, Mobil Research &
Development Corporation, Pfizer Inc., The Proctor & Gamble Company, Schering
Corporation, SmithKline Beckman Corporation, Triton Biosciences Inc. and the
Upjohn Company. The opinions expressed within respective chapters reflect
the views of the authors, which may not be necessarily those of the sponsor-
ing agencies. The articles and discussion comments have been edited for com-
pleteness and clarity.

CONTENTS

INTRODUCTION: IS THERE A BOUNDARY BETWEEN PROMOTION AND PROGRESSION?

Oscar Sudilovsky

Case Western Reserve University
School of Medicine
Cleveland, OH

The need for a meeting that would discuss the poorly defined areas of promotion and progression, and the impetus to promote it, came as a result of a Gordon Conference on Cancer in August, 1987. It had become evident to me and other participants that the concepts of promotion and progression as two separate stages in a multi-step model of carcinogenesis required further evaluation.

Promotion can be defined operationally as the development of tumors by non-carcinogenic agents in tissues previously treated with an incomplete carcinogen or with low doses of a complete carcinogen. On the other hand, tumor progression, as first defined by Foulds in 1949 (1), is the development of tumors by way of permanent, irreversible qualitative changes in one or more of its characters. The name progression is usually reserved for the acquisition of more and more aberrant characteristics by cells in benign or malignant tumors and is commonly considered as a third stage in the natural history of neoplasia, after promotion. Following exposure to a carcinogen, the initiated cells replicate and are promoted until a visible tumor develops. The cell population at that time consists of euploid cells. Progression occurs later on in these tumors and results, among other things, in increasing aneuploidy and metastasis.

The distinction between promotion and progression, however, is not always possible because these two processes are very frequently linked together. Their "boundaries" are ill-defined inasmuch as the irreversible changes required by Foulds occur frequently during promotion. They involve modifications of gene expression, DNA strand breakage, the formation of chromosomal abnormalities, or the development of aneuploidy. A case in point is probably that of circulating myelodysplastic cells, in which chromosomal aberrations or deletion of genes takes place sometimes before the appearance of malignancy. Similar situations obtain in solid tumors. For example, aneuploid DNA patterns might occur in dysplasias of the uterine cervix prior to the appearance of tumors (2). In experimental animals, very high doses of radiation result in chromosomal damage, and progression ensues in the viable cells since the promotion stage is no longer identifiable. In the rat liver, we have found that after initiation with diethylnitrosamine and promotion with a diet deficient in choline supplemented with phenobarbital for 16 weeks, aneuploidy occurs in the enzyme altered foci before a visible tumor can be recognized (3).

Boundaries between Promotion and Progression during Carcinogenesis
Edited by O. Sudilovsky *et al.*, Plenum Press, New York, 1991

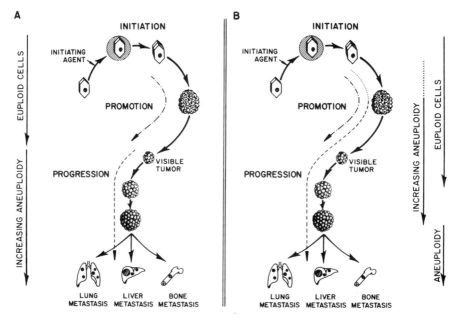

Figure 1. The natural history of neoplasia. Common conception of promotion and progression (A). Proposed conception showing overlapping of promotion and progression (B).

In view that any addition or deletion of general information could place the cells in the stage of progression, I would like to propose that there are instances in which the classical scheme should be modified, as shown in Figure 1. In our schema, the left part shows a sketch similar to the one of Henry Pitot (4). The right part of the diagram purports our conception that promotion overlaps progression while the cell population is composed of both euploid and aneuploid cells, with progressively increasing aneuploidy as time passes. Some of the speakers or participants may not accept this notion, but this symposium, in the end, should generate debates and new ideas about "The Boundaries."

REFERENCES

1. L. Foulds, Mammary tumors in hybrid mice: growth and progression of spontaneous tumors, Brit J Cancer 3:345-375 (1949).
2. Y.S. Fu, J.W. Reagan, and R.M. Richart, Definition of precursors. Gynecol Oncol 12:S220-S231 (1981).
3. O. Sudilovsky, T.K. Hei, Aneuploid nuclear DNA content in some enzyme-altered rat liver foci during tumor promotion. Fed Proc 42:2225 (1983).
4. H.C. Pitot, "Fundamentals of Oncology," 3rd edition, p. 190, Marcel Dekker, Inc., New York (1986).

CHARACTERIZATION OF THE STAGE OF PROGRESSION

IN HEPATOCARCINOGENESIS IN THE RAT

Henry C. Pitot

McArdle Laboratory for Cancer Research
Departments of Oncology and Pathology
The Medical School
University of Wisconsin-Madison
Madison, Wisconsin 53706

INTRODUCTION--BASIC CONCEPTS AND DEFINITIONS

The concept of neoplastic progression as a stage in the natural history of neoplastic development was first enunciated by Foulds[1]. He distinguished the stage of initiation as that which established "a persistent region of <u>incipient neoplasia</u> whence tumors of varied kinds emerge at a later time"[2]. All of the remainder of neoplastic development he termed progression. In a sense, such a concept was quite analogous to that proposed earlier by several authors as the stages of initiation and promotion[3]. Interestingly, the basis for the development of these two concepts of multistage carcinogenesis were, respectively, experimental mammary adenocarcinoma and epidermal carcinoma in mice. In retrospect, more emphasis was placed on the development of malignant lesions in the concept of progression, whereas early benign and preneoplastic lesions were emphasized as characteristic of the stage of promotion during multistage epidermal carcinogenesis. It is only during the last decade that these two general concepts have been reconciled on the basis of an increased understanding and characterization of the early stages of development, initiation and promotion, with the relegation of the term progression to the final stage of neoplastic development, in which the malignant characteristics and genetic heterogeneity of neoplasms appear[4].

Tumor Progression Today

In his discussions of progression, Foulds[2] emphasized the "characters" of a neoplasm, such characters being growth rate, invasiveness, potential and actual metastases, hormone responsiveness, and histomorphology. Furthermore, he pointed out that any one of these characters could undergo progression independent of the other characters, and each neoplasm within any one host may progress independently of others in that host[2]. Today it is apparent that many of the characters described by Foulds are a direct function of, or closely associated with, measurable changes in the genome of the cell. Karyotypic alterations in neoplasms are directly correlated with increased growth rate[5], invasiveness[6], metastatic potential and capability[7,8], hormone responsiveness[9], and morphologic characteristics[10]. The genetic heterogeneity characteristically seen in the stage of progression[7,11] by both karyotypic analyses and more detailed molecular studies[12] is thus reflected

Boundaries between Promotion and Progression during Carcinogenesis
Edited by O. Sudilovsky *et al.*, Plenum Press, New York, 1991

3

Table 1. Biological Characteristics of the Stage of Progression,
         the Final Stage in Neoplastic Development

Irreversible - demonstrable alterations in cell genome
Evolving karyotypic instability
Relatively autonomous malignant neoplasia
Induction by progressor agents and/or complete carcinogens
Spontaneous (fortuitous) progression of cells in the stage of promotion

---

in the "characters" of the stage of progression described by Foulds (see
above). Furthermore, although significant phenotypic heterogeneity has been
described during the stage of promotion in hepatocarcinogenesis in the
rat[13,14], significant biochemical homogeneity[15] and a lack of demonstrable genetic heterogeneity and instability characterize the stages of initiation and promotion. Unlike the succinct and limited nature of the stages of
initiation and promotion, the stage of progression involves a continued evolution of neoplastic cells towards increased autonomy from host influences.
The continued evolution of karyotypic changes that accompany the evolution of
the stage of progession has been described in a variety of systems, both
experimental[16,17] and in the human[11].

On the basis that the stage of progression is distinguishable from the
stages of initiation and promotion in the multistage development of neoplasia, it should be possible to delineate specific characteristics of this
stage that distinguish it from the earlier two stages. Such a set of characteristics is listed in Table 1. The irreversibility of this stage is emphasized by the obvious alterations in the cell genome accompanying this stage
and distinguishing it from the reversible preceding stage of promotion. However, it is clear that, under certain circumstances, cells in the stage of
progression may be induced, by treatment with specific chemicals, into terminal differentiation, thereby removing them from continued progression to a
more malignant state[18]. That readily demonstrable, irreversible alterations in the genome are in turn associated with genetic instability of neoplastic cells in the stage of progression has been recognized by a number of
investigators in the field[7,19]. Karyotypic instability has a variety of
consequences for the neoplastic cell in the stage of progression. These
include gene amplification[20], gene and chromosomal translocations and
rearrangements[21], gene deletions[22], proto-oncogene activation[23,24], and
more efficient transfection of genes into neoplastic cells in the stage of
progression[25]. In fact, of all of the characteristics seen in Table 1,
karyotypic instability is unique to the stage of progression and is probably
the major factor distinguishing this stage from the stages of initiation and
promotion and is also the basis for the malignant transformation of cells in
the stage of progression.

Neoplasia has been defined as a "heritably altered, relatively autonomous growth of tissue"[26]. Relative autonomy in the stage of progression
tends toward complete autonomy of cells from environmental regulatory factors
such as hormones, growth factors, and other regulatory environmental components. Complete dependence of neoplasms on the presence of hormones and/or
other regulatory factors is characteristic of the stage of promotion[4]. It
is in the stage of progression that relative autonomy of neoplastic cells
with respect to such agents is noted. This is especially true of the responsiveness and autonomy of neoplasms in the stage of progression to hormones,
as has been described by many investigators[27]. Another potential mechanism
for the increasing relative autonomy of neoplastic cells in the stage of progression is alteration of the methylation of DNA, a process shown to be
important in the regulation of gene expression[28]. In particular, admini-

stration of the "demethylating" agent, 5-azacytidine, to neoplastic cells in culture altered their growth rates and morphologic characteristics[29,30].

Agents that act only to induce the entrance of cells into the stage of progression have not been definitively characterized, as have promoting agents and initiating agents. Perhaps the best example of a "progressor" agent is the free radical generator, benzoylperoxide, an agent active in inducing the stage of progression in experimental epidermal carcinogenesis[31]. Obviously, by definition complete carcinogens will also have "progressor" agent activity. Theoretically, such progressor agents should be capable of inducing the genetic changes characteristic of the stage of progression. Examples of such agents would include clastogens and similar agents effecting major chromosomal alterations. That such agents may act in this manner has been demonstrated with an "initiation-promotion-initiation" format (I-P-I) proposed earlier by Potter[32] and later experimentally demonstrated in multistage epidermal carcinogenesis in the mouse by Hennings and his associates[33], who found that when the usual initiation-promotion format was subsequently followed by the application of a second complete carcinogen, such as an alkylating agent, a rapid high incidence of carcinomas resulted. This was in contrast to the standard initiation-promotion format in this tissue, which resulted primarily in benign neoplasms during the time span of the experiment. A similar regimen has been employed in rat liver by Scherer and his associates[34] in multistage hepatocarcinogenesis in the rat. In this instance, monitoring of the effect of the regimen was carried out by evaluating the appearance of "foci-within-foci", which Scherer has proposed as representing the earliest morphologic appearance of cells in the stage of progression. Since foci-within-foci do occur spontaneously in multistage hepatocarcinogenesis protocols in which progression to malignancy is an extremely infrequent event[13], and since carcinomas are seen infrequently in the standard initiation-promotion protocol of multistage epidermal carcinogenesis[33], spontaneous or fortuitous progression, in analogy to spontaneous initiation[4], must occur.

A Definition of the Stage of Tumor Progression

On the basis of the characteristics seen in Table 1, one may propose a working definition of the stage of progression as evidenced from the numerous studies quoted, as well as many others. The proposed definition, which is somewhat different from those previously advanced[35,36], is as follows:

> Progression is that stage of carcinogenesis exhibiting readily measurable (by molecular biological and/or microscopic means) changes in the structure of the cell genome. The evolution of such changes is directly related to the evolution of increasing growth rate, invasiveness, metastatic capability, and the increasing autonomy of the neoplastic cell during this stage.

Although, as emphasized earlier, it is likely that karyotypic instability forms the basis for the evolution of the characters during the stage of progression, the definition implies rather than explicitly stating the critical nature of karyotypic instability for this stage, since as yet we have no definition of the molecular mechanisms of this process. The remainder of the discussion in this text will be related to the experimental characterization of the stage of progression during multistage hepatocarcinogenesis.

MORPHOLOGIC, KARYOLOGIC, AND MOLECULAR ASPECTS OF PROGRESSION IN MULTISTAGE CARCINOGENESIS

In the "classical" two-stage carcinogenesis of mouse epidermis, tumor promoters were effective only when chronic treatment resulted in the ap-

Table 2. Preneoplastic Lesions in the Human
and Their Counterparts in Rodents

Tissue	Human	Rodent
Skin	Keratoacanthoma[49]	Papilloma[37]
Tracheobronchial epithelium	Atypical metaplasia[50]	Atypical metaplasia[51]
Esophagus	Moderate to severe dysplasia[52]	Moderate to severe dysplasia[53]
Stomach	Intestinal metaplasia[54]	Glandular dysplasia[55]
Colon	Polyp[44]	Polyp[56]
Pancreas	Focal acinar cell dysplasia[57]	Atypical acinar cell foci[58]
Liver	Liver cell dysplasia[59] Focal nodular hyperplasia[60]	Altered hepatic foci[38] "Neoplastic" nodules[61]
Bladder	Moderate to severe dysplasia[62]	Papillary hyperplasia[63]
Adrenal gland	Adrenocortical nodules[64]	Adrenocortical hyperplasia[65]
Mammary gland	Atypical lobule type A[66]	Hyperplastic terminal end buds[67]

pearance of malignant neoplasms. This was felt to be true despite the fact, at least in the epidermal system, that benign tumors, papillomas, were the endpoint in most experiments[37]. Therefore, it is not surprising that the same rationale was applied to analogous stages in hepatocarcinogenesis in the rat, again regardless of the fact that nodules of relatively benign-appearing hepatocytes were monitored as endpoints in several such models[38]. Furthermore, although regulatory agencies in the United States tend to equate benign and malignant neoplasms as endpoints in chronic bioassays, the use of intermediate lesions, such as papillomas in epidermal carcinogenesis and altered hepatic foci[38] or nodules in liver carcinogenesis, as endpoints in carcinogen bioassays has not met with general acceptance.

Despite resistance in academic and regulatory circles to the use of intermediate lesions as endpoints in bioassays of carcinogenic substances, it is becoming increasingly evident that such lesions represent transient, unstable cell populations derived from initiated cells from which carcinomas arise, at least during hepatocarcinogenesis, with a frequency greater than in uninitiated cells. Since cells of these intermediate lesions, which are in the stage of promotion, do have an increased risk for the development of malignancy, it is important to determine the mechanism for this increased risk and its role in the genesis of malignancy.

Morphologic Evidence

It is now generally accepted that tumor promotion is a reversible process[4,37] and that transient and unstable intermediate lesions can be considered as the cellular expression of this stage[26,39]. The genesis of the major neoplasms of the human -lung, breast, prostate, and uterus- probably involves promotion, exogenous or endogenous, as the principal stage in their development[26]. Many lesions have been described in the human which have many of the biological characteristics of the intermediate, transient lesions seen during multistage carcinogenesis in the rodent; some examples are seen in Table 2. This is not to say that carcinomas may not arise in the absence of any evidence of such intermediate lesions, both in the human[40,41] and in the experimental animal[42,43]. Presumably, in these latter instances, the initial carcinogenic insult is sufficient to convert the target cell to one in the stage of progression from the onset of carcinogenesis. However, in many instances one can demonstrate morphologic evidence that a secondary change occurs in a group of cells within the intermediate lesion, such that these cells assume the histologic characteristics of malignancy. In the human, numerous examples of carcinomas arise within benign[44] or preneoplastic lesions[45]. Carcinoma in situ of the uterine cervix can arise within areas of preneoplastic cervical epithelium[46]. In experimental systems, carcinomas can be seen to arise in areas of preneoplasia[47] as well as in benign neoplastic lesions[48].

As indicated above, one likely explanation of the morphologic changes seen when carcinomas arise in areas of preneoplasia or benign neoplasia is that further genetic alterations have taken place in one or more cells of the initial lesion. Model systems that can mimic this phenomenon of malignancy arising in preneoplasia are those employing the "initiation-promotion-initiation" format described above[32]. Scherer and his associates have developed an experimental model system in rat liver in which focal carcinomas can be induced to arise in pre-existing preneoplastic altered hepatic foci and/or nodules[34,47]. He and his colleagues have designated such lesions as "foci-in-foci". Using a system slightly different from that described by Scherer but based on the protocol proposed by Potter[32], we have utilized quantitative techniques developed in our laboratory[66] for the quantification of focal lesions to monitor the "foci-in-foci" as heterogeneous foci seen in livers of animals undergoing such protocols. A typical example of such a computer-derived plot can be seen in Figure 1. In this figure, the number of foci-in-foci in two dimensions (i.e., per cm^2) is 10, whereas

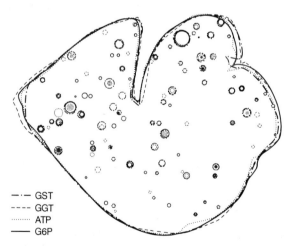

—·— GST
---- GGT
········ ATP
—— G6P

Fig. 1. Computer-driven overlay of plot of four serial sections stained respectively for the placental isozymic form of glutathione S-transferase (GST), γ-glutamyl-transpeptidase (GGT), canalicular ATPase (ATP), and glucose-6-phosphatase (G6P). See references 68 and 105 for details of the methods utilized. Foci-in-foci can readily be seen on the plot and their phenotypes identified.

that of all altered hepatic foci in the section is 60/cm². It has not been possible thus far to develop techniques to quantitate the number of foci-in-foci in three dimensions, as is done by quantitative stereology of the homogeneous foci[68]. Despite this barrier, it has been possible to obtain a crude estimate of the induction of foci-in-foci resulting from specific agents given as the second initiating agent in the protocol modeled after Potter. A sample of such data utilizing ethylnitrosourea or hydroxyurea as the second initiating agent, termed a "progressor" agent, is seen in Table 3. Since ethylnitrosourea is a complete carcinogen, a significant increase in the total number of altered hepatic foci is seen following a second initiation in the presence of the promoting agent phenobarbital when compared with animals receiving only a single initiation shortly after birth. In contrast, administration of hydroxyurea as the progressor agent shows no significant change in the total number of altered hepatic foci, but an approximately threefold increase in the foci-in-foci (fif) resulting from the I-P-I protocol, when the number of foci-in-foci are normalized for the total number of altered hepatic foci seen in the section (fif/AHF/cm²). Expressed in this way, the progressor activities of ethylnitrosourea and hydroxyurea are essentially the same, although it is clear that their initiating activities (AHF/liver) are quite different, hydroxyurea exhibiting essentially no initiating action. The initiating action of ethylnitrosourea at the time of partial hepatectomy is also noted by the significantly lower mean volume of the altered hepatic foci, since those foci initiated by ethylnitrosourea had a much shorter period to develop than those initiated by diethylnitrosamine at the time of birth. The chronic administration of phenobarbital following diethylnitrosamine administration as an initiating agent, but with no second initiation, also resulted in a significant number of foci-in-foci. Since in preliminary experiments the number of foci-in-foci following diethylnitrosamine initiation alone was very low (less than 1/cm²), it would appear that phenobarbital also has some progressor action of itself. Since this agent has been shown to be clastogenic, at least in lower life forms[69], and its metabolism involves the formation of active oxygen radicals[70] which themselves are clastogenic[71], some degree of progressor activity would be expected for this agent. However, until better quantitative methods are available, it will be difficult to ascertain estimates of the effectiveness of chemicals as progressor agents.

These experiments thus make use of the morphologic change apparently resulting from the administration of a second agent, in this case termed the progressor agent, acting primarily on cells within already formed focal lesions in the stage of promotion. Such a system can be utilized to identify agents that act specifically at the interface of the stages of promotion and progression. Such agents may be termed progressor agents. These agents may be extremely important in the genesis of human cancer since, as pointed out above, the stage of promotion is probably the most important factor in the genesis of the most frequent human neoplasms[26].

Karyologic Evidence

Boveri[72] was one of the first to recognize that significant karyotypic abnormalities are seen in most human and animal malignant neoplasms. Today it is clear that essentially all malignant neoplasms in the human exhibit some degree of karyotypic abnormality if one uses the modern techniques of banding and/or premature chromosome condensation[73]. Such malignancies in the human as acute nonlymphocytic leukemia[74] and in the animal as the "minimal deviation" hepatomas were found to exhibit abnormalities even in apparently diploid neoplasms when banding techniques were used[75].

Recently, in multistage epidermal carcinogenesis in the mouse, Aldaz and his associates[16] demonstrated that early appearing papillomas after initiation and promotion with TPA showed normal banded karyotypes. However, as

Table 3. Stereologic Parameters and Foci-in-Foci/cm² (fif/cm²) in Livers of Female Rats Subjected to I-P-I Protocols[a]

Protocol	Focal volume (% of liver)	AHF/liver	fif/cm²	AHF/cm²	fif/AHF/cm² x 100	Mean volume of AHF (mm³)
DEN/PH/HU & PB	11.6±0.9	14,000±1,000	10.3±0.5	59.5±4.1	17.1±0.7	0.075±.005
DEN/PH/ENU & PB	16.9±2.8	30,000±2,000	15.5±1.2	96.8±3.3	16.0±1.0	0.060±.015
DEN/PH & PB	14.2±0.9	13,950±600	4.2±0.4	57.4±2.0	6.0±1.0	0.101±.008

[a]In the protocols listed, diethylnitrosamine (DEN) was administered at a dose of 15 µg/kg 5 days after birth by intraperitoneal injection. At weaning, 10 female Sprague-Dawley rats were placed on a laboratory chow diet containing 0.05% phenobarbital (PB), and 12 weeks later they were subjected to a 70% partial (PH) hepatectomy under ether anesthesia. Twenty-four hours later, one group of animals received an intraperitoneal injection of ethylnitrosourea (ENU), which had been dissolved in water just prior to injection at a dose of 100 µg/kg. Another group of 10 rats were given an aqueous solution of hydroxyurea (HU) by intraperitoneal injection at a dose of 60 mg/kg. This dose was repeated again 26 and 32 hours after partial hepatectomy (PH). All animals were sacrificed 12 weeks following the operation. Sections of liver were taken, frozen on solid CO_2, and serial sections prepared and stained for the four histochemical markers, (γ-glutamyl transpeptidase, canalicular adenosine triphosphatase, glucose 6-phosphatase, and the placental form of glutathione-S-transferase) by techniques previously described in this laboratory[105]. The focal volume percentage and altered hepatic foci (AHF) per liver were determined by quantitative stereology[68]. All values are the mean ± standard error of the mean. See text for further details.

9

promotion continued with frequent applications of TPA, karyotypic abnormalities appeared with increasing complexity as the time of promotion was extended. Earlier studies on mammary carcinogenesis[5] showed a very similar
effect, and recent studies in our laboratory[76] have revealed the fact that
cells isolated from altered hepatic foci in the Peraino protocol[14] exhibit
essentially no chromosomal abnormalities. However, hepatocytes from altered
hepatic foci induced by the Solt-Farber protocol[77] demonstrated that the
majority of hepatocytes from the foci exhibited significant chromosomal abnormalities, indicating that the toxicity resulting from the extremely high
dose of diethylnitrosamine and the selection procedure employed in this protocol caused significant karyotypic damage. This is in line with the fact
that the Solt-Farber protocol rapidly induces hepatocellular carcinomas when
animals are promoted with PB[78], whereas the Peraino protocol takes considerably longer[14]. Since earlier studies did show that virtually all hepatocellular carcinomas in the rat exhibited chromosomal abnormalities[79,80],
it is reasonable to propose that the develpoment of such changes is the
result of genetic alteration, different from those that produced the initiated cells and their progeny in the stage of promotion.

Furthermore, cells that continue in the stage of progression exhibit
more and more chromosomal abnormalities[9,11], exhibiting greater and greater
degrees of karyotypic instability. The experiment of Kraemer[81] showed con-
clusively that ultimately even clones of neoplastic cells in the stage of
progression develop a variety of karyotypes, but maintain the same DNA content. Therefore, it is quite possible that the earliest karyotypic changes
represent the earliest expression of karyotypic instability, which can proceed to the extreme described by Kraemer and his associates.

Molecular Evidence

With the exception of the aberrant regulation of genetic expression, no
ubiquitous biochemical or molecular abnormality present in all neoplasms has
been described. Considerable evidence has recently developed to support the
proposition that malignant neoplasms may all exhibit an abnormal expression
of one or more proto-oncogenes; however, no single proto-oncogene is affected
in all malignant neoplasias. Therefore, just as with the induction of neoplasia by a variety of retroviruses, many pathways may lead to chemical and
radiation carcinogenesis. In these studies the "activation" of proto-oncogenes occurs by a variety of mechanisms, both mutational and transcriptional. The major known methods of activation of such genes are seen in Table
4. With relatively few exceptions, the activation of proto-oncogenes is a
phenomenon characteristic of the stage of progression in both human and
animal neoplasia. The exceptions, however, pose interesting questions. The
mutational activation of the Ha-ras proto-oncogene in mouse papillomas of the
skin[95] and adenomas of the mouse liver[96] suggests that this activation
occurs during the stage of initiation in these two tissues of this rodent.
However, no such mutational activation of this or any other proto-oncogene is
seen at the stage of initiation and/or promotion in multistage hepatocarcinogenesis in the rat. On the other hand, transcriptional activation of proto-
oncogenes in the mouse, rat, human, and other species during the stage of
progression is quite common[97-99].

In multistage hepatocarcinogenesis in the rat, transcriptional activation of proto-oncogenes is not found in altered hepatic foci during the stage
of promotion, although one report has presented evidence that foci resulting
from initiation by a necrogenic dose of diethylnitrosamine did show some
transcriptional activation of this proto-oncogene[100]. On the other hand,
when initiation was carried out with a non-necrogenic dose of diethylnitrosamine, the resulting foci actually showed lower levels of the protein products of a number of proto-oncogenes in the rat (M. Neveu and H. Pitot,

Table 4. Potential Mechanisms of Proto-Oncogene Activation

Event	Consequence	Example
Base mutation in coding sequences	New gene product with altered activity	Rous sarcoma[82] and other v-onc genes[83] Bladder carcinoma[84]
Deletion in noncoding sequences	Altered regulation of normal gene product	3T3 cell transformation in vitro[85]
Altered promotion for RNA polymerase	Increased transcription of mRNA (normal gene product)	Murine plasmacytoma[86] Lymphoma in chickens[87]
Insertion of substitution with repetitive DNA elements ("transposons")	Altered regulation of gene product (?normal)	Canine venereal tumor[88] Mouse myeloma[89]
Chromosomal translocation	Altered mRNA, new gene product Altered expression of gene	Non-Hodgkin's lymphoma[90] Mouse plasmacytoma[91]
Gene amplification	Increased expression of normal gene	Human colon carcinoma[92] Human neuroblastomas[93]
Hypomethylation of c-onc gene	Altered regulation of gene expression, normal gene product	Human colon and lung cancer[94]

unpublished observations). However, occasional foci, especially those seen in the I-P-I experiment, did show foci-in-foci with an increased expression of the fos or raf proto-oncogene in the internal or new focus (M. Neveu and H. Pitot, unpublished observations). Therefore, it is possible that transcriptional and/or mutational activation of proto-oncogenes, growth factors, nuclear DNA-binding proteins, and other as yet unknown "proto-oncogenes" may occur very early during the stage of progression and be mechanistically associated with its development, as might be predicted from carcinogenesis by the acutely oncogenic retroviruses. Since karyotypic changes are among the major mechanisms for transcriptional activation and mutational activation of proto-oncogenes (Table 4), karyotypic instability is again shown to be critical in the genesis and maintenance of the stage of progression.

CONCLUSIONS

There is considerable evidence, some of which has been discussed in this paper, to distinguish the stage of progression as a distinct phase in multi-stage carcinogenesis. Morphologic and karyotypic changes, as well as alterations in gene expression that are unique to this stage and distinguish it from the first two stages, have been described in one or more systems both in the animal and in the human. As pointed out earlier, high doses of complete carcinogens, acutely transforming retroviruses, and high doses of ionizing radiation may convert normal cells to cells in the stage of progression without any demonstrable intervening stage of promotion. However, where promotion occurs, progression follows, either spontaneously or induced.

The relationship of the three-stage concept of carcinogenesis to the requirement for two genetic changes proposed by Knudson[101] becomes obvious. Although the actions of proto-oncogenes are dominant, it is only with acutely transforming oncogenic viruses that such dominance is apparent without any known further genetic change. Even in artificial transfection experiments with activated proto-oncogenes, it appears that in most instances more than one genetic change is necessary[102,103]. In the genetic systems originally forming the basis for Knudson's hypothesis, such as retinoblastoma, the second genetic change was presumed to be in the same allele as the first. Subsequent recent findings have borne out this hypothesis[104]. Obviously, more gross chromosomal changes such as are seen in the stage of progression would give a greater chance for the second mutational change, especially if this had to occur in the same allele. Furthermore, it is of importance in this respect that cells of altered hepatic foci are largely diploid[76], in contrast to normal hepatocytes, which are largely tetraploid. Knudson's hypothesis could not function appropriately in a tetraploid cell, thus suggesting that the progenitors of carcinomas in the liver are this diploid population.

Finally, the preliminary studies described in this paper (see above), as well as studies in epidermal carcinogenesis[31], suggest that agents specifically acting to convert cells from the stage of promotion to that of progression actually exist. Such progressor agents may be very important in human health considerations, especially since humans exist in an environment, largely self-made, in which tumor promotion is predominant, especially in relation to the major human cancers (see above). Therefore, an understanding of the stage of progression, both its cellular and molecular biology, will probably be very important in the rational prevention and therapy of human cancer in the future.

REFERENCES

1. L. Foulds, The experimental study of tumor progression: a review, Cancer Res. 14:327-339 (1954).

2. L. Foulds, Multiple etiologic factors in neoplastic development, Cancer Res. 25:1339-1347 (1965).

3. P. Shubik, Progression and promotion, J. Natl. Cancer Inst. 73:1005-1011 (1984).

4. H. C. Pitot, D. Beer, and S. Hendrich, Multistage carcinogenesis: the phenomenon underlying the theories, in: "Theories of Carcinogenesis", O. Iversen, ed., Hemisphere Press, Washington, D.C. (1988).

5. E. R. Fisher, R. H. Shoemaker, and A. Sabnis, Relationship of hyperplasia to cancer in 3-methylcholanthrene-induced mammary tumorigenesis, Lab. Invest. 33:33-42 (1975).

6. S. J. Bevacqua, C. W. Greeff, and M. J. C. Hendrix, Cytogenetic evidence of gene amplification as a mechanism for tumor cell invasion, Somatic Cell Mol. Genet. 14:83-91 (1988).

7. G. L. Nicolson, Tumor cell instability, diversification, and progression to the metastatic phenotype: from oncogene to oncofetal expression, Cancer Res. 47:1473-1487 (1987).

8. P. Frost, R. S. Kerbel, B. Hunt, S. Man, and S. Pathak, Selection of metastatic variants with identifiable karyotypic changes from a non-metastatic murine tumor after treatment with 2'-deoxy-5-azacytidine or hydroxyurea: implications for the mechanisms of tumor progression, Cancer Res. 47:2690-2695 (1987).

9. S. R. Wolman, Karyotypic progression in human tumors, Cancer Metast. Rev. 2:257-293 (1983).

10. A. C. Ritchie, The classification, morphology, and behaviour of tumours, in: "General Pathology", H. W. Florey, ed., W. B. Saunders Co., Philadelphia (1970).

11. P. C. Nowell, Mechanisms of tumor progression, Cancer Res. 46:2203-2207 (1986).

12. A. P. Feinberg and D. S. Coffey, The concept of DNA rearrangement in carcinogenesis and development of tumor cell heterogeneity, in: "Tumor Cell Heterogeneity", A. H. Owens, D. S. Coffey, and S. B. Baylin, eds., Academic Press, Inc., New York (1982).

13. H. C. Pitot, L. Barsness, T. Goldsworthy, and T. Kitagawa, Biochemical characterization of stages of hepatocarcinogenesis after a single dose of diethylnitrosamine, Nature 271:456-458 (1978).

14. C. Peraino, E. F. Staffeldt, B. A. Carnes, V. A. Ludeman, J. A. Blomquist, and S. D. Vesselinovitch, Characterization of histochemically detectable altered hepatocyte foci and their relationship to hepatic tumorigenesis in rats treated once with diethylnitrosamine or benzo(a)pyrene within one day after birth, Cancer Res. 44:3340-3347 (1984).

15. L. Eriksson, M. Ahluwalia, J. Spiewak, G. Lee, D. S. R. Sarma, M. J. Roomi, and E. Farber, Distinctive biochemical pattern associated with resistance of hepatocytes in hepatocyte nodules during liver carcinogenesis, Environ. Health Persp. 49:171-174 (1983).

16 C. M. Aldaz, C. J. Conti, A. J. P. Klein-Szanto, and T. J. Slaga, Progressive dysplasia and aneuploidy are hallmarks of mouse skin papillomas: relevance to malignancy, Proc. Natl. Acad. Sci, U.S.A., 84:2029-2032 (1987).

17. T. H. Yosida, Karyotype evolution and tumor development, Cancer Genet. Cytogenet. 8:153-179 (1983).

18. G. B. Pierce and W. C. Speers, Tumors as caricatures of the process of tissue renewal: prospects for therapy by directing differentiation, Cancer Res. 48:1996-2004 (1988).

19. P. C. Nowell, Genetic instability in cancer cells: relationship to tumor cell heterogeneity, in: "Tumor Cell Heterogeneity," A. H. Owens, D. S. Coffey, and S. B. Baylin, eds., Academic Press, Inc., New York (1982).

20. R. Sager, I. K. Gadi, L. Stephens, and C. T. Grabowy, Gene amplification: an example of accelerated evolution in tumorigenic cells, Proc. Natl. Acad. Sci, U.S.A. 82:7015-7019 (1985).

21. M. Chorazy, Sequence rearrangements and genome instability, J. Cancer Res. Clin. Oncol. 109:159-172 (1985).

22. D. R. Welch, and S. P. Tomasovic, Implications of tumor progression on clinical oncology, Clin. Exp. Metastasis 3:151-188 (1985).

23. H. C. Pitot, Oncogenes and human neoplasia, Clin. Lab. Med. 6:167-179 (1986).

24. G. Klein, and E. Klein, Conditioned tumorigenicity of activated onco-genes, Cancer Res. 46:3211-3224 (1986).

25. M. G. Parker, and M. J. Page, Use of gene transfer to study expression of steroid-responsive genes, Mol. Cell. Endocrinol. 34:159-168 (1984).

26. H. C. Pitot, "Fundamentals of Oncology," 3rd ed., Marcel Dekker, New York (1986).

27. A. A. Sinha, Hormone sensitivity and autonomy of tumours, in: "Hormonal Management of Endocrine-Related Cancer," B. A. Stoll, ed., Lloyd-Luke, Ltd., London (1981).

28. H. Cedar, DNA methylation and gene activity, Cell 53:3-4 (1988).

29. L. E. Babiss, S. G. Zimmer, and P. B. Fisher, Reversibility of progres-sion of the transformed phenotype in Ad5-transformed rat embryo cells, Science 228:1099-1101 (1985).

30. R. S. Kerbel, P. Frost, R. Liteplo, D. A. Carlow, and B. E. Elliott, Possible epigenetic mechanisms of tumor progression: induction of high-frequency heritable but phenotypically unstable changes in the tumorigenic and metastatic properties of tumor cell populations by 5-azacytidine treatment, J. Cell. Physiol. Suppl. 3:87-97 (1984).

31. J. F. O'Connell, A. J. P. Klein-Szanto, D. M. DiGiovanni, J. W. Fries, and T. J. Slaga, Enhanced malignant progression of mouse skin tumors by the free-radical generator benzoyl peroxide, Cancer Res. 46:2863-2865 (1986).

32. V. R. Potter, A new protocol and its rationale for the study of initia-tion and promotion of carcinogenesis in rat liver, Carcinogenesis 2:1375-1379 (1981).

33. H. Hennings, R. Shores, M. L. Wenk, E. F. Spangler, R. Tarone, and S. H. Yuspa, Malignant conversion of mouse skin tumours is increased by tumour initiators and unaffected by tumour promoters, Nature 304:67-69 (1983).

34. E. Scherer, Neoplastic progression in experimental hepatocarcinogenesis, Biochim. Biophys. Acta 738:219-236 (1984).

35. R. Schulte-Hermann, Tumor promotion in the liver, Arch. Toxicol. 57:147-158 (1985).

36. E. Farber and D. S. R. Sarma, Hepatocarcinogenesis: a dynamic cellular perspective, Lab. Invest. 56:4-22 (1987).

37. R. K. Boutwell, Some biological aspects of skin carcinogenesis, Progr. Exp. Tumor Res. 4:207-250 (1964).

38. T. L. Goldsworthy, M. H. Hanigan, and H. C. Pitot, Models of hepato-carcinogenesis in the rat-contrasts and comparisons, CRC Crit. Rev. Toxicol. 17:61-89 (1986).

39. C. B. Wigley, Experimental approaches to the analysis of precancer, Cancer Surv. 2:495-515 (1983).

40. D. Y. Lin, Y.-F. Liaw, C. M. Chu, C. S. Chang-Chien, C. S. Wu, P. C. Chen, and I. S. Sheen, Hepatocellular carcinoma in noncirrhotic patients, Cancer 54:1466-1468 (1984).

41. S. Kuramoto and T. Oohara, Minute cancers arising de novo in the human large intestine, Cancer 61:829-834 (1988).

42. G. M. Williams, The pathogenesis of rat liver cancer caused by chemical carcinogens, Biochim. Biophys. Acta 605:167-189 (1980).

43. A. P. Maskens and R.-M. Dujardin-Loits, Experimental adenomas and car-cinomas of the large intestine behave as distinct entities: most carcinomas arise de novo in flat mucosa, Cancer 47:81-89 (1981).

44. D. W. Day, The adenoma-carcinoma sequence, Scand. J. Gastroenterol. 19 (suppl. 104):99-107 (1984).

45. Y. Nakanuma, G. Ohta, H. Sugiura, K. Watanabe, and K. Doishita, Incidental solitary hepatocellular carcinomas smaller than 1 cm in size found at autopsy: a morphologic study, Hepatology 6:631-635 (1986).

46. W. M. Christopherson, Dysplasia, carcinoma in situ, and microinvasive carcinoma of the uterine cervix, Human Pathol. 8:489-501 (1977).

47. E. Scherer, Relationship among histochemically distinguishable early lesions in multistep-multistage hepatocarcinogenesis, Arch. Toxicol. Suppl. 10:81-94 (1987).

48. P. J. Hermanek and J. Giedl, The adenoma-carcinoma sequence in AMMN-induced colonic tumors of the rat, Pathol. Res. Pract. 178:548-554 (1984).

49. H. Haber, The skin, in: "Systemic Pathology, Vol. II," G. P. Wright and W. St. C. Symmers, eds., American Elsevier Publishing Company, New York (1966).

50. B. F. Trump, E. M. McDowell, F. Glavin, L. A. Barrett, P. J. Becci, W. Schürch, H. E. Kaiser, and C. C. Harris, The respiratory epithelium. III. Histogenesis of epidermoid metaplasia and carcinoma in situ in the human, J. Natl. Cancer Inst. 61:563-575 (1978).

51. E. M. McDowell and B. F. Trump, Histogenesis of preneoplastic and neoplastic lesions in tracheobronchial epithelium, Surv. Synth. Pathol. Res. 2:235-279 (1983).

52. P. Correa, Precursors of gastric and esophageal cancer, Cancer 50:2554-2565 (1982).

53. C. A. Rubio, Epithelial lesions antedating oesophageal carcinoma. I. Histologic study in mice, Pathol. Res. Pract. 176:269-275 (1983).

54. K. Saito, and T. Shimoda, The histogenesis and early invasion of gastric cancer, Acta Pathol. Jpn. 36(9):1307-1318 (1986).

55. D. Tsiftsis, J. R. Jass, M. I. Filipe, and C. Wastell, Altered patterns of mucin secretion in precancerous lesions induced in the glandular part of the rat stomach by the carcinogen N-methyl-N'-nitro-N-nitrosoguanidine, Invest. Cell Pathol. 3:399-408 (1980).

56. J. L. Madara, P. Harte, J. Deasy, D. Ross, S. Lahey, and G. Steele, Jr., Evidence for an adenoma-carcinoma sequence in dimethylhydrazine-induced neoplasms of rat intestinal epithelium, Am. J. Pathol. 110:230-235 (1983).

57. D. S. Longnecker, H. Shinozuka, and A. Dekker, Focal acinar cell dysplasia in human pancreas, Cancer Res. 45:534-540 (1980).

58. D. S. Longnecker and T. J. Curphey, Adenocarcinoma of the pancreas in azaserine-treated rats, Cancer 35:2249-2257 (1975).

59. S. Watanabe, K. Okita, T. Harada, T. Kodama, Y. Numa, T. Takemoto, and M. Takahashi, Morphologic studies of the liver cell dysplasia, Cancer 51:2197-2205 (1983).

60. D. M. Knowles and M. Wolff, Focal nodular hyperplasia of the liver, Human Pathol. 7:533-545 (1976).

61. R. A. Squire and M. H. Levitt, Report of a workshop on classification of specific hepatocellular lesions in rats, Cancer Res. 35:3214-3223 (1975).

62. W. M. Murphy and M. S. Soloway, Urothelial dysplasia, J. Urology 127:849-854 (1982).

63. E. Kunze, A. Schauer, and S. Schatt, Stages of transformation in the development of N-butyl-N-(4-hydroxybutyl)-nitrosamine-induced transitional cell carcinomas in the urinary bladder of rats, Z. Krebsforsch. 87:139-160 (1976).

64. R. B. Cohen, Observations on cortical nodules in human adrenal glands. Their relationship to neoplasia, Cancer 19:552-556 (1966).

65. T. B. Dunn, Normal and pathologic anatomy of the adrenal gland of the mouse, including neoplasms, J. Natl. Cancer Inst. 44:1323-1389 (1970).

66. L.-J. van Bogaert, Mammary hyperplastic and preneoplastic changes: taxonomy and grading, Breast Cancer Res. Treat. 4:315-322 (1984).

67. D. M. Purnell, The relationship of terminal duct hyperplasia to mammary carcinoma in 7,12-dimethylbenz(a)anthracene-treated LEW/Mai rats, Am. J. Pathol. 98:311-322 (1980).

68. H. A. Campbell, Y.-D. Xu, M. H. Hanigan, and H. C. Pitot, Application of quantitative stereology to the evaluation of phenotypically hetero-geneous enzyme-altered foci in the rat liver, J. Natl. Cancer Inst. 76:751-767 (1986).

69. S. Albertini, U. Friederich, U. Gröschel-Stewart, F. K. Zimmermann, and F. E. Würgler, Phenobarbital induces aneuploidy in Saccharomyces cerevisiae and stimulates the assembly of porcine brain tubulin, Mutat. Res. 144:67-71 (1985).

70. P. J. O'Brien, Hydroperoxides and superoxides in microsomal oxidations, Pharmac. Ther. A. 2:517-536 (1978).

71. M. L. Cunningham, J. G. Peak, and M. J. Peak, Single-strand DNA breaks in rodent and human cells produced by superoxide anion or its reduc-tion products, Mutat. Res. 184:217-222.

72. T. Boveri, Zur Frage der Entstehung maligner Tumoren, Gustav Fischer, Jena (1914).

73. J. J. Yunis, The chromosomal basis of human neoplasia, Science 221:227-236 (1983).

74. J. J. Yunis, C. D. Bloomfield, and K. Ensrud, All patients with acute nonlymphocytic leukemia may have a chromosomal defect, N. Engl. J. Med. 305:135-139 (1981).

75. S. R. Wolman, A. A. Horland, and F. F. Becker, Altered karyotypes of transplantable "diploid" tumors, J. Natl. Cancer Inst. 51:1909-1914 (1973).

76. L. Sargent, Y.-H. Xu, G. L. Sattler, L. Meisner, and H. C. Pitot, Ploidy and karyotype of hepatocytes isolated from enzyme-altered foci in two different protocols of multistage hepatocarcinogenesis in the rat, Carcinogenesis, 10:387-391 (1989).

77. D. Solt, and E. Farber, New principle for the analysis of chemical car-cinogenesis, Nature 263:701-703 (1976).

78. V. Préat, J. de Gerlache, M. Lans, H. Taper, and M. Roberfroid, Compara-tive analysis of the effect of phenobarbital, dichlorodiphenyltri-chloroethane, butylated hydroxytoluene and nafenopin on rat hepato-carcinogenesis, Carcinogenesis 7:1025-1028 (1986).

79. P. C. Nowell and H. P. Morris, Chromosomes of "minimal deviation" hepatomas: a further report on diploid tumors, Cancer Res. 29:969-970 (1969).

80. E. Kovi and H. P. Morris, Chromosome binding studies of several trans-plantable hepatomas, Adv. Enz. Reg. 14:139-162 (1976).

81. P. M. Kraemer, L. L. Deaven, H. A. Grissman, and M. A. Van Dilla, DNA constancy despite variability in chromosome number, Adv. Cell Mol. Biol. 2:47-108 (1972).

82. J.-Y. Kato, T. Takeya, C. Grandori, H. Iba, J. B. Levy, and H. Hanafusa, Amino acid substitutions sufficient to convert the nontransforming $p60^{c-src}$ protein to a transforming protein, Mol. Cell. Biol. 6:4155-4160 (1986).

83. J. M. Bishop, The molecular genetics of cancer, Science 235:305-311 (1987).

84. C. J. Tabin, S. M. Bradley, C. I. Bargmann, R. A. Weinberg, A. G. Papageorge, E. M. Scolnick, R. Dhar, D. R. Lowy, and E. H. Chang, Mechanism of activation of a human oncogene, Nature 300:143-149 (1982).

85. F. Meijlink, T. Curran, A. D. Miller, and I. M. Verma, Removal of a 67-base-pair sequence in the noncoding region of proto-oncogene fos converts it to a transforming gene, Proc. Natl. Acad. Sci., U.S.A. 82:4987-4991 (1985).

86. J.-Q. Yang, S. R. Bauer, J. F. Mushinski, and K. B. Marcu, Chromosome translocations clustered 5' of the murine c-_myc_ gene qualitatively affect promoter usage: implications for the site of normal c-_myc_ regulation, _EMBO J._ 4:1441-1447 (1985).

87. W. S. Hayward, B. G. Neel, and S. M. Astrin, Activation of a cellular _onc_ gene by promoter insertion in ALV-induced lymphoid leukosis, _Nature_ 290:475-480 (1981).

88. N. Katzir, G. Rechavi, J. B. Cohen, T. Unger, F. Simoni, S. Segal, D. Cohen, and D. Givol, "Retroposon" insertion into the cellular oncogene c-_myc_ in canine transmissible veneral tumor, _Proc. Natl. Acad. Sci., U.S.A._ 82:1054-1058 (1985).

89. G. Rechavi, D. Givol, and E. Canaani, Activation of a cellular oncogene by DNA rearrangement: possible involvement of an IS-like element, _Nature_ 300:607-611 (1982).

90. A. C. Hayday, S. D. Gillies, H. Saito, C. Wood, K. Wiman, W. S. Hayward, and S. Tonegawa, Activation of a translocated human c-_myc_ gene by an enhancer in the immunoglobulin heavy-chain locus, _Nature_ 307:334-340 (1984).

91. P. D. Fahrlander, J. Sümegi, J.-Q. Yang, F. Wiener, K. B. Marcu, and G. Klein, Activation of the c-_myc_ oncogene by the immunoglobulin heavy-chain gene enhancer after multiple switch region-mediated chromosome rearrangements in a murine plasmacytoma, _Proc. Natl. Acad. Sci., U.S.A._ 82:3746-3750 (1985).

92. C. C. Lin, K. Alitalo, M. Schwab, D. George, H. E. Varmus, and J. M. Bishop, Evolution of karyotypic abnormalities and c-_myc_ oncogene amplification in human colonic carcinoma cell lines, _Chromosoma_ 92:11-15 (1985).

93. G. M. Brodeur and R. C. Seeger, Gene amplification in human neuroblastomas: basic mechanisms and clinical implications, _Cancer Genet. Cytogenet._ 19:101-111 (1986).

94. A. P. Feinberg and B. Vogelstein, Hypomethylation of _ras_ oncogenes in primary human cancers, _Biochem. Biophys. Res. Commun._ 111:47-54 (1983).

95. A. Balmain, M. Ramsden, G. T. Bowden, and J. Smith, Activation of the mouse cellular Harvey-_ras_ gene in chemically induced benign skin papillomas, _Nature_ 307:658-660 (1984).

96. S. H. Reynolds, S. J. Stowers, R. R. Maronpot, M. W. Anderson, and S. A. Aaronson, Detection and identification of activated oncogenes in spontaneously occurring benign and malignant hepatocellular tumors of the B6C3F1 mouse, _Proc. Natl. Acad. Sci., U.S.A._, 83:33-37 (1986).

97. J. G. Guillem, L. L. Hsieh, K. M. O'Toole, K. A. Forde, P. LoGerfo, and I. B. Weinstein, Changes in expresion of oncogenes and endogenous retroviral-like sequences during colon carcinogenesis, _Cancer Res._ 48:3964-3971 (1988).

98. M. D. Erisman, P. G. Rothberg, R. E. Diehl, C. C. Morse, J. M. Spandorfer, and S. M. Astrin, Deregulation of c-_myc_ gene expression in human colon carcinoma is not accompanied by amplification or rearrangement of the gene, _Mol. Cell. Biol._ 5:1969-1976 (1985).

99. T. Tanaka, D. J. Slamon, H. Battifora, and M. J. Cline, Expression of p21 _ras_ oncoproteins in human cancers, _Cancer Res._ 46:1465-1470 (1986).

100. P. Galand, D. Jacobovitz, and K. Alexandre, Immunohistochemical detection of c-Ha-_ras_ oncogene p21 product in pre-neoplastic and neoplastic lesions during hepatocarcinogenesis in rats, _Int. J. Cancer_ 41:155-161 (1988).

101. A. G. Knudson, Jr., Genetics and the etiology of childhood cancer, _Pediat. Res._ 10:513-517 (1976).

102. D. G. Thomassen, T. M. Gilmer, L. A. Annab, and J. C. Barrett, Evidence for multiple steps in neoplastic transformation of normal and preneoplastic Syrian hamster embryo cells following transfection with Harvey murine sarcoma virus oncogene (v-Ha-ras), Cancer Res. 45:726-732 (1985).

103. H. Land, L. F. Parada, and R. A. Weinberg, Tumorigenic conversion of primary embryo fibroblasts requires at least two cooperating oncogenes, Nature 304:596-602 (1983).

104. M. F. Hansen and W. K. Cavenee, Genetics of cancer predisposition, Cancer Res. 47:5518-5527 (1987).

105. S. Hendrich, H. A. Campbell, and H. C. Pitot, Quantitative stereological evaluation of four histochemical markers of altered foci in multistage hepatocarcinogenesis in the rat, Carcinogenesis 8:1245-1250 (1987).

CRITICAL EVENTS IN SKIN TUMOR

PROMOTION AND PROGRESSION

Thomas J. Slaga

The University of Texas
M. D. Anderson Cancer Center
Science Park-Research Division
Department of Carcinogenesis
Smithville TX 78957

INTRODUCTION

Carcinogenesis in skin as well as other target tissues in a number of species has been shown to be a multistage process which can be divided into at least three major stages, initiation, promotion and progression. An important aspect of the multistage skin carcinogenesis is that it has suggested that both genetic and epigenetic mechanisms are important. Altered growth control and differentiation leading to a more embryonic phenotype appear to be critical consequences of the genetic and epigenetic changes.

The sequential application of a subthreshold dose of a carcinogen (initiation stage) followed by repetitive treatment with a noncarcinogenic promoter (promotion stage) will induce skin tumors. The initiation phase requires only a single application of either a direct or an indirect carcinogen and is essentially an irreversible step, while the promotion phase is initially reversible later becoming irreversible[1]. The progression stage is defined as those events occurring after the initial appearance of skin tumor and represents the transition from a benign (papilloma) to malignant (squamous cell carcinoma) and finally to metastatic tumors.

CRITICAL TARGETS AND EVENTS
IN SKIN TUMOR INITIATION

Skin tumor initiation in mouse skin appears to be an irreversible stage that probably involves a somatic mutation in some aspect of epidermal growth control and/or epidermal differentiation. Extensive data has revealed a good correlation between the carcinogenicity of many chemical carcinogens and their mutagenic activities. Most tumor initiating agents either generate or are metabolically converted to electrophilic reactants, which bind covalently to cellular DNA and other macromolecules. Previous studies have demonstrated a good correlation between the skin tumor initiating activities of several polycyclic aromatic hydrocarbons (PAH) and their abilities to bind covalently to DNA.

Data suggests that skin tumor initiation probably occurs in dark basal keratinocytes since a good correlation exists between the degree of tumor

Boundaries between Promotion and Progression during Carcinogenesis
Edited by O. Sudilovsky *et al.*, Plenum Press, New York, 1991

19

initiation and the number of dark basal keratinocytes present in the skin[1].
The dark basal keratinocytes are present in the skin in large numbers during
embryogenesis, in moderate numbers in newborns, in low numbers in young
adults, and in very low numbers in old adults which suggests that these cells
may be epidermal stem cells[1,2]. The initiating potential of mouse skin
decreases with the age of the mouse to the point that it is very difficult to
initiate mice greater than one year of age when the number of dark basal
keratinocytes are extremely rare. These data suggest that the number of dark
basal cells correlate with the degree of tumor initiating. The dark basal
cells are very dense cells which do not enter into cell division very often
unless wounding occurs or when they are stimulated by tumor promoters[2].

There is substantial evidence suggesting that normal basal keratinocytes
within epidermal proliferative units and hair follicles form a maturation
series of slowly cycling, self-renewing stem cells, proliferative but non-
renewing transit (amplifying) cells, and postmitotic maturing keratinocytes
in mice[3,4]. The slowly cycling self-renewing stem cells are also very
dense cells[6]. Tumor initiator label-retaining cells were found by Morris
and coworkers[6] to have characteristics of the slowly cycling cells:
i) most of the carcinogen labeled nuclei were found in the central regions of
the epidermal proliferative units; ii) treatment of the carcinogen label-
retaining cells with 2 μg of 12-0-tetradecanoylphorbol-13-acetate (TPA)
elicited labeled mitosis within 1 day and a general decrease in grain density
over basal nuclei. In contrast, "maturing" basal cells 4 days after a single
injection of [³H]thymidine were found at the periphery of the epidermal proliferative units. Within 1 day after treatment with 2 μg of TPA, "maturing"
basal cells were displaced to the suprabasal layers. Double isotope-double
emulsion autoradiographs demonstrated doubly labeled cells 1 month after continuous labeling with [³H]thymidine and [¹⁴C]benzo[a]pyrene and provide
evidence that the radioactive carcinogen is retained by the slowly cycling
[³H]thymidine label-retaining cells. These observations suggest that a
slowly cycling population of epidermal cells may be relevant to the initiation phase of two-stage carcinogenesis[6].

CRITICAL EVENTS IN SKIN TUMOR PROMOTION

In general, skin tumor promoters do not bind covalently to DNA and are
not mutagenic but bring about a number of important epigenetic changes[1,7].
In addition to causing inflammation and epidermal hyperplasia, the phorbol
ester and other tumor promoters produce many other morphological and biochemical changes in skin. Of the observed promoter related effects on the
skin, the induction of epidermal cell proliferation, ornithine decarboxylase
(ODC) and subsequent polyamines, prostaglandins and dark basal keratinocytes
have the best correlation with promoting activity[1,8]. In addition to the
induction of dark cells, which are normally present in large numbers in embryonic skin, many other embryonic conditions appear in adult skin after
treatment with tumor promoters. Most of them occur after tumor promotion
treatment and may be a consequence of the alteration in differentiation.

It is difficult to determine which of the many effects associated with
tumor promotion are in fact essential components of the promotion process. A
good correlation appears to exist between promotion and epidermal hyperplasia[8]. However, some agents that induce epidermal cell proliferation do not
promote carcinogenesis[1]. Nevertheless it should be emphasized that all
known skin tumor promoters do induce epidermal hyperplasia[3]. O'Brien et.
al.[9] have reported an excellent correlation between the tumor-promoting
ability of various compounds and the induction of ODC activity in mouse
skin. However, mezerein (a diterpene similar to TPA but with weak promoting
activity) induced ODC to levels that were comparable to those of TPA[10].

Raick[11] found that phorbol ester tumor promoters caused the appearance of "dark basal cells" in the epidermis, whereas ethylphenylpropiolate (EPP), a nonpromoting epidermal hyperplastic agent, did not. A weak promoter[11], like wounding, induced a few dark cells. In addition, a large number of these dark cells are found in papillomas and carcinomas[1]. Klein-Szanto et. al.[12] reported that TPA induced about 3 to 5 times the number of dark cells as mezerein, which was the first major difference between these compounds.

There are a number of other important epigenetic changes in the skin such as membrane and differentiation alterations and an increase in protease activity, cAMP independent protein kinase activity and phospholipid synthesis have been caused by promoters[1]. In addition, the skin tumor promoters cause a decrease in epidermal superoxide dismutase and catalase activity as well as a decrease in the number of glucocorticoid receptors[13]. The phorbol ester tumor promoters and teleocidin induced changes appear to be mediated by their interaction with specific membrane receptors whereas many of the other promoters such as benzoyl peroxide and anthralin do not act through this receptor but may involve a free radical mechanism[13].

Altered differentiation appears to play a critical role in tumor promotion and carcinogenesis in general. Tumor promoters transiently induce, in epidermal and other cells, a set of phenotypic changes which resemble those found in embryonic cells as well as malignant cells. Raick[11] found that tumor promoting agents induced in basal keratinocytes certain morphologic changes resembling those found in embryonic, papillomas and carcinoma cells. Klein-Szanto et. al.[12] found that tumor promoters increased the number of dark basal keratinocytes in adult skin, which were normally in high numbers in embryonic skin, as well as in papillomas and carcinomas. Theoretically, modulation of the commitment of differentiation potential of subpopulations of keratinocytes could result in the accumulation of subpopulations of initiated cells. Reiners and Slaga[14] found that tumor promoters induce a subpopulation of basal cells to commit to terminal differentiation, and to acelerate the rate of differentiation of committed cells. Yuspa et. al.[15] also found that tumor promoters, in culture, can induce subpopulations of basal cells to differentiate. This could be an important mechanism in the expansion of the initiated cell population. In this regard, Yuspa and Morgan[16] found that initiated epidermal cells in culture do not change under high calcium, a physiological stimulus to differentiate.

Skin tumor promoters may also have an effect on the genetic material of cells. Free radicals may be the candidates for the many genetic effects. Some promoters such as benzoyl peroxide spontaneously give rise to free radicals whereas others such as phorbol ester and teleocidin type promoters may give rise to free radicals by their clastogenic effect[17]. Oxidized lipids and oxygen radicals are likely candidates induced by the clastogenic effect of TPA which could have a direct effect on the genetic material. Tumor promoters have been shown to cause gene amplification[18], mitotic aneuploidy in yeast[19], synergistic interactions with viruses in enhancing cell transformation[20], enhancement of irreversible anchorage-independent growth in mouse epidermal cell lines[21], and sister chromatid exchange[22]. The epigenetic effects of the tumor promoters are reversible and thus may be more important in the earlier stages of promotion since promotion is reversible for a reasonable period of time. On the other hand, the genetic effects of the tumor promoters may be responsible for the irreversible portion or late state of promotion.

Overall, the critical aspect of skin tumor promotion and possibly tumor promotion in other systems is the selective expansion of initiated cells by the tumor promoter. Table 1 summarizes various possible mechanisms of selection of initiated cells.

Table 1. Mechanisms of Selection of Initiated Epidermal Stem Cell
by Skin Tumor Promoters

1. Some tumor promoters may have direct effect on initiated stem cells (dark cells?) causing them to divide and expand in number.

2. Tumor promoters convert some basal keratinocytes to an embryonic phenotype similar to the dark cells, thereby supplying a positive environment for the initiated dark cells to expand in number.

3. Tumor promoters stimulate terminal differentiation of some epidermal cells and thus decrease a negative feedback mechanism on cell proliferation.

4. Some tumor promoters have a selective cytotoxic effect which may cause initiated cells to expand in number.

CRITICAL EVENTS IN SKIN TUMOR PROGRESSION

Although many studies have been directed toward understanding the mechanisms involved in skin tumor initiation and promotion, only a few studies have been performed on the progression stage of skin carcinogenesis. Table 2 summarizes some of the important characteristics of skin papillomas and carcinomas in order to emphasize those events which appear to be critical in the conversion of papillomas to carcinomas. As can be seen there are several events which appeared during tumor promotion that are continued or even exaggerated during tumor progression such as an increase in dark cells, loss of glucocorticoid receptors and an increase in polyamines and prostaglandins[13].

There are also a number of changes that occur very late in the carcinogenesis process which are related to the conversion of benign to malignant tumors. We have found that all squamous cell carcinomas lack several differentiation product proteins such as high molecular weight keratins (60,000-62,000) and filaggrin, but are positive for gamma glutamyltransferase (GGT). Only about 20% of the papillomas generated by an initiation-promotion protocol exhibit a similar condition[23,24]. Before visible tumors are observed using the initiation-promotion protocol, these conditions appear normal suggesting that the changes are very late responses[24].

Balmain and coworkers[25] have recently found that a percentage of papillomas and carcinomas induced by 9,10-dimethyl-1,2-benzanthracene (DMBA)-TPA contained elevated levels of Ha-ras transcripts compared with normal epidermis. Furthermore, the tumor DNA was capable of malignantly transforming NIH 3T3 cells in DNA transfection studies[26]. Studies in our laboratory[27] indicate that initiation alone or repetitive TPA treatments are insufficient to turn on the expression of the Ha-ras oncogene in adult SENCAR mouse epidermis. Initiation followed by either one or six weeks of TPA treatment also failed to activate Ha-ras expression. Like Balmain, we observed elevated levels of Ha-ras RNA in a percentage of papillomas and carcinomas tested. We have also found that the expression of c-src and c-abl are increased in the majority of carcinomas examined[28]. Presently, it still remains to be determined whether oncogene activation plays a critical role in multistage skin carcinogenesis.

Hennings and coworkers[29] recently reported that if mice with papillomas are treated repetitively with N-methyl-N-nitrosoguanidine (MNNG), a significant increase in the conversion of papillomas to carcinomas occurs. We have also found similar results with limited treatment of MNNG as well as with

Table 2. Characteristics of Skin Tumors

BENIGN PAPILLOMAS

1. Large number of dark cells.

2. Loss of glucocorticoid receptors.

3. High level of polyamines and prostaglandins.

4. Approximately 80% of the papillomas induced by two stage protocol have molecular weight keratins and filaggrin and are negative for GGT. 20% have reverse conditions.

5. Approximately 50% of papillomas induced by the two stage protocol express Ha-ras RNA.

6. Some papillomas are reversible while others are irreversible.

7. Early papillomas (10 wks) during promotion are well differentiated hyperplastic lesions with mild or no cellular atypia, whereas late ones (40 wks) are dysplastic, show atypia and are aneuploid.

8. Sequential appearance of trisomy of chromosome 6 followed by trisomy of chromosome 7.

CARCINOMAS

1. Large number of dark cells.

2. All lack glucocorticoid receptors.

3. High level of polyamines and prostaglandins.

4. All lack high molecular weight keratins and filaggrin.

5. All positive for GGT.

6. Approximately 67% of carcinomas induced by two-stage protocol express Ha-ras RNA. Complete carcinogenesis protocol by MNNG does not give increase in expression of Ha-ras RNA but increase expression of src and abl.

7. All are aneuploid with some non-random chromosomal changes such as trisomy in chromosome 6, 7, and 2.

---

ethylnitrosourea (ENU) and benzoyl peroxide and hydrogen peroxide[30,31]. This type of treatment (initiation-promotion-initiation) produces a carcinoma response similar to complete carcinogenesis, i.e., the repetitive application of a carcinogen such as DMBA or MNNG probably supplies both initiating and promoting influences continuously. The reason why a different type of promoter, like benzoyl peroxide, or a non-promoter, like hydrogen peroxide, can increase the conversion of papillomas to carcinomas is presently not known[32] Table 3 shows that the potency of a progressing agent does not necessarily parallel its ability to promote.

We also investigated the possible role of free radicals during progression by utilizing free radical scavengers and antioxidants to inhibit tumor

Table 3. Activity of Promoting Agents throughout Carcinogenesis

Chemical	Complete Carcinogenesis	Initiation	Promotion	Progression
Acetone	-[a]	-	-	-
TPA	-	-	Strong	Weak
Mezerein	-	-	Moderate (2nd Stage)	Weak
Benzoyl Peroxide	-	-	Moderate	Strong
Hydrogen Peroxide	-	-	Weak	Strong
Acetic Acid	-	-	Weak	Strong
Calcium Ionophore A23187	-	-	Moderate (1st Stage)	-
Diethylhexyl-phthalate	-	-	Moderate	Weak

[a]Indicates no activity.

progression. Sodium benzoate, copper diisopropylsalicylate, vitamin E, buty-lated hydroxyanisole, glutathione (GSH), or acetone was applied topically to papillomatous mice. The doses used were similar to those that had been demonstrated to be effective in decreasing papilloma formation during TPA-induced promotion. These findings are further evidence that free radicals may play a role in tumor progression, since topical application of a free radical scavenger like GSH inhibited cancer formation[33,34]. We presently do not know why the other agents did not inhibit tumor progression. Higher doses of these compounds are currently being tested because of the possible penetration problem into the papillomas[33]. These results are summarized in Table 4.

Another approach to test the role of free radicals during progression uses chemicals which deplete or overwhelm free radical defense mechanisms. In this study, papilloma-bearing mice were treated with one of the following: diethylmaleate (DEM), which decreases glutathione levels; acetic acid (AA), which would overwhelm free radical defenses by cytotoxicity; aminotriazole (AT), which would decrease catalase activity; diethylhexylphthalate (DEHP), a peroxisome proliferator which would generate excess H_2O_2 within the cell and overcome free radical defenses; or acetone. The results are summarized in Table 5. The data shows that DEM and AA are effective enhancers of pro-gression whereas DEHP and AT are not effective at the dose given[34]. This experiment suggests that at least one inhibitor of the free radical defense system, DEM, can enhance tumor progression and that by overwhelming these defenses, as in the case with AA, cancer formation can be increased[33,34].

The mechanisms involved in progression in the mouse skin system are unclear. The carcinogens, ENU and MNNG and the peroxides are all genotoxic compounds. Chromosomal studies have shown that squamous cell carcinomas are

Table 4. Inhibitors of Promotion Tested for Activity
During Progression[a]

Chemical	Promotion	Progression
Butylated hydroxyanisole	Moderate	-[b]
Vitamin E	Moderate	-
Copper Diisopropylsalicylate	Strong	-
Sodium Benzoate	Moderate	-
Glutathione	Moderate	Moderate
Disulfiram	Moderate	Moderate

[a]See references 33 and 34 for details concerning
these experiments.
[b]Indicates no activity.

Table 5. Effect of Modifiers of Free Radical Defenses
During Progression[a]

Chemical	Effect	Progression
Diethylmaleate hydroxyanisole	↓Glutathione	Strong
Aminotriazole	↓Catalase	Weak
Diethylhexyl-phthalate	Peroxisome proliferator	Weak
Acetic Acid	Toxicity	Strong

[a]See references 33 and 34 for details concerning
these experiments.

highly aneuploid lesions often exhibiting hyperdiploid stem cell lines[35]. Although early papillomas (10 wks of promotion) are diploid, they progressively show chromosomal changes and eventually become all aneuploid after 30 to 40 weeks of promotion[35]. Using a direct cytogenetic technique specific chromosome alterations were found[36,37]. Aldaz and coworkers[36] identified a non-random trisomy of chromosome 6 in 100% of aneuploid mouse skin papillomas and in 10 of 11 squamous cell carcinomas induced by chemical carcinogenesis. The second most common abnormality observed was trisomy of chromosome 7 found in most dysplastic papillomas and 9 of 11 carcinomas[36]. Trisomy of chromosome 6 occurred before trisomy of chromosome 7[36]. Both trisomies were the only abnormalities found in all aneuploid papillomas and in several carcinomas[36]. More progressed carcinomas also had trisomies of chromosomes 2 and 13. Whether the genotoxic effects of the agents used in progression experiments are able to induce such specific alterations is presently unknown.

In addition to chromosomal alterations, squamous cell carcinomas exhibit a number of changes in protein expression, including the lack of high molecular weight keratins[38], filaggrin[39], and the presence of GGT[23]. Possibly these phenotypic changes are the result of the gene alterations and rearrangements and can be induced by genotoxic agents. Histological and cytochemical studies of keratoacanthomas induced by both ENU and benzoyl peroxide when these agents were used as "progressors" showed a high percentage of GGT-positive tumors possibly reflecting a novel expression of this enzyme in benign lesions[30].

A different mechanism of genetic alteration which could be relevant to progression are changes in the methylation state of DNA. Preliminary evidence from this laboratory has indicated that a gradient in DNA methylation exists from normal mouse epidermis to papillomas to carcinomas, with carcinomas being highly under-methylated (La Peyre et. al., unpublished results). With regard to the agents used as "progressors" in these studies, both ENU and MNNG have been shown to inhibit methylation by blocking DNA methyltransferase activity[40].

An alternate hypothesis for the action of "progressor" agents is related to their high degree of selective cytotoxicity. In this model for progression, highly cytotoxic agents may 1) selectively kill cells within a tumor, allowing the growth of more malignant cells and/or 2) kill normal cells, reducing the constraints against expansion along the border between normal and tumor tissue. Both alternatives assume that cells capable of invasion pre-exist within the benign tumor or that expansion of tumor clones increases the chance of the natural progression of cells toward malignancy. We are currently testing a number of cytotoxic agents for activity during the progression stage.

ACKNOWLEDGEMENTS

The research was supported by Public Health Service grant CA-43278 from the National Cancer Institute and the Sid W. Richardson Foundation.

REFERENCES

1. T. J. Slaga, S. M. Fischer, C. E. Weeks, A. J. P. Klein-Szanto, and J. Reiners, Studies on the mechanisms involved in multistage carcinogenesis in mouse skin, J. Cellular Bioch. 18:99-119 (1982).
2. A. J. P. Klein-Szanto and T. J. Slaga, Numerical variation of dark cells in normal and chemically induced hyperplastic epidermis with age of animal and efficiency of tumor promoter, Cancer Res. 41:4437-4440 (1981).

3. E. Christophers, Cellular architecture of the stratum corneum, J. Invest. Dermatol. 56:165-169 (1981).

4. I. C. Mackenzie, Relationship between mitosis and the ordered structure of the stratum corneum in mouse epidermis, Nature 226:653-655 (1970).

5. F. Marks, Epidermal growth control mechanisms, hyperplasia, and tumor promotion in the skin, Cancer Res. 36:2636-2643 (1976).

6. R. J. Morris, S. M. Fischer, and T. J. Slaga, Evidence that a slowly cycling subpopulation of adult murine epidermal cells retains carcinogen, Cancer Res. 46:3061-3066 (1986).

7. T. J. Slaga and A. J. P. Klein-Szanto, Initiation-promotion versus complete skin carcinogenesis in mice: Importance of dark basal keratinocytes (stem cells), Cancer Investigation 1:425-436 (1983).

8. T. J. Slaga, S. M. Fischer, C. E. Weeks, and A. J. P. Klein-Szanto, Cellular and biochemical mechanism of mouse skin tumor promoters, in: "Reviews in Biochemical Toxicology," E. Hodgson, J. Bend, R. M. Philpot, eds., Elsevier North-Holland, Inc., New York 3:231-281 (1981).

9. T. G. O'Brien, R. C. Simsiman, and R. K. Boutwell, Induction of the polyamine biosynthetic enzymes in mouse epidermis by tumor promoting agents, Cancer Res. 35:1662-1670 (1975).

10. R. A. Mufson, S. M. Fischer, A. K. Verma, G. L. Gleason, T. J. Slaga, and R. K. Boutwell, Effects of 12-0-tetradecanoylphorbol-13-acetate and mezerein on epidermal ornithine decarboxylase activity, isoproterenol-stimulated levels of cyclic adenosine 3',5'-monophosphate, and induction of mouse skin tumors, Cancer Res. 39:4791-4795 (1979).

11. A. N. Raick, Cell proliferation and promoting action in skin carcinoenesis, Cancer Res. 34:920-926 (1974).

12. A. J. P. Klein-Szanto, S. M. Major, and T. J. Slaga, Induction of dark keratinocytes by 12-0-tetradecanoylphorbol-13-acetate and mezerein as an indicator of tumor promoting efficiency, Carcinogenesis 1:399-406 (1980).

13. T. J. Slaga, Cellular and molecular mechanisms of tumour promotion, Cancer Surveys 2:595-612 (1983).

14. J. J. Reiners and T. J. Slaga, Effects of tumor promoters on the rate and commitment to terminal differentiation of subpopulations of murine keratinocytes, Cell 32:247-255 (1983).

15. S. H. Yuspa, H. Hennings, M. Kulsease-Martin, and U. Lichti, The study of tumor promotion in a cell culture model for mouse skin, in: "Cocarcinogenesis and Biological Effects of Tumor Promoters," pp. 217-230. E. Hecker, ed., Raven Press, New York (1982).

16. S. H. Yuspa and D. L. Morgan, Mouse skin cells resistant to terminal differentiation associated with initiation of carcinogenesis, Nature 293:72-74 (1981).

17. P. Cerutti, I. Emerit, and P. Amstad, Membrane-mediated chromosomal damage, in: "Proc. of P and S Biomedical Sciences Symposium," I. B. Weinstein and H. Vogel, eds., Academic Press, New York (in press).

18. A. Varshavsky, Phorbol ester dramatically increases incidence of methotrexate-resistant mouse cells: Possible mechanisms and relevance to tumor promotion, Cell 25:561-572 (1981).

19. J. M. Parry, E. M. Parry, and J. C. Barrett, Tumor promoters induce mitotic aneuploidy in yeast, Nature 294:263-265 (1981).

20. P. B. Fischer and I. B. Weinstein, Chemical viral interactions and multistep aspects of cell transformation, in: "Molecular and Cellular Aspects of Carcinogen Screening Tests," R. Montesano, H. Bartsh, and L. Tomatis, eds., IARC Scientific Publications, Lyon, France (1980).

21. N. H. Colburn, B. F. Former, K. A. Nelson, and S. H. Yuspa, Tumor promoter induces anchorage independence irreversibly, Nature 281:589-591 (1979).

22. A. R. Kinsella and M. Radman, Tumor promoter induces sister chromatid exchanges: Relevance to mechanisms of carcinogenesis, Proc. Natl. Acad. Sci. USA 75:6149-6153 (1978).

23. A. J. P. Klein-Szanto, R. G. Nelson, Y. Shah, and T. J. Slaga, Keratin modifications and GGT activity as indication of tumor progression in skin papillomas, J. Natl. Cancer Inst. 70:161-168 (1983).

24. K. G. Nelson, K. B. Stephenson, and T. J. Slaga, Protein modification induced in mouse epidermis by potent and weak tumor-promoting hyperplasiogenic agents, Cancer Res. 42:4164-4174 (1982).

25. A. Balmain, M. Ramsden, G. T. Bowden, and J. Smith, Activation of the mouse cellular Harvey-ras gene in chemically induced benign skin papillomas, Nature 307:658-660 (1984).

26. A. Balmain, and I. D. Pragnell, Mouse skin carcinomas induced in vivo by chemical carcinogens have a transforming Harvey-ras oncogene, Nature 303:72-74 (1983).

27. J. C. Pelling, D. C. Hixson, R. S. Nairn, and T. J. Slaga, Altered gene expression during two-stage tumorigenesis in SENCAR mouse skin (abstract), Proc. Amer. Assoc. Cancer Res. 25:78 (1984).

28. G. J. Patskan, J. C. Pelling, R. S. Nairn, and T. J. Slaga, Altered oncogene expression in mouse skin squamous cell carcinomas, presented at the Pennsylvania State University Fourth Summer Symposium in Molecular Biology (1985).

29. H. Hennings, R. Shores, M. L. Wenk, E. F. Spangler, R. Tarone, and S. H. Yuspa, Malignant conversion of mouse skin tumours is increased by tumour initiators and unaffected by tumour promoters, Nature 304:67-69 (1983).

30. J. F. O'Connell, A. J. P. Klein-Szanto, D. M. DiGiovanni, J. W. Fries, and T. J. Slaga, Malignant progression of mouse skin papillomas treatd with ethylnitrosourea, N-methyl-N'-nitrosoguanidine or 12-0-tetradecanoylphorbol-13-acetate, Cancer Letters 30:269 (1986).

31. J. F. O'Connell, A. J. P. Klein-Szanto, J. M. DiGiovanni, J. W. Fries, and T. J. Slaga, Enhanced malignant progression of mouse skin tumors by the free-radical generator benzoyl peroxide, Cancer Res. 46:2863-2865 (1986).

32. J. B. Rotstein, J. F. O'Connell, and T. J. Slaga, The enhanced progression of papillomas to carcinomas by peroxides in the 2-stage mouse skin model, Proc. Am. Assoc. Cancer Res. 27:143 (1986).

33. J. B. Rotstein and T. J. Slaga, Effect of exogenous glutathione on tumor progression in the murine skin multistage carcinogenesis model, Carcinogenesis 9:1547-1551 (1988).

34. J. B. Rotstein, J. F. O'Connell, and T. J. Slaga, A possible role for free radicals in tumor progression, in: "Anticarcinogenesis and Radiation Protection," pp. 211-219. P.A. Cerrutti, O. F. Nygaard, and M. G. Simic, Plenum Publishing Corp, New York (1987).

35. C. M. Aldaz, C. J. Conti, A. J. P. Klein-Szanto, and T. J. Slaga, Progressive dysplasia and aneuploidy are hallmarks of mouse skin papillomas: Relevance to malignancy, Proc. Natl. Acad. Sci. USA 84:2029-2032 (1987).

36. C. M. Aldaz, D. Trono, F. Larcher, T. J. Slaga, and C. Conti, Sequential trisomization of chromosomes 6 and 7 in mouse skin premalignant lesions, Molecular Carcinogenesis (in press) (1989).

37. C. J. Conti, C. M. Aldaz, J. O'Connell, A. J. P. Klein-Szanto, and T. J. Slaga, Aneuploidy, an early event in mouse skin tumor development, Carcinogenesis 7:1845-1848 (1986).

38. A. J. P. Klein-Szanto, Morphological evaluation of tumor promoter effects on mammalian skin, in: "Mechanisms of Tumor Promotion," Vol. II, T. J. Slaga, ed., CRC Press, Boca Raton (1984).

39. M. D. Mamrack, A. J. P. Klein-Szanto, J. J. Reiners, Jr., and T. J. Slaga, Alteration in the distribution of the epidermal protein filaggrin during two stage chemical carcinogenesis in the SENCAR mouse skin, Cancer Res. 44:2634-2641 (1984).
40. V. L. Wilson, and P. A. Jones, Inhibition of DNA methylation by chemical carcinogens in vitro, Cell 32:239-246 (1983).

DISCUSSION

Moderator (Oscar Sudilovsky): There is time for only a few questions for Dr. Slaga.

Hazan Mukhtar: Do you have any data on DSTP activity in this initiated tumor?

Slaga: No.

Farber: You talk about stem cells being initiated, but what evidence do you have that they are stem cells?

Slaga: We have published a couple of papers showing that there are clonogenic cells in the epidermis.

Farber: (Interrupting) Well, that doesn't make them stem cells.

Slaga: Well, you wanted a definition ...

Farber: (Interrupting) You take normal epithelium and it grows nicely.

Slaga: Not all the cells in the epidermis are clonogenic.

Farber: (Interrupting) You know that the current idea is all cancers come from stem cells. But the data isn't very convincing, at least not to me, anyway. Maybe it is to everybody else, so I shouldn't say it.

Slaga: I'm not trying to relate it to all systems...

Farber: (Interrupting) I think we have to be very careful and humble because the use of "stem cell" has enormous implications which I don't think are justified at this stage of the game.

Slaga: I think they're more justified in the skin than in liver, for example. I don't think I would even come up with a stem cell relation in the liver, but in the skin there is enough data to lead one to think...

Farber: (Interrupting) I believe that normal skin has stem cells, but how do you know that the cancer or the papilloma comes from them?

Sudilovsky: Can I interrupt now? Since the question doesn't have a simple answer, I will use my prerogative as moderator and ask Dr. Hennings to begin his presentation.

MALIGNANT CONVERSION, THE FIRST STAGE IN PROGRESSION,

IS DISTINCT FROM PHORBOL ESTER PROMOTION IN MOUSE SKIN

Henry Hennings

Laboratory of Cellular Carcinogenesis
and Tumor Promotion
National Cancer Institute
National Institutes of Health
Bethesda, MD

Multiple stages in the induction of benign and malignant tumors have
been demonstrated experimentally in the mouse skin model system of chemical
carcinogenesis[1,2]. Although skin tumors can be induced by repeated topical
applications of a carcinogen[3], protocols have been developed which define
at least three distinct stages: initiation, promotion and malignant conversion[4-6]. In a typical experiment, the first stage, initiation, is accomplished by a single exposure to a low dose of a mutagenic carcinogen. Initiation, which may represent a single mutational event[7], causes a heritable
change in some epidermal cells, which are termed "initiated." Without subsequent treatment, the initiated cells do not develop into tumors. Repeated
topical treatment of initiated mice with a tumor promoter allows the expression of the neoplastic change resulting in the formation of benign squamous
papillomas. The second stage, promotion, is effective even when promoter
treatments are delayed for several months after initiation, indicating the
irreversibility of the initiating mutation. In contrast, the promoting effects of individual TPA applications are reversible since papillomas do not
develop after insufficient exposure of initiated skin to promoters or when
the interval between individual promoter applications is increased. The
reversibility of promotion suggests an epigenetic mechanism. Promotion can
be defined as the selective clonal expansion of initiated cells.

Although large numbers of papillomas can be induced in sensitive mice by
initiation with 7,12-dimethylbenz[a]anthracene (DMBA) and promotion with
12-0-tetradecanoylphorbol-13-acetate (TPA), few of these benign tumors progress to malignancy. The rate of malignant conversion can be significantly
increased by the repeated treatment of papilloma-bearing mice with genotoxic
agents such as 4-nitroquinoline-N-oxide (4-NQO) or urethane, but is not affected by continued treatment with TPA[5]. The agents active in this third
stage of epidermal carcinogenesis (malignant conversion) are likely to act
via a genetic mechanism. Further stages in progression, characterized by
increased independent growth and metastasis to the lungs and lymph nodes,
have also been demonstrated in the mouse skin model[6]. This manuscript reviews the experimental evidence supporting a malignant conversion stage, and
describes the heterogeneity among individual papillomas which complicates the
study of the conversion of papillomas to malignancy.

Boundaries between Promotion and Progression during Carcinogenesis
Edited by O. Sudilovsky *et al.*, Plenum Press, New York, 1991

MATERIALS AND METHODS

Chemicals were purchased from the following sources: DMBA from Eastman (Rochester, NY), TPA from LC Services (Woburn, MA), 4-NQO from Sigma (St. Louis, MO), and urethane from MCB (Cincinnati, OH). SENCAR female mice, 4-5 weeks old, were obtained from the NCI-DCT Animal Program (Frederick, MD); CD-1 female mice were purchased from Charles River Laboratories, Kingston, NY. Seven to eight week old mice in the resting phase of the hair growth cycle were initiated with a single topical application of DMBA. Once-weekly promotion by topical application of TPA was begun one week later. The duration of promotion varied among experiments; weekly treatments with the converting agents 4-NQO or urethane were begun one week after the last TPA treatment. Papilloma and carcinoma counts were recorded bi-weekly and mice were weighed once per month. Noninvasive, raised lesions were classified as papillomas when their diameter exceeded 1 mm and were present for at least two weeks. Suspected carcinomas were verified by pathological evaluation of tumor histology. Carcinomas which result from initiation-promotion protocols generally progress from papillomas[5]. The percent conversion of papillomas to carcinomas was calculated using the following formula: Percent Conversion = (Total Carcinomas/Total Papillomas) x 100. Since all papillomas were not evaluated histologically, the number of carcinomas (as well as the percent conversion) is a minimal estimate.

HETEROGENEITY OF PAPILLOMAS

Most of the benign lesions which develop as the result of DMBA initiation-TPA promotion protocols are papillomas. Initiation of papillomas is characterized by a one-hit dose-response pattern[8], and the papillomas formed in initiation-promotion experiments are monoclonal[9]. Thus, papillomas apparently result from the clonal expansion of individual initiated cells. At least two types of initiated cells, differing in their potential for conversion to malignancy, have been hypothesized[10] as indicated in Figure 1. The majority of papillomas (the benign lesions which develop from promotion of Type A initiated cells) do not convert to malignant carcinomas, and may not have the potential for conversion to either a persistent benign lesion or to malignancy. In an experiment in CD-1 mice[5,6], two groups of animals were initiated with DMBA and promoted with TPA for 12 weeks, resulting in 14-15 papillomas per mouse. Beginning at week 13, one group continued to receive weekly applications of TPA; the second was treated with acetone

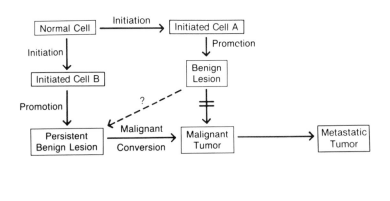

Fig. 1. Cell lineage in tumor development and progression.

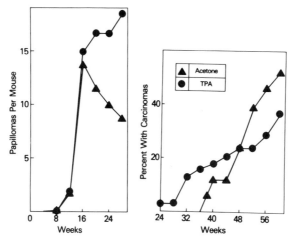

Fig. 2. Papilloma and carcinoma incidence in
papilloma-bearing CD-1 mice treated
with TPA or acetone solvent. Two
groups of 40 mice were initiated with
50 μg DMBA and promoted with 12 weekly
applications of 10 μg TPA. Beginning
at week 13, one group was treated once
per week with 0.2 ml acetone (▲) and
the other was treated once per week
with 10 μg TPA/0.2 ml acetone (●).

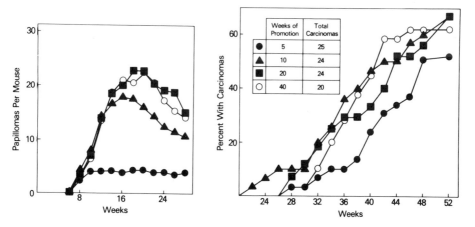

Fig. 3. Effect of varying the duration of TPA promotion on papilloma and
carcinoma development. Groups of 30 SENCAR mice were initiated by
20 μg DMBA and promoted once weekly with 2 μg TPA for either
5 (●), 10 (▲), 20 (■) or 40 (○) weeks.

33

solvent. As shown in Figure 2, the papilloma yield was maintained in TPA-treated mice, but nearly half of the papillomas regressed in the acetone-treated mice. All of the carcinomas apparently progress from the TPA-independent papillomas (developing from Type B initiated cells) since the final carcinoma yield was similar in the two groups (Figure 2). Obviously, the carcinomas develop from the population of papillomas which persist after termination of TPA promotion. Papilloma heterogeneity appears to result from changes induced by initiation rather than by promotion since the potential for progression to malignancy depends on the dose of DMBA used for initiation[11] or on the initiator used[12].

In SENCAR mice, all of the papillomas which spontaneously convert to malignancy (developing from promotion of Type B initiated cells) are promoted by the first few applications of the promoter TPA[5]. Promotion for 5 weeks resulted in only about 30% of the maximum number of papillomas inducible by promotion for 10-40 weeks, but the number of carcinomas was identical regardless of the duration of promotion (Figure 3). The non-progressing papillomas (Type A) are those which appear later as a result of continued TPA promotion. Further support for the existence of subclasses of papillomas comes from the study of weak promoters such as mezerein[5] and chrysarobin[13], which are much less effective than TPA as promoters of papillomas, but which produce as many carcinomas as TPA. Mezerein-promoted papillomas, like those promoted by short-term TPA (Figure 3), do not regress when promoter treatments are stopped (not shown). Thus, with several promotion protocols, persistence characterizes that subpopulation of papillomas with a high spontaneous rate of conversion to malignancy.

INDUCTION OF MALIGNANT CONVERSION

The rate of spontaneous conversion of papillomas to carcinomas can be increased by treating papilloma-bearing mice with 4-NQO topically or urethane intraperitoneally[4,6]. After DMBA initiation of CD-1 mice and 12 weeks of TPA promotion, weekly treatments with malignant converting agents for 10-40 weeks were necessary to increase the rate of conversion to malignancy (Figure 4). The mechanisms involved in the malignant conversion stage must differ from those involved with TPA promotion since continued TPA treatment did not increase carcinoma formation (Figure 4). However, another promoter, benzoyl peroxide, is active as a converting agent[14]. This result emphasizes the importance of comparing the mechanisms of action of these promoters, as well as chrysarobin[13], teleocidin[15], mezerein[5,16] and others, since the ratio of carcinomas to papillomas is much lower with TPA than it is for the other promoters. A promoter effective in the production of carcinomas could act by either the selective induction of those papillomas likely to progress to carcinomas or by its additional activity in the malignant conversion stage.

In experiments in which individual tumors have been followed, all carcinomas appeared to arise in tumors with the gross appearance of papillomas[5]. Foci of carcinoma in situ have also been observed in initiation-promotion experiments[17,18]. In a group of 40 DMBA-initiated CD-1 mice in which acetone solvent was substituted for TPA treatment after initiation, only 3 papillomas and no carcinomas developed in response to 40 weeks of urethane treatment[6]. In a similar group of mice promoted with TPA for 12 weeks prior to urethane treatment, over 500 papillomas and 28 carcinomas developed[6]. Thus, a TPA-promoted papilloma stage appears necessary for carcinoma development as a result of this treatment protocol. The TPA-induced clonal expansion of the population of initiated cells (with perhaps one critical mutation) to form a papilloma may be required to provide a large target population for a second genetic change induced by the converting agent.

Table 1. Lack of Effect of Fluocinolone Acetonide on Malignant Conversion[a]

Stage of FA Treatment	Total Papillomas	Total Carcinomas	Percent Conversion
None	480	24	5.0
Promotion	64	4	6.3
Malignant Conversion	392	23	5.9

[a]Groups of 40 Charles River CD-1 mice were initiated with 50 µg DMBA and promoted with 10 µg TPA once per week for 12 weeks. Malignant conversion was accomplished by weekly i.p. injections of 20 mg urethane from week 13-40. FA (1 µg/0.2 ml acetone) was applied topically 30 minutes before each weekly TPA treatment or 30 minutes before each urethane injection. The cumulative papilloma and carcinoma incidences are shown at the time the experiment ended at week 52.

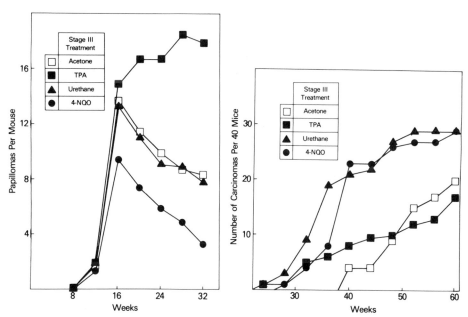

Fig. 4. Papilloma and carcinoma incidence in an initiation-promotion-conversion experiment. Groups of 40 CD-1 mice were initiated with 50 µg DMBA (stage I) and promoted with 12 weekly applications of 10 µg TPA (stage II). Stage III treatments once weekly from weeks 13 to 52 were as follows: 0.2 ml acetone (□), 10 µg TPA (■), 20 mg urethane i.p. (▲), 250 µg 4-NQO (●).

To further establish that the properties of malignant conversion differ from those of promotion by TPA, fluocinolone acetonide (FA), an inhibitor of promotion, was tested for its effect on malignant conversion[2]. After initiation by DMBA and promotion by TPA, treatment with 1 ug FA 30 minutes before each injection of urethane did not affect either the number of carcinomas or the percent conversion (Table 1). However, when FA was given 30 minutes before each promoting treatment of TPA, both papilloma and carcinoma formation were inhibited by more than 80%. The rate of malignant conversion was unaffected by FA application during the promotion stage, suggesting an equal inhibition of formation of papillomas with a high or low probability of conversion to carcinomas.

As more agents are tested for activity in the malignant conversion stage of carcinogenesis, the mechanism of action of converting agents can be better defined. In addition to 4-NQO, urethane and benzoyl peroxide, hydrogen peroxide[19] and ionizing radiation[20] are active as converting agents. Mechanistic studies will be facilitated if conversion can be accomplished by fewer treatments with converting agents. We have recently found that a single injection of the chemotherapeutic agent cisplatin, a tumor initiator for mouse skin[21], increases the rate of malignant conversion by about 2-fold (Hennings, H. and Yuspa, S. H., unpublished results). In addition, work is underway on a cell culture model to study the possible cooperation between initiation by an activated ras[Ha] oncogene and malignant conversion by chemical treatment (Yuspa, S. H., unpublished observations).

METASTASIS IN THE MOUSE SKIN MODEL

In the experiment shown in Figure 4, possible metastases were noted in the lymph nodes and lungs of several animals bearing squamous cell carcinomas. Histological examination of these lesions indicated metastases in about 20% of the control animals initiated with DMBA, promoted with TPA for 12 weeks, then treated for 40 weeks with either acetone or TPA (Table 2). When carcinomas were induced by treating papilloma-bearing mice for 40 weeks with 4-NQO, the frequency of metastases was not altered. In contrast, the carcinomas resulting from a DMBA-TPA-urethane protocol metastasized to the lungs more frequently than in the mice in the other three groups (Table 2). Six of the 24 mice in this group bore lung metastases, compared to only 3 out of 51 mice in the other groups shown in Table 2. These urethane-induced lung metastases were not primary urethane-induced tumors since they were observed only in mice bearing skin carcinomas; identical urethane injections in groups of mice which were either not initiated or initiated but not promoted did not develop squamous cell carcinomas in the lungs. However, lung adenomas were found in about half of the urethane-treated mice. The possible effect of these benign tumors in facilitating metastases in the lung is unknown. This result emphasizes that the experimental protocol used can affect the metastatic potential of the carcinomas that develop.

SUMMARY AND CONCLUSIONS

Papillomas induced by DMBA initiation-TPA promotion protocols are necessary precursor lesions of squamous cell carcinomas. The papillomas are heterogeneous in their potential for progression to carcinomas, a property apparently induced at the time of initiation. The probability of conversion to malignancy is highest for the papillomas most easily promoted, by either the first few TPA treatments or by "weak" promoters such as mezerein or chrysarobin. The conversion frequency is lowest for TPA-dependent papillomas and those papillomas which appear late in a TPA promotion protocol. The spontaneous rate of malignant conversion is not altered by continued TPA treatment; TPA promotion may simply expand clones of initiated cells which are already

Table 2. Metastases from Epidermal Squamous Cell Carcinomas[a]

Treatment	Number of Mice with Squamous Cell Carcinomas	Number of Carcinoma-bearing Mice with Metastases	Percent Metastases	Site of Metastases	
				Lymph Node	Lung
TPA	13	3	23	3	0
Acetone	18	3	17	3	2
4-NQO	20	3	15	3	1
Urethane	24	8	33	3	6

[a]Groups of 40 female CD-1 mice were initiated with 50 μg DMBA and promoted with 12 weekly applications of 10 μg TPA. Once weekly treatments from week 13 to 52 were as follows: TPA, 10 μg; acetone, 0.2 ml; 4-NQO, 250 μg; urethane, 20 mg i.p. Mice were sacrificed at week 60. In some mice, metastases developed in both lymph nodes and lung.

programmed with a given probability of conversion. Treatment of papilloma-bearing mice with a genotoxic agent, such as 4-NQO, urethane or cisplatin, increases the rate of malignant conversion. The properties of active converting agents differ markedly from those of the phorbol ester promoting agents, suggesting differences in the mechanisms of action of these two classes of compounds. A genetic mechanism appears likely to explain conversion. The differences between the two stages are further emphasized by the finding that inhibitors of tumor promotion are not inhibitory when given during malignant conversion. The converting agent urethane also affects the subsequent discrete stage in tumor progression, tumor metastasis. Differences in metastatic potential have been found between carcinomas which progress spontaneously after TPA promotion and carcinomas induced in TPA-promoted papillomas by urethane. The multistage nature of experimental epidermal carcinogenesis is well established. The mouse skin model will continue to be valuable for mechanistic studies since similar stages have been described in other tissues[22] as well as in man[23].

REFERENCES

1. T. J. Slaga, Mechanisms involved in two-stage carcinogenesis in mouse skin, in: "Mechanisms of Tumor Promotion, Vol. II, Tumor Promotion and Skin Carcinogenesis," T. J. Slaga, ed., CRC Press, Inc., Boca Raton (1984), pp. 1-16.

2. H. Hennings, Tumor promotion and progression in mouse skin, in: "Mechanisms of Environmental Carcinogenesis," Vol. II. Multistep Models of Carcinogenesis, J. C. Barrett, ed., CRC Press, Inc., Boca Raton (1987), pp. 59-71.

3. P. Shubik, The growth potentialities of induced skin tumors in mice. The effects of different methods of chemical carcinogenesis, Cancer Res. 10:713-717 (1950).

4. H. Hennings, R. Shores, M. L. Wenk, E. F. Spangler, R. Tarone, and S. H. Yuspa, Malignant conversion of mouse skin tumors is increased by tumour initiators and unaffected by tumour promoters, Nature 304:67-69 (1983).

5. H. Hennings, R. Shores, P. Mitchell, E. F. Spangler, and S. H. Yuspa, Induction of papillomas with a high probability of conversion to malignancy, Carcinogenesis 6:1607-1610 (1985).

6. H. Hennings, E. F. Spangler, R. Shores, P. Mitchell, D. Devor, A. K. M., Shamsuddin, K. M. Elgjo, and S. H. Yuspa, Malignant conversion and metastasis of mouse skin tumors: A comparison of SENCAR and CD-1 mice, Env. Health Perspec. 68:69-74 (1986).

7. M. Quintanilla, K. Brown, M. Ramsden, and A. Balmain, Carcinogen-specific mutation and amplification of Ha-ras during mouse skin carcinogenesis, Nature 322:78-80 (1986).

8. F. Burns, R. Albert, B. Altshuler, and E. Morris, Approach to risk assessment for genotoxic carcinogens based on data from the mouse skin initiation-promotion model, Environ. Health Perspect. 50:309-320 (1983).

9. A. L. Reddy, and P. J. Fialkow, Papillomas induced by initiation-promotion differ from those induced by carcinogen alone, Nature 304:69-71 (1983).

10. J. D. Scribner, N. K. Scribner, B. McKnight, N. K. Mottet, Evidence for a new model of tumor progression from carcinogenesis and tumor promotion studies with 7-bromomethylbenz[a]anthracene, Cancer Res. 43:2034-2041 (1983).

11. A. L. Reddy, P. J. Fialkow, Influence of dose of initiator on two-stage skin carcinogenesis in BALB/c mice with cellular mosaicism, Carcinogenesis 9:751-754 (1988).

12. H. Hennings, D. Devor, M. L. Wenk, T. J. Slaga, B. Former, N. H. Colburn, G. T. Bowden, K. Elgjo, and S. H. Yuspa, Comparison of two-stage epidermal carcinogenesis initiated by 7,12-dimethylbenz[a]anthracene or N-methyl-N'-nitro-N-nitrosoguanidine in newborn and adult SENCAR and BALB/c mice, Cancer Res. 41:773-779 (1981).

13. F. H. Kruszewski, C. J. Conti, and J. DiGiovanni, Characterization of skin tumor promotion and progression in SENCAR mice, Cancer Res. 47:3783-3790 (1987).

14. J. F. O'Connell, A. J. P. Klein-Szanto, D. M. DiGiovanni, J. W. Fries, and T. J. Slaga, Enhanced malignant progression of mouse skin tumors by the free-radical generator benzoyl peroxide, Cancer Res. 46:2863-2865 (1986).

15. H. Fujiki, M. Suganuma, N. Matsukura, T. Sugimura, and S. Takayama, Teleocidin from Streptomyces is a potent promoter of mouse skin carcinogenesis, Carcinogenesis 3:895-898 (1982).

16. H. Hennings, and S. H. Yuspa, Two-stage tumor promotion in mouse skin: an alternative explanation, J. Natl. Cancer Inst. 74:735-740 (1985).

17. G. L. Knutsen, R. M. Kovatch, and M. Robinson, Gross and microscopic lesions in the female SENCAR mouse skin and lung in tumor initiation and promotion studies, Env. Health Perspec. 68:91-104 (1986).

18. J. M. Ward, S. Rehm, D. Devor, H. Hennings, and M. L. Wenk, Differential carcinogenic effects of intraperitoneal initiation with 7,12-dimethyl-benz[a]anthracene or urethane and topical promotion with 12-0-tetra-decanoylphorbol-13-acetate in skin and internal tissues of female SENCAR and BALB/c mice, Env. Health Perspec. 68:61-68 (1986).

19. J. Rotstein, J. O'Connell, and T. Slaga, The enhanced progression of papillomas to carcinomas by peroxides in the 2-stage mouse skin model, Proc. Amer. Assn. Cancer Res. 27:143 (1986).

20. D. R. Jaffe, J. F. Williamson, and G. T. Bowden, Ionizing radiation enhances malignant progression of mouse skin tumors, Carcinogenesis 8:1753-1755 (1987).

21. K. M. Barnhart, and G. T. Bowden, Cisplatin as an initiating agent in two-stage mouse skin carcinogenesis, Cancer Letters 29:101-105 (1985).

22. H. C. Pitot, and H. A. Campbell, Quantitative studies on multistage hepatocarcinogenesis in the rat, in: "Tumor Promoters: Biological Approaches for Mechanistic Studies and Assay Systems," R. Langenbach, E. Elmore, and J. C. Barrett, eds., Raven Press, New York (1988), pp. 79-95.

23. S. H. Moolgavkar, and A. G. Knudson, Jr., Mutation and cancer: a model for human carcinogenesis, J. Natl. Cancer Inst. 66:1037-1052 (1981).

DISCUSSION

Harry Rubin: What you showed is that you have a lot of regressions among your papillomas. My first question is how do you square this with the view of a classical mutation, which you wouldn't think would result in regression. Do you just consider the term mutation to be a very general event that could include a differentiation type of change? This is partly also a question for Dr. Manny Farber, because as you know he shows that many of the hepatocyte nodules that arise, most of them ultimately regress. He characterized that as a physiological adaptation rather than a mutation.

My second question has to do with the differences among your papillomas. I know from work we've done on spontaneous transformation in culture, that every transformant is different from every other one. If you want to talk about "mechanisms", then the mechanism is actually different in every one of these individual cases.

Hennings: Yes, I think the mechanisms could be quite different. Considering that, I think it was surprising that Alan Balmain (6) found an activated

Harvey ras in such a high percentage of papillomas and carcinomas. Apparently the heterogeneity is not directly related to ras activation.

Henry C. Pitot: It doesn't demonstrate causality.

Emmanuel Farber: The mutagen induces mutations. Okay, that's nice, but it's not a very revolutionary discovery.

Hennings: I think most people would agree that initiation can be explained most simply by a mutation, and I'm not sure that a mutation has to give you a persistent tumor. Why couldn't it give you a regressable neoplasm? You just have papillomas with different properties or you have foci with different properties. I think they could still be caused by a mutation.

Rubin: The mutation then could back mutation, in all the cells where it has initially mutated, or why else should it revert.

Hennings: I think Dr. Pitot showed that the tumor could regress to a size below which you're not going to count it as a tumor; but then if you promote again, that same tumor may very well come back.

Rubin: I don't know what the answer is. I don't have an answer, but I think one should consider the kind of interpretation that Manny Farber did make for hepatocyte nodules, that it is not a classical kind of mutation. In fact, it is like a physiological adaptation, whether you want to call it induction as a counterweight to mutation, but at least that kind of possibility ought to be considered, and it isn't often considered.

Pitot: Henry, in many of your slides you were expressing papillomas per mouse; but then when you expressed carcinomas I thought it was percent animals with carcinomas. What happens if you express it in carcinomas per mouse? What is the relationship?

Hennings: You don't get very many carcinomas per mouse because the first carcinoma usually kills the mouse. In our experience, we've probably had as many as three or four on a mouse, but as a rule you don't get more than one, and you don't get them in more than about half the animals. So as far as carcinomas per mouse, it would be something less than one.

Unidentified speaker: What about malignant conversion?

Hennings: When we express malignant conversion, we took the total number of carcinomas in the mice in that group and divided it by the total number of papillomas that had developed in that group.

Dan Branstetter: In the different types of papillomas that you get, is it due to the nature of the mutation or initiating event that occurs in a cell that distinguishes the two? Is it a difference in the type of cell that is initiated, or is it a stem cell that distinguishes the two? Is there an experiment that you could think of that would distinguish those two possibilities?

Hennings: I can't really think of a way of getting at that. That is certainly a possibility. I don't really mean to imply that there are only two types of initiated cells. I mean, there is probably a whole spectrum of different types of initiated cells.

Farber: Not in liver. In liver, you take six months of nothing before you start to get cancer from nodules if you use a synchronized system. I presume that your system is pretty synchronous, that all papillomas come up pretty much together. Do they come up...?

Hennings: They do come up within an eight or ten week period.

Farber: No, that's not synchronous. I mean do they come up within an eight or ten _day_ period?

Hennings: No, they are not synchronous.

Farber: So you have quite an asynchronous development of the papillomas. They could all be at different phases of certain kinds of development, differentiation, and so on.

Hennings: Here I'm really just showing what we do with our experiments. There could be several steps in each of these stages.

George Yoakum: First of all, Henry, I'd like to congratulate you on what I thought was a clear talk. If you look at the metabolism of the drugs you're using, TPA, Ionomycin and Mezerin, can you look at the metabolic effects that you would expect from those applications and discern any kind of mechanistic rationale for your observations? Do you foresee or anticipate any kind of an interaction with the local immunity in the animal?

Hennings: Are you asking what's the rationale for the Ionomycin?

Yoakum: If you have three different drugs, you have three different metabolic sets of events that can go on. If you compare pathways and outcomes, what sorts of mechanistic rationale could you draw or sketch up from what you see?

Hennings: Whatever the initiating change is, the cells no longer terminally differentiate in response to the calcium signal or in response to TPA. Ionomycin, which brings in a lot of calcium into the cell, apparently allows the cell to respond to TPA. The differentiative response to TPA, we think, requires perhaps something to do with protein kinase C and something to do with a certain level of calcium in the cell. That's the rationale for using Ionomycin along with TPA. As far as your second question, I don't have any information.

G. Bowden: Henry, I found very interesting your studies on Ionomycin's effects on the 308 cells. They do, in fact, have an activated _ras_ gene. As you are aware, Yuspa has shown that introduction of an activated _ras_ gene in the primary epidermal cells blocked the response to calcium in terms of terminal differentiation. Have you looked at other papilloma producing cell lines that don't have an activated _ras_ gene? Do they still also respond to Ionomycin in terms of terminal differentiation?

Hennings: We've looked at three different lines, but the lines we've looked at all have an activated _ras_. We do, however, have lines available to do the experiment you suggest.

Farber: Henry, am I correct in interpreting that you think that whatever the final step is, the first step is malignant? That you get cancer right away? Is that what you're saying?

Hennings: You certainly can see something like carcinoma in situ in papillomas, very similar to the foci within foci which Dr. Pitot was showing.   Further stages in progression, including metastasis, come later.

BOUNDARIES IN MAMMARY CARCINOGENESIS

Jose Russo and Irma H. Russo

Department of Pathology
Michigan Cancer Foundation
Detroit, MI 48201

INTRODUCTION

Although there are many questions to be answered in order to understand the biology of breast cancer[1], the four basic ones that will be the subject of this work are: what causes breast cancer? how does breast cancer start? what determines the susceptibility of the gland to undergo carcinogenesis? what makes an initiated cell progress to a fully malignant one?

We do not know what the etiological factors are, but we do know that several risk factors are related to breast cancer[1-4]. Breast cancer is diagnosed more frequently in women over 40 years of age[1,4] even though a trend to increased incidence in younger patients has been observed[5]. Women with family history of breast cancer have a 9:1 greater chance to develop the disease[4,6]. The presence of premalignant lesions is also associated with higher risk[1-4] and lastly, it has been shown that reproductive history has a significant weight in the risk to develop breast cancer[1,3,6-8].

To understand how these factors affect the natural history of the disease and how they can be manipulated, an adequate experimental system is needed[1]. The one that we have used in our study is that developed by Huggins[9], and consists in the inoculation of the polycyclic hydrocarbon 7,12-dimethylbenz(a)anthracene (DMBA) to virgin Sprague-Dawley rats, which is a strain highly susceptible strain to induced breast carcinomas. In this model, carcinogen dose and route of administration, susceptibility of the strain of animal used, endocrine manipulation and diet are all factors that play an important role in carcinogenesis[9-12].

PATHOGENESIS OF MAMMARY CARCINOMA

Mammary cancer results from the interaction of two basic components, the carcinogen and the target organ[1]. The carcinogen in our experimental system is known, and the target organ is obviously the mammary gland. The mammary gland, however, does not respond as a unit to the carcinogenic insult. In this complex branching system we have been able to identify the terminal end bud (TEB) as the site of origin of mammary carcinomas[13-15]. The TEB is the primitive element of the gland; at the time of puberal growth in the rat (at about 25-35 days of age) it starts to bifurcate into alveolar buds (ABs); these, with successive estrous cycles progress to the virginal

Boundaries between Promotion and Progression during Carcinogenesis
Edited by O. Sudilovsky *et al.*, Plenum Press, New York, 1991

43

lobules (14-16). If the carcinogen is given at the time of active differentiation of TEBs to ABs, the first lesions are observed at the level of one
or various adjacent TEBs. Within 14 days of carcinogen administration there
is enlargment of TEBs and a darker appearance in the whole mount prepara
tion[14-16]. These lesions are called intraductal proliferations or IDPs[14,16]
(Fig. 1). IDPs are easily distinguished from the gland's normal ductal
structures by the presence of a multilayered epithelium[14,16,17]. Lesions
arising in adjacent TEBs tend to coalesce forming microtumors, which become
evident after 20 days of DMBA administration[16]. These are in general intraductal carcinomas, composed of epithelial cells with nuclear pleomorphism,
increased nuclear size and moderate mitotic activity[16]. The IDPs that
evolve to intraductal carcinomas are also characterized by stromal reaction
with desmoplasia and mast cell and lymphocytic infiltration[17,18]. From
these structures the fully developed or palpable tumors grow into cribiform,
comedo or papillary patterns, alone or in combination (Fig. 1). As the
tumors progress, they invade the stroma, forming the typical infiltrating
ductal carcinoma[17]. If the animal is allowed to live long enough metastases occur, mainly to the lungs[17]. In Figure 1 is depicted the pathogenesis
of chemically induced rat mammary cancer, indicating that the TEB is the site
of origin of malignancies, whereas benign lesions such as cysts, adenomas,
alveolar hyperplasias and fibroadenomas are originated from the ABs. These
observations indicate that there are two different pathogenetic pathways:
one for the malignant and another for the benign lesions. In addition,
benign lesions tend to appear later than the malignant ones, indicating that
the former are not precursors of the latter[1,11,12].

Once the target structure of the carcinogen was identified, it remains
to be elucidated which is the stem or the target cell of the carcinogen; what
are the cell kinetics in the different compartments of the mammary gland, and
whether the carcinogen acts by increasing cell proliferation or by altering
the balance between different cell types. The rat mammary gland parenchyma
is composed of three cell types, myoepithelial, dark and intermediate
cells[1,11] (Fig. 2). In the TEBs, ducts and ABs of the resting mammary
gland the most abundant is the dark cell type that comprises 77% of them,

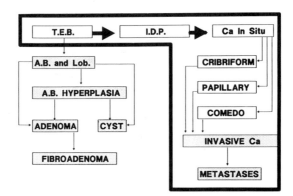

Fig. 1. Chart representing two different
        pathogenetic pathways for benign
        and malignant lesions. Malignant
        lesions originate from TEB and
        appear earlier than benign lesions
        originated from AB. The earliest
        lesion detected after DMBA adminis-
        tration is the IDP.

followed by the intermediate and myoepithelial cells that represent 12 and 11% respectively[1,11]. The distribution of cell populations in the mammary gland during carcinogenesis varies in TEBs, but not in ABs or lobules. There is a progressive shift due to a diminution in number of dark cells and an increase in number of intermediate cells. Myoepithelial cells are not affected. When the tumors become palpable, around 40 days after DMBA administration, intermediate cells comprise 65% of the total cell population; 100 day-old tumors are dominated by intermediate cells, which comprise nearly 90% of the total[1,11]. We have shown[1,11] that even though in the resting gland both TEBs and ABs contain the same proportion of the three cell types, they have different proliferative rates. Depending upon their location in either TEBs or in ABs, the length of the cell cycle (Tc) in intermediate cells is 13 hours when they are located in the TEB, but it lengthens to 34 hours when this cell type is located in AB[1]. This difference explains the higher susceptibility of the intermediate cell of the TEBs to the effects of the carcinogen. Initiation of this stimulus causes further expansion of the proliferative compartment of the intermediate cells and depression in the dark cell population[1].

FACTORS THAT REGULATE THE SUSCEPTIBILITY OF THE MAMMARY GLAND TO CARCINOGENESIS

During the postnatal development of the mammary gland, the number of TEBs reaches a peak in the 21 day-old female, after which they reduce both in size and number due to their differentiation into ABs and lobules[14,15]. Since the TEB is the site of origin of mammary carcinomas, it is possible to predict that carcinogen inoculation at various times during the lifespan of a virgin animal will result in an incidence of carcinomas proportional to the number of TEBs, as depicted in Figure 3. The older the animal at the time of carcinogen administration, the fewer the numbers of tumors developed[1,13]. It was also observed that not all the mammary glands responded to the administration of the carcinogen in the same fashion; tumor incidence in thoracic mammary glands is higher than in the abdominal mammary glands[1,19,20]. This different carcinogenic response is due to the asynchronous development of glands in different topographic areas; thoracic glands lag behind in development and retain a higher concentration of TEBs[1,20]. This asynchrony reduces with aging, and the thoracic glands approach the degree of development of glands seen in other locations; by the age of 330 days tumor incidence is similar in both glands located in either a thoracic or abdominal region (Fig. 3)[20].

A further evidence in support of the TEB as the site of origin of mammary carcinomas has been obtained by plotting the incidence of adenocarcinomas,

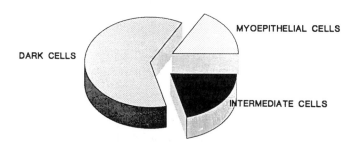

Fig. 2. Distribution of dark, intermediate and myoepithelial cells on mammary gland.

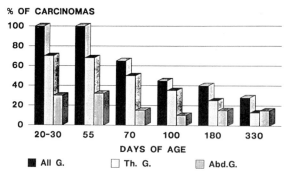

% OF CARCINOMAS

Fig. 3. Histogram showing the percentage of
animals bearing carcinomas induced
by DMBA administered at various
ages (abscissa) in all mammary
glands (All G.), thoracic glands
(Th. G.) and abdominal glands (Abd.
G.).

or percentage of animals bearing this tumor type against the percentage of
TEBs percent in the mammary gland at the time of carcinogen administration.
A high correlation coefficient between these two parameters has been observed
(Fig. 4). However, there is no correlation between tumor incidence and the
number of other terminal structures such as ABs or lobules[20].

Further evidence that it is the number of TEBs which affects the susceptibility of the mammary gland to carcinogenesis is obtained through the study
of pregnancy. Full term pregnancy, that completely eliminates TEBs, also results in inhibition of tumor development in parous animals[1,11,21,22,23].
If pregnancy is interrupted, however, this protection is minimized or nul-
lified[24]. Hormonal manipulation with estrogenic compounds[20,25-28] or
with chorionic gonadotropin[1,20,29] demonstrates that reductions in number
of TEBs similar to those occurring after pregnancy can be induced by these
treatments, thus resulting in different degrees of refractoriness[1]

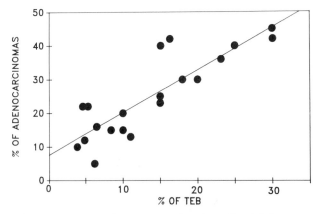

Fig. 4. Regression curve showing the high correlation (cc 0.87 p = 0.001) between
percentage of adenocarcinomas and percentage of TEB.

One of the elements that influence the susceptibility of the TEB to carcinogenesis is the high proliferative activity of its epithelium[1,30,31]. We have measured the growth fraction of the TEBs and found that it contains the highest proliferative compartment. Fifty-five percent of the cells are proliferating in the TEB of virgin rats whereas only 23% are in the proliferative pool for the ABs and lobules[31]. The growth fraction decreases with both aging and differentiation[31]. In the gland of parous animals, which is refractory to carcinogenesis, the growth fraction is only 1%[31].

Cells of the TEBs cycle approximately every 10 hours, with an S phase of about 7 hours and a G1 or resting phase of one hour[31]. When the mammary gland is differentiated by pregnancy, the growth fraction is reduced about 60 times and the cells that are proliferating have a cell cycle lengthened to about 35 hours. This increase occurs at the expense of the G1 phase of the cell cycle[1,31]. Therefore, pregnancy shifts the cells to a resting compartment.

Carcinogenic initiation requires the stable alteration of DNA molecules[1,11]. For this the carcinogen has to bind to the DNA, and maximal DNA binding occurs when a peak of maximal DNA synthesis is taking place in the organ. Carcinogens damage DNA mostly during the S-phase[32-35]. If the damage is not repaired during the G1 phase, the damaged DNA is transmitted to the daughter cells and it becomes fixed during successive S-phases of the cycle[32-35]. We have demonstrated that the uptake of tritiated DMBA is selectively higher in TEBs than in other structures of the mammary gland; this uptake, expressed as the number of grains per nucleus, highly correlates with the DNA synthetic activity of the cells or DNA-LI[11].

DMBA is metabolized by the mammary epithelium to both polar and phenolic metabolites (Fig. 5). The metabolic pathway is similar in both the TEBs of virgin rats and the lobules of parous animals; however, the formation of polar metabolites is higher in the TEB epithelial cells and the binding of the carcinogen to DNA is also higher in these cells than in lobular cells[1,36,37]. Removal of adducts from the DNA has also a different pattern (Fig. 5). TEBs have a very low rate of adduct removal, whereas lobules are more efficient, indicating that these latter structures repair the damage induced by the carcinogen more efficiently (Fig. 5)[29,37].

All these data allowed us to conclude that the susceptibility of the mammary gland to carcinogenesis is modulated by the following parameters: 1) The presence of terminal end buds; 2) the size of the proliferative compartment; 3) the amount of binding of the carcinogen to the DNA, and 4) the ability of the cells to repair the DNA damaged by the carcinogen[1].

FACTORS THAT MODULATE THE PROGRESSION
OF THE INITIATED CELL TO FULL MALIGNANCY

In the preceding sections we have shown that mammary carcinogenesis induced in Sprague-Dawley rats by administration of DMBA is the result of the interaction of the carcinogen with the TEB; when damaged, this structure further evolves to IDP and this to carcinoma (Fig. 1).

IDPs are morphologically distinguishable from TEBs by their size, which is more than twice that of the TEB and by the homogenous cell composition, which consists preponderantly of intermediate cells[1,11,20]. As it is depicted in Figure 6 the induction of IDPs is by no means a rare event. Within three weeks of DMBA administration there are between 10 and 20 IDPs per mammary gland, and this number increases with time, such that by 6 to 10 weeks after treatment there are approximately 30 lesions per gland, and around 200 per animal. Although IDPs occur in large numbers following DMBA administra-

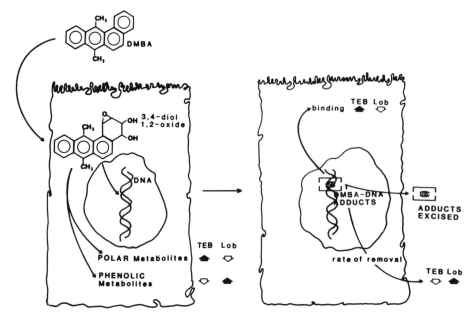

Fig. 5. Schematic representation of DMBA metabolism in a mammary epithelial cell. DMBA is metabolized to the active compound 3,4 diol-1,2-oxide, that interacts with cellular DNA, and to polar and phenolic metabolites. Polar metabolites are increased in TEBs, but are lower in the lobular compartment, whereas the phenolic metabolites are lower in TEBs and increased in the lobular structures. DMBA binding to DNA is higher in TEBs and lower in lobular structures. After binding DMBA, adduct removal is higher in lobular and lower in TEB structures. From: Russo, J. and Russo, I. H., Lab. Invest. 57:112, 1987 (with permission from the publisher).

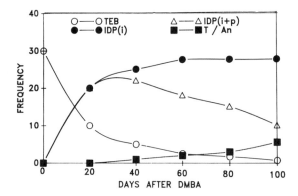

Fig. 6. Emergence of preneoplastic lesions IDP (i); IDP (i+p) and carcinomas (T/An) as the number of TEB decreases with time in the mammary gland at different periods after DMBA administration.

tion, the likelihood of any one IDP of progressing to carcinoma in the intact mammary gland is lower than that, since the maximal tumorigenic response rarely goes beyond 5 to 6 adenocarcinomas per animal (Fig. 6). This is attributed to the fact that there are two different types of IDP. The one that we call "initiated" IDP [IDP (i)], increase in numbers steadily with a concomitant decrease in the number of TEBs; they reach a plateau by 60 days post-DMBA (Fig. 6). The IDP (i) are characterized by having a diameter larger than that of the TEB, and are composed of a greater number of epithelial cells. They do not elicit a response in the surrounding stroma, and remain unchanged during the whole post carcinogen observation period. A second type of IDP, which we call "initiated and promoted" [IDP (i+p)], arises at the same time and reaches the same level as IDPs (i); by 20 to 30 days post-DMBA they are also characterized by having a larger diameter and a greater cell number than TEBs. They elicit a marked stromal reaction, consisting in collagen deposition and infiltration by mast cells and lymphocytes (Fig. 7). IDPs (i+p) progress to carcinoma in situ and to invasive carcinoma. The fact that not all the IDPs progress to carcinoma indicates that although both IDPs (i) and (i+p) are preneoplastic lesions, there are factors that regulate the progression of initiated cells, which affect differently IDPs (i) and IDPs (i+p).

The factors that regulate the progression of an initiated cell to preneoplasia and to neoplasia are unknown. We have been able to identify, however, a host response elicited by the initiated cells that might play a role in this mechanism.

Figure 7 depicts graphically the two types of IDPs: IDP (i), which does not elicit stromal reaction and fails to progress to carcinoma in situ (C.I.S.), and IDP (i+p), which is surrounded by numerous mast cells and lymphocytes and originates malignant lesions. The number of mast cells around the IDP (i+p) is three times higher than in TEBs and IDP (i) (Fig. 8). This

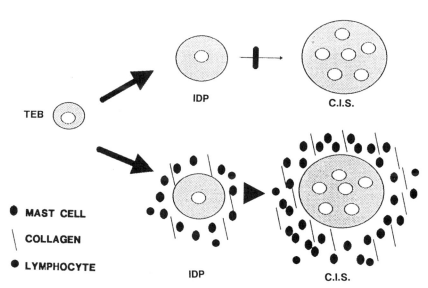

Fig. 7. Chart representing the evolution of TEB to IDP and carcinoma in situ (C.I.S.). Those IDP that elicit a host response with attraction of mast cells, lymphocytes and collagen deposition, are the ones that evolve to carcinoma in situ.

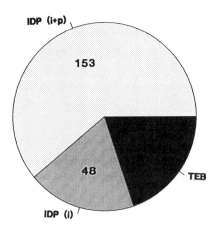

Fig. 8. Distribution of mast cells in
TEBs, IDPs (i) and IDP (i+p).
The values represent the number
of mast cells per 100 μm².

increase in mast cells is accompanied by an increase in lymphocytes,
fibroblasts, collagen fibers and proteoglycans.

Mast cells are found in different parts of the body, and the mammary
gland is not an exception. They contain in their cytoplasm numerous granules
measuring up to 0.8 μm in diameter which stain metachromatically with
toluidine blue or alcian blue[38]. Mast cells have membrane receptors to IgE
which participate in the immediate and delayed type of hypersensitivity (Fig.
9)[39]. When an antigen combines with IgE and bind the cell receptor, the
cell degranulates releasing histamine and heparin. Heparin is a heparan sulfate that has been shown to stimulate cell proliferation (Fig. 9)[40]. It
has been shown that transplanted tumors in the chick embryo may increase by
40-fold the number of mast cells around the tumor implant before new capilla-
ries arise[41]. Mast cell lysates or mast cell-conditioned medium stimulate
locomotion of capillary endothelial cells in vitro[42,43], an effect that is
attributed to heparin. It has been postulated that heparin or fragments of
heparin on the surface of endothelial cells may selectively bind endothelial
cell mitogens that are also angiogenic[40]. Interestingly enough, there are
several growth factors that have great affinity for heparin[44,45]. It has

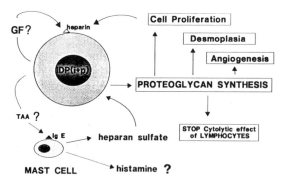

Fig. 9. Chart indicating the interaction
between IDP (i+p) and the host,
namely presence of mast cells, and
local regulatory factors such as
secretion of proteoglycans that
affect cell proliferation, desmoplasia and angiogenesis.

50

also been shown that heparin-coated tumor cells exhibit altered transplantation and cytotoxicity reaction[46,47], presumably due to blockage of cell surface antigens by heparin. Thus heparan sulfate proteoglycans also may be deposited in an annulus around the tumor and may modulate the immunoreactivity or accessibility of the enclosed cells (Fig. 9).

An early change observed during the process of transformation is the synthesis of a large amount of proteoglycans by IDP (i+p) which is evidenced by the deposition of an electron dense material on the cell surface of the epithelial cells, and by an increased reactivity with alcian blue pH 2.7 and PAS (Figs. 10-13). This is accompanied by an increase in uptake of ^{3}H-fucose and ^{3}H-glucosamine (Fig. 14). The number of cells uptaking these precursors is almost three times the number found in TEBs and IDPs (i) (Fig. 14). All these data clearly indicate that the initiated cells that are progressing to malignancy are secreting proteoglycans which accumulate in the stroma. We do not know whether these proteoglycans are influencing the response of the host by eliciting a higher mobilization of mast cells and inhibiting the cytotoxic effect of lymphocytes, or by inducing angiogenesis, desmoplasia and cell proliferation. Some of these proteoglycans, such as heparan sulfate, act as receptors for growth factors, which in turn initiate an autocrine response.

Proteoglycans occur on the plasma membranes of mammalian cells and in the extracellular matrix[48-50]. They are neutral glycoproteins, stained by PAS, which are negative with alcian blue pH 2.7 and contain fucose residues; and acid mucopolysaccharides, which react positively with alcian blue pH 2.7 and negatively with PAS. There are two types of glycosaminoglycans: glycosaminoglycans of condroitin sulfate, containing residues of D-glucuronic acid, and 2) N-acetyl-D-galactosamine and glycosaminoglycans of heparan sulfate, which contain alternating residues of D-glucuronic acid or L-iduronic acid and N-acetyl-D-glucosamine.

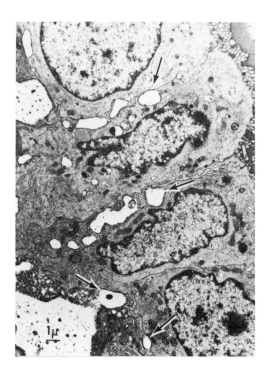

Fig. 10. Formation of intercellular spaces (arrows) in the IDP (i+p) of rat mammary gland treated with DMBA. The electron dense material are proteoglycans.

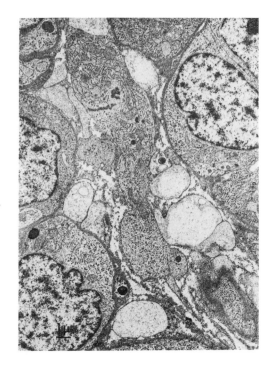

Fig. 11. Deposition of proteoglycan in interstitial
spaces of an IDP (i+p).

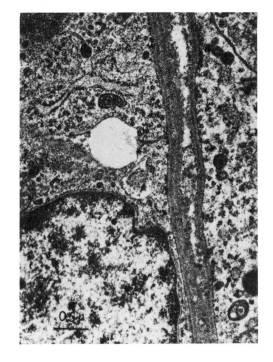

Fig. 12. Deposition of proteoglycans between clus-
ters of neoplastic
cells in C.I.S.

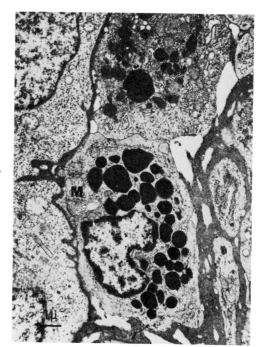

Fig. 13. Deposition of proteo-
        glycans around a carci-
        noma in situ, and
        around mast cells (M).

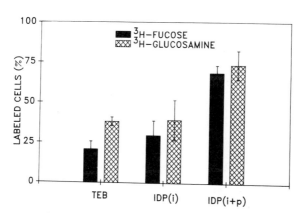

Fig. 14. Histogram showing the percentage of
         cells labeled with ^{3}H-fucose and
         ^{3}H-glucosamine in TEB, IDP (i)
         and IDP (i+p).

Neoplastic transformation of cells dramatically alters proteoglycan synthesis both in the tumor and in the surrounding tissues[51]. This is thought to stimulate tumorigenic growth by decreasing the adhesion of transformed cells to the extracellular matrix[51].

Based upon our own data and those reported in the literature, it is possible to speculate that the production of proteoglycans allows the IDP to progress to carcinoma in situ by stimlation of cell proliferation and by interference with an immune reaction toward the cells. Of these newly synthesized proteoglycans, both those that incorporate ^{3}H-D-glucosamine and stain with alcian blue pH 2.7 and those that uptake ^{3}H-fucose and stain with PAS may act like epiglycanin, a high molecular weight sialoglycoprotein present in mouse mammary carcinoma (Ta3) cells, which is though to mask histocompatibility antigens. These, in turn, can prevent the generation and penetration of cytolytic lymphocytes (Fig. 9)[46].

CONCLUSIONS

Using the DMBA rat mammary model we have been able to demonstrate that the carcinogen acts on the intermediate cell of the TEB, and that this structure is the one that evolves to IDP and carcinoma in situ. There are several factors that regulate the susceptibility of the TEB; some of them are: a) topographic location of the mammary gland, b) age of the animal, and c) reproductive history.

The high proliferative activity of the TEB is associated with higher binding of the carcinogen, and its short cell cycle makes them less likely to repair the DNA damaged by the carcinogen.

Even though most of the TEBs are transformed to IDPs, not all of them evolve to carcinomas. The regulatory mechanism of this process is more complex due to intrinsic properties of the TEB. The fact that IDPs progress to carcinomas, secrete proteoglycans and attract lymphocytes and mast cells, emphasize the importance of the interaction of the initiated cells with the host as a mechanism in the progression of the disease.

It is clear that the understanding of the mechanisms that modulate the progression of an IDP to a carcinoma will not only further our knowledge and understanding of carcinogenesis, but will also provide the tools for the prevention of the disease, as a result of the development of strategies for stopping the progression of initiated cells to fully manifested malignancy.

REFERENCES

1. J. Russo and I. H. Russo, Biological and molecular bases of mammary carcinogenesis, Lab. Invest. 57:112-137 (1987).
2. I. H. Russo and J. Russo, From pathogenesis to hormone prevention of mammary carcinogenesis, Cancer Surveys 5:649-670 (1986).
3. V. G. Valaoras, B. MacMahon, D. Trichopoulus, and A. Polychronopoulou, Lactation and reproductive histories of breast cancer patients in greater Athens, 1965-1967;, Int. J. Cancer 4:350-363 (1969).
4. B. MacMahon, Etiology of human breast cancer, J. Natl. Cancer Inst. 50:21-42 (1973).
5. N. Krieger, Rising incidence of breast cancer, J. Natl. Cancer Inst. 30:2-3 (1988).
6. B. MacMahon, P. Cole, M. Liu, C. R. Lowe, A. P. Mirra, B. Ravinihar, E. J. Salber, V. G. Valaoras, and S. Yuasa, Age at first birth and breast cancer risk, Bull. WHO 34:209-221 (1970).

7. S. Yuasa and B. MacMahon, Lactation and reproductive histories of breast cancer patients in Tokyo, Japan, Bull. WHO 42:195-204 (1970).

8. E. J. Salber, D. Trichopoulo, and B. MacMahon, Lactation and reproductive histories of breast cancer patients in Boston, J. Natl. Cancer Inst. 43:1013-1014 (1969).

9. C. B. Huggins, L. C. Grand, and F. P. Brillantes, Mammary cancer induced by a single feeding of polynuclear hydrocarbons, and its suppression, Nature 189:204-207 (1961).

10. T. L. Dao, F. G. Bock, and M. J. Greiner, Mammary carcinogenesis by 3-Methylcholanthrene. II. Inhibitory effect of pregnancy and lactation on tumor induction, J. Natl. Cancer Inst. 25:991-1003 (1960).

11. J. Russo, L. K. Tay, and I. H. Russo, Differentiation of the mammary gland and susceptibililty to carcinogenesis, Breast Cancer Res. Treat. 2:5-73 (1982).

12. J. Russo, Basis of cellular autonomy in susceptibility to carcinogenesis, Toxicologic Pathol. 11:149-166 (1983).

13. J. Russo, G. Wilgus, and I. H. Russo, Susceptibility of the mammary gland to carcinogenesis. I. Differentiation of the mammary gland as determinant of tumor incidence and type of lesion, Am. J. Pathol. 96:721-735 (1979).

14. I. H. Russo and J. Russo, Developmental stage of the rat mammary gland as determinant of its susceptibility to 7,12-dimethylbenz(a)anthracene, J. Natl. Cancer Inst. 61:1439-1449 (1978).

15. J. Russo and I. H. Russo, DNA-labeling index and structure of the rat mammary gland as determinants of its susceptibility to carcinogenesis, J. Natl. Cancer Inst. 61:1451-1459 (1978).

16. J. Russo, J. Saby, W. M. Isenberg, and I. H. Russo, Pathogenesis of mammary carcinomas induced by 7,12-dimethylbenz(a)anthracene, J. Natl. Cancer Inst. 59:435-445 (1977).

17. J. Russo, I. H. Russo, M. J. van Zwieten, A. E. Robers, and B. Gusterson, Classification of neoplastic and non-neoplastic lesions of the rat mammary gland, in: "Integument and Mammary Gland, Monograph series on the Pathology of Laboratory Animals," T. C. Jones, Y. Konishi, and U. Mohr, eds., Springer-Verlag, Berlin (1988) (In press).

18. I. H. Russo, J. Saby, and W. Isenberg, Early signs of malignant transformation in rat mammary carcinoma, Proc. Am. Assoc. Cancer Res. 17:463a (1976).

19. I. H. Russo, M. Tewari, and J. Russo, Morphology and development of the rat mammary gland, in: "Integument and Mammary Gland of Laboratory Animals," T. C. Jones, Y. Konishi, and U. Mohr, eds., Springer-Verlag, Berlin (1988) (In press).

20. I. H. Russo and J. Russo, Hormone prevention of mammary carcinogenesis: a new approach in anticancer research, Anticancer Res. 8:1-18 (1988).

21. J. Russo, I.H. Russo, M. Ireland, and J. Saby, Increased resistance of multiparous rat mammary gland to neoplastic transformation by 7,12-dimethylbenz(a)anthracene, Proc. Am. Assoc. Cancer Res. 18:149a (1977).

22. J. Russo, J. Miller, and I. H. Russo, Hormonal treatment prevents DMBA-induced rat mammary carcinoma, Proc. Am. Assoc. Cancer Res. 23:348a (1982).

23. J. Russo and I. H. Russo, Is differentiation the answer in breast cancer prevention? IRCS Med. Sci. 10:877-880 (1982).

24. J. Russo and I. H. Russo, Susceptibility of the mammary gland to carcinogenesis. II. Pregnancy interruption as a risk factor in tumor incidence, Am. J. Pathol. 100:497-512 (1980).

25. I. H. Russo, M. Al-Rayess, and S. Sabharwal, Effect of contraceptive agents on mammary gland structure and susceptibility to carcinogenesis, Proc. Am. Assoc. Cancer Res. 26:460a (1985).

26. I. H. Russo, M. Al-Rayess, and J. Russo, Role of contraceptive agents in breast cancer prevention, Biennial Intl. Breast Cancer Res. Conf. p 1;187 (1985).

27. I. H. Russo, T. Pokorzynski, and J. Russo, Contraceptives as hormone-preventive agents in mammary carcinogenesis, Proc. Am. Assoc. Cancer Res. 27:912a (1986).

28. I. H. Russo, J. Frederick, and J. Russo, Hormone prevention of mammary carcinogenesis by Norethyndrel-Mestranol, Breast Cancer Res. and Treat. (In press).

29. L. K. Tay and J. Russo, Effect of human chorionic gonadotropin on 7,12-dimethylbenz(a)anthracene induced DNA binding and repair synthesis by rat mammary epithelial cells, Chem. Biol. Interact. 55:13-21 (1985).

30. D. R. Ciocca, A. Parente, and J. Russo, Endocrinologic milieu and susceptibility of the rat mammary gland to carcinogenesis, Am J. Pathol. 109:47-56 (1982).

31. J. Russo and I. H. Russo, Influence of differentiation and cell kinetics on the susceptibility of the mammary gland carcinogenesis, Cancer Res. 40:2671-2687 (1980).

32. I. Berenblum, A speculative review: The probable nature of promoting action, its significance in the understanding of the mechanism of carcinogenesis, Cancer Res. 14:471-476 (1976).

33. J. V. Frei and T. Harsano, Increased susceptibility to low doses of carcinogen of epidermal cells in stimulated DNA synthesis, Cancer Res. 27:1482-1491 (1967).

34. T. Kakunaga, The role of cell division in the malignant transformation of mouse cells treated with 3-Methylcholanthrene, Cancer Res. 35:1637-1642 (1975).

35. H. Marquardt, S. Baker, B. Tierney, P. L. Grover, and P. Sims, Comparison of mutagenesis and malignant transformation by dihydrodiols of 7,12-dimethyl-benz(a)anthracene, Br. J. Cancer 39:540-547 (1979).

36. L. K. Tay and J. Russo, 7,12-Dimethylbenz(a)anthracene-induced DNA binding and repair synthesis in susceptible and nonsusceptible mammary epithelial cells in culture, J. Natl. Cancer Inst. 67:155-161 (1981).

37. L. K. Tay and J. Russo, Formation and removal of 7,12-dimethylbenz(a)-anthracene nucleic acid adducts in rat mammary epithelial cells with different susceptibility to carcinogenesis, Carcinogenesis 2:1327-1333 (1981).

38. K. Hashimoto, W. M. Tarnowski, W. F. Lever, Reifung und degranulierung der Mastzellen in der mersch Lichen Haut, Hautarzt 18:318-324 (1967).

39. H. F. Dvorak and A. M. Dvorak, Basophils, mast cells and cellular immunity in animals and man, Human Pathol. 3:454-456 (1972).

40. J. Folkman, How is blood vessel growth regulated in normal and neoplastic tissue?, Cancer Res. 46:467-473 (1986).

41. D. A. Kessler, R. S. Langer, N. A. Pless, and J. Folkman, Mast cells and tumor angiogenesis, Int. J. Cancer 18:703-709 (1976).

42. B. R. Zetter, Migration of capillary endothelial cells is stimulated by tumour-derived factors, Nature 285:41-43 (1980).

43. R. G. Azizkhan, J. C. Azizkhan, B. R. Zetter, and J. Folkman, Mast cell heparin stimulates migration of capillary endothelial cells in vitro, J. Exp. Med. 152:931-944 (1980).

44. D. Gospardorowicz, J. Cheng, G. M. Lui, A. Baird, and P. Bohlent, Isolation of brain fibroblast growth factor by heparin-sepharose affinity chromatography: identify with pituitary fibroblast growth factor, Proc. Natl Acad. Sci. U.S.A. 81:6963-6967 (1984).

45. R. R. Lobb and J. N. Fett, Purification of two distinct growth factors from bovine neural tissue by heparin affinity chromatography, Biochemistry 23:6295-6299 (1984).

46. M. Lippman, Transplantation and cytotoxicity changes induced by acid mucopolysaccharides, Nature 219:33-36 (1968).

47. W. M. McBride and J. B. L. Bard, Hyaluronidase-sensitive halos around adherent cells, J. Exp. Med. 149:507-515 (1979).

48. I. Ito, Radioactive labeling of the surface coat on enteric microvilli, Anat. Res. 151:489a (1965).

49. J. G. Bekesi and R. J. Winzler, The metabolism of plasma glycoproteins: Studies on the incorporation of L-fucose-1-^{14}C into tissue and serum in the normal rat, J. Biol. Chem. 242:3873-3879 (1967).

50. H. B. Bossmann, A. Hagopian, and E. H. Eylar, Cellular membranes: the biosynthesis of glycoprotein and glycolipids in the HeLa cell membranes, Arch. Biochem. Biophys. 130:573-583 (1969).

51. J. D. Esko, K. S. Rostand and J. L. Weinke, Tumor formation dependent on proteoglycan biosynthesis, Science 241:1092-1096 (1988).

DISCUSSION

Emmanuel Farber: That was obviously very nice, very enjoyable. But I am not clear now about "initiation". Let's say you have the carcinogen acting on the epithelium of the terminal bud. Presumably it has been initiated; however you have no index of initiation. How can you tell that initiation has occurred? Is there anything that you can see?

Jose Russo: We know that the TEB is initiated because the increase in cell number and proliferation evolves in the formation of intraductal proliferation.

Farber: But after all, in many systems there is a focal proliferation at some stage during the so-called promotion. Do you see a focal proliferation of a few cells, or is the whole bud responding? It is not clear, at least to me. Do you have a picture of what is going on at that particular step, before you get the obvious differentiation along two different pathways?

Russo: The mammary gland has basically three cell types. One is the myoepithelial cell, in the basal portion adjacent to the basement membrane. The other two are epithelial cells that can be divided into two subtypes: dark and intermediate cells. One of the first changes observed is a decrease in the number of dark cells and an increase in the number of intermediate cells. Quantitatively, the carcinogen inhibits the proliferation of the dark cells producing an imbalance. This imbalance causes a shifting of the number of intermediate cells because they keep dividing, whereas the dark cells have stopped replicating. The number of intermediate cells increases steadily in the fully developed tumor, thus making them the preponderant cell type of the tumor.

Farber: It seems that this is very special to the mammary gland. Can you think of another system that has any kind of analogous biological history, or is this unique?

Russo: Dr. Slaga mentioned that in the skin, dark cells are the proliferating cells. They could be the equivalent to intermediate cells found in the mammary gland. In mammary glands we don't have the different steps found in the liver or in the skin. The fact that we can identify TEBs that have been initiated, and others that are also promoted, may indicate that there are different steps in carcinogenesis. In humans there is a significant amount of preneoplastic lesions, but the final number of carcinomas is very low. This indicates a similarity between the rat mammary gland and the human breast.

Joseph Locker: You raised a point that I found very interesting about the relationship between excision repair and the S phase of the cell cycle. Could you review what is the evidence for that? I have the impression that several experimental systems have been shown in which inducing excision repair blocks entry into the S phase. Is there actual experimental data?

Russo: Mammary epithelial cells from terminal and lobular structures treated with a carcinogen in vitro were studied to determine their ability to remove the adducts formed. The cells that are able to remove the adducts more efficiently are the ones from the more differentiated structures such as lobules. The cells that remove the adduct less efficiently are derived from terminal end bud structures. We correlated this data with their cell cycle in the terminal end buds and in the lobules, and concluded that because the cells of TEBs are proliferating with a shorter cell cycle than the ones in the lobules, they have less chance to repair the damaged DNA. We do not know if the cells are able to carry on the repair during the S phase of the cell cycle.

Locker: So you can say that the cells proliferate fast and have a decreased ability to carry out excision repair, but you can't actually conclude that it is because they carry it into the S phase. That is only a hypothesis.

Russo: It means that is a property of the differentiated cells to repair the damage more efficiently, and this correlates with the longer cell cycle.

George Michalopoulos: I was very interested in your presentation on the role of the mast cells presumably helping these cells into progression. I'd just like to point out that there is evidence that heparin activates certain growth factors. Tom Masiac showed that with the endothelial cell growth factor. In the liver we found that heparin will stimulate the effect of endothelial cell growth factor on proliferation of hepatocytes. At the same time, it will inhibit totally the effect of hepatopoietin A on hepatocytes. So heparin is playing a very serious growth modulatory role vis a vis two heparin binding growth factors on the same cell target. The role of heparin in your system and in the liver system needs to be examined, actually, as a potential growth modulatory substance.

Russo: Yes, I agree with you on the need to explore the subject more in detail.

Michael Lieberman: Let me ask a focus question by way of opening up another area of investigation. It has to do with your observation on the proteoglycans and desmoplasia and angiogenesis. As we all know, transforming growth factor beta is an agent that can produce both of these effects and is well known to do so. The specific question is, have you looked at the role of this factor in any of your lesions, and more specifically, have you thought about using these powerful agents to dissect progression in this system?

Russo: Yes, we have some preliminary data that I didn't present for reason of time. We have inserted pellets of TGF-beta and TGF-alpha near the terminal end buds. The TGF-alpha does not produce any effect, but the TGF-beta produces a 30% decrease in the proliferative activity of the terminal end buds. Therefore, it is quite possible that the reason why some IDPs progress to carcinomas is because they are synthesizing TGF-beta that, when released into the stroma, stop the proliferation of other IDPs. But, we don't know how this or other growth factors are acting in the initiated or progressed IDPs.

Farber: Since this was raised, I think it's worth perhaps pointing out to the few people that might not be aware of this, that the role of the mast cell is very puzzling in carcinogenesis. I want to just note that Wally Clark, who's done an enormous amount of work on the pathogenesis of melanoma, points out that at a certain stage in human malignant melanoma you start to see quite an inflammatory infiltrate and mast cells begin to appear in larger numbers. You can make some kind of correlation between "progression," (it's a little bit rough and vague) and the presence of mast cells. So perhaps this could be a more general phenomenon than just simply in the mammary gland in the breast.

Russo: The literature shows that in many tumoral processes, there is accumulation of mast cells. In most cases they don't show up because they don't stain well with conventional methods. But, there is a relation between tumorigenic response and mast cell proliferation. Heparin synthesized by mast cells may play an important role in the process of cell proliferation.

Yoakum: To me it seems as though you have a very beautifully dissected system here for staging the interaction with several well defined carcinogens, and I'm wondering, in the light of Barbacid and Balmain's work on the ras gene, whether you have looked to see if ras mutation occurs in your system, firstly, and secondly, if you can temporally identify the point at which ras mutation may play a role in the generation of these types of carcinomas.

Russo: We are in the process of doing this kind of work. If the data of Barbacid are correct, meaning that mammary carcinoma is produced by a point mutation, a good way to corroborate it is to do a sequential study during all the different steps in carcinogenesis. We are doing some experiments in this direction but we don't have any data yet.

DISCUSSION OF SESSION

Moderator (Oscar Sudilovsky): After such interesting presentations I think it is time to concentrate on the mechanistic similarities between tumor promotion and progression. Dr. Pitot, if I interpret him correctly, said that the effect of promoting agents can extend to cells in the stage of progression, while Dr. Slaga expressed the opinion that both processes exist as a continuum. So I would like to start the discussion by asking Dr. Pitot how he would contrast his belief with those of Dr. Slaga.

Henry C. Pitot: In multistage carcinogenesis in the skin there is no way as yet to either identify initiated cells or cells that are on the boundary between the stage of progression and promotion except, by carcinoma "in situ" in the later case. In the liver we feel that we can identify early lesions of progression as the foci-in-foci. We haven't proven it, obviously, but at least that's a potential. In any event, if in the liver one examines all the focal lesions and combines them together, one would get the same sort of result that Dr. Slaga gets. Therefore, I think that depending upon how one can dissect that critical boundary between promotion and progression, one may interpret the results in one of several ways. At the moment in the skin there is no readily available method to clearly distinguish the boundary between the two, whereas in the liver I think there is, at least potentially.

Emmanuel Farber: I think I have to disagree with Dr. Pitot. In Dr. Pitot's system with phenobarbital, and with many other systems in the liver, the foci, nodules, nodules in nodules and cancer develop asynchronously. It is axiomatic that in any multi-step process, be it molecular, biochemical or cellular, synchronous appearance of each step is essential for sequential analysis. So I think the question you ask is a very important one, but without discrete steps with synchrony there are no boundaries. This lack of synchrony seems also to be the case with the development of papillomas and carcinomas in the skin model, according to what we have just heard from Dr. Hennings. So I don't know how one can approach your question unless you have a system where one can say this is the end of promotion and this is the beginning of progression.

Sudilovsky: I do not entirely agree with everything that Dr. Pitot said. But I also have problems in understanding why synchronism is so critical in any experimental design of liver carcinogenesis (or for that matter of any other organ). I assume that the value of synchrony is that similar alterations would occur simultaneously in every focus or nodule, so that they can be analyzed collectively. But even in the Resistant Hepatocyte Model no two cells in coetaneous foci are alike! On the other hand, although a number of us have, as Henry indicated, examined all the focal lesions and combined them together, I am not certain either that that is the best approach. Let's assume, Manny, that you analyze not the entire liver but a single, given focus. If you were able to chose a specific parameter or parameters, such as

Boundaries between Promotion and Progression during Carcinogenesis
Edited by O. Sudilovsky *et al.*, Plenum Press, New York, 1991

a genetic alteration, you could analyze each individual focus without the
need of synchronism in all foci, as required by your postulation.

Farber: But you can't ask the question, Oscar, because if you have a
multiple lesions system without synchrony you have no way to tell what the
fate of any single lesion is.

Sudilovsky: That may not always be the case. For example, we have done
experiments in the rat in which we examined the liver after initiation with
diethylnitrosamine and promotion with a choline deficient diet supplemented
with phenobarbital (1, 2). It turned out that while the majority of the foci
analyzed cytospectrophotometrically were euploid, a minority of them had an
aneuploid modal DNA content. When we scrutinized further our raw data, we
discovered that only a variable proportion of the total number of cells in
each focus was aneuploid. The balance of cells in any such foci were
euploid. The implication was that since there is genomic heterogeneity among
the cells of individual foci or nodules, each of them is a world unto
itself! This is a most relevant inference, because it denotes that what is
important is what happens to each particular focus, not what occurs in a
cross section of synchronized nodules. For the purposes of the question I
asked Henry Pitot initially, if it means that there is such a possibility as
the simultaneous appearance of promotion and progression in different cells
within an individual focus.

Farber: That's the worst possible way you can go. Instead of having to
determine one you're going to have 1,000. Everything is magnified
one-thousand fold!

Sudilovsky: I don't think that what is important is to answer the question
of simultaneity of promotion and progression in cells of different foci. If
we estimate that the rat liver has 3×10^8 hepatocytes and that 1 per 10^5
or 10^6 hepatocytes are altered after exposure to chemical carcinogens, from
300 to 3000 foci will be obtained following initiation and promotion. According to your own data only 1% of them, or 3 to 30, will develop into persistent nodules (which are the ones that may acquire cancerous properties).
Therefore, the amount of determinations may not be an insurmountable obstacle
when the proper technology becomes available. It is possible that a methodology not yet developed will allow us to trace a single focus to cancer or vice
versa; probably newer methodologies in molecular biology could eventually
allow us to do that. With regards to the aneuploidy we observed, it is
probably that most of those genomic changes will turn out to be epiphenomena;
however the possibility exists that such alterations are risk factors for
malignant transformation. Presumably other event(s) must occur in addition
to, or independent of, a modification in chromosomal DNA for the accrual of
malignant properties.

Daniel Longnecker: I'd like to ask Drs. Sudilovsky or Pitot since aneuploidy
is being used as a criterion of progression, if there can be a euploid malignant tumor. I'd like to know how aneuploidy is defined. Is karyotype required or is flow cytometry an adequate way to measure?

Pitot: I think it is an open question as to whether one can identify a malignant neoplasm with a perfectly normal karyotype. In the past it was thought,
for example, that human acute lymphocytic leukemia (ALL) would be a good example of this. On ranking the chromosomes of ALL cells, as Dr. Yunis has now
shown that 98% or more will show abnormalities. Thus, a karyotypic abnormality is something that one can demonstrate either by standard cytogenetic techniques or by in situ hybridization. In some translocations, such as those
Dr. Nowell will be discussing, there is such a tiny amount of a chromosome

translocated that it can not be easily identified under the microscope, but by means of recombinant DNA techniques such as Restriction Fragment Length Polymorphism one may see such changes. Thus, although the observation of aneuploidy cytologically is very straightforward and can be seen in greater than 95-98% of cancers, it is likely that more refined techniques would recognize one or more major structural changes in the DNA of all cancers.

Sudilovsky: Aneuploidy can be analyzed by karyotyping or by flow cytometry and cytospectrophotometry. These methods evaluate different parameters that should not be confused: the first one measures metaphase cells, the other two assess cells at any phase of the mitotic cycle. Dr. Pitot has just answered the question regarding karyotypes; I will address the other. Both flow cytometry and cytospectrophotometry are adequate for the measurement of mitotic and non-mitotic cells. The accuracy of the determinations, however, depends on the sensitivity of the instrument and on procedural techniques -all of them measured by the coefficient of variation. Most of these methods have a coefficient of variation of 5% or greater, but some refinements are now possible to decrease those values. The smaller the coefficient of variation, the smaller the amount of chromosomal DNA required for a diagnosis of aneuploidy.

George Michalopoulos: Since this is a conference on the boundaries between promotion and progression, I'd like to throw something to the floor, either as a consensus building platform or as a balloon to be punctured. That is, are we correct in assuming that promotion is based on processes which do not alter the genome, whereas progression is always based on mutations, chromosomal rearrangements, or some kind of a genomic structural alterations which bring about changes from a benign to a malignant cell? Is this a fundamental difference between the two processes? Is there a firm mechanistic boundary between those two processes?

Michael Lieberman: I'd like to comment on George's point. In a sense I think the question and some of the premises that we're discussing here this morning are artificial. My view is that progression is what happens to cells, that is, they acquire increasingly aggressive phenotypic properties; promotion is what one does to cells or animals to get changes. In fact, I view progression as a biological property and promotion as a property of the investigator. I'm not sure that we couldn't in another way view even these early "hyperplastic changes" as a form of progression. I wonder if we haven't derived an artificial schema on which to hang our experimental data.

George Yoakum: I'd like to address a question to both Dr. Farber and Dr. Pitot. There has been an implication and almost a direct statement that the focus that progresses to cancer is somehow different from the other foci and I'm not clear why that focus has to be in any way different. If, for example, in a mutagenic model a thousand foci represent a thousand targets for another event, the number of foci will determine the probability that that event will happen in one of them. It is not clear to me that there has to be any distinction. It makes this an extremely difficult problem for investigation.

Pitot: At our present state of knowledge there is no obvious pattern as to where the second focus, that is, the focus-in-focus arises in a particular set of phenotypes. In fact, the phenotype heterogeneity of altered hepatic foci is extensive, as several of us have shown, including Dr. Farber, Dr. Peraino and others. As a matter of fact, one could make the argument, if one used enough markers, that no two foci are identical phenotypically. Thus a focus arising within an established focus is the result of chance similar to what is seen when one initiates cells. A focus-in-focus may occur spontaneously, or as a result of the action of a known progressor agent. As Dr. Farber indicated, we do not agree on the subject of the importance of the synchrony of the foci. In fact, one can study mitosis in the cell cycle in an asynchronous population just as well as in a synchronous population and obtain the same result. In either case, certain assumptions as to mechanisms will have to be made. Some experiments to test the assumptions can be formulated now while others, as Dr. Sudilovsky says, cannot.

Harry Rubin: I partly wanted to respond to what Dr. Lieberman said, and I agree with him. I think what we're talking about is the tyranny of words. We're dealing with complex biological processes and we invent words. I have the same reaction as he did, actually, that progression is a biological process and promotion is an operation that we do to the cells; it's like comparing apples and oranges. Maybe I'm wrong about that. The other problem I have is that the term heterogeneity keeps coming up. I've had a fair amount of experience with it in culture. If you look carefully enough, almost every transformed cell differs from every other one. They share certain properties. They behave in a certain way. They continue to grow when they are not supposed to grow. When you get down to detailed characterization of fine characters, they all differ from one another. I think this elusive quest, which I've been witnessing for the last 35 years or so is for a mechanism that really doesn't exist. What we have is almost an infinite variety of changes that are possible in cells, some of which lead to neoplastic behavior, most of which do not. If we continue to always look for a mechanism, we're going to find lots of them, but I think we really miss the point. I think there's really a fundamental way of looking at this, which is basically probabilistic. We have to adopt some basic reorientation in thinking similar to the one that physicists had to go through with the advent of quantum mechanics. I believe a probabilistic way of thinking about the problem, rather than a mechanistic one could be a very profitable way of attacking some of these problems.

Charles Boone: I would like to refer to the landmark observations on human cancer biology described by Lester Foulds in his book, "Neoplastic Development" (Academic Press, New York, 1975). He was the first to use the term "progression", and meant it to apply to the entire neoplastic process, from its earliest beginning at the first mutation and clonal overgrowth in grossly normal appearing tissue, to an invading and metastasizing cancer. Only later was the term "progression" preempted by others for use with more restrictive meaning as the name of a "step" in the neoplastic process. As Foulds describes it, there are no "steps" or "boundaries" during neoplastic development, but rather one single continuum of mutations and clonal overgrowths. And I think he commented on promotion as being a superimposed co-phenomenon to this basic mutational clonal expansional process of progression. So I'd like to get comments on that.

Farber: I'm not sure I understand you, because to me promotion offers no conceptual problem. Its as stressed by Dr. Hennings and by Dr. Pitot, it's simply clonal expansion. You get a few altered cells and you have to expand those, otherwise they're not going to go anywhere. It's clonal expansion. Whether there is, in addition to clonal expansion something else, nobody knows. I don't think I would agree with Dr. Foulds. It is not justified to say promotion isn't a real phenomenon. It's a real phenomenon, because if you don't promote you don't go anywhere.

Michalopoulos: I raised the question about the boundaries between the two processes, and of the answers that came out, some were that the interpretation of processes depends upon the investigator whereas others stated that promotion may not be relevant. I would just like to bring up one point which brings out again the boundary between the two processes. All the substances which induce promotion in the liver, and maybe in the skin, induce a hyperplastic response in the target tissue while all the conversion or the progressor agents would score positive with the Ames' test. I think this is a fundamental difference (between the two processes) which needs to be examined very critically, and either accepted or rejected as a possibility. Is promotion essentially a process which does not proceed through genetic alterations, whereas progression or conversion unavoidably go through a restructuring of the cellular genome to achieve the malignant phenotype?

Peter Duesberg: I have a question on Dr. Pitot's talk. He mentioned, it seemed to me in passing, that normal liver cells are tetraploid and that tumor cells become diploid. Did I understand that correctly?

Pitot: The normal adult liver in the rat exhibits about 80% tetraploidy in the hepatocytes. The remaining hepatocytes are mostly diploid but there is a small number of some higher ploidy hepatocytes, octoploid and above, the number of which increases with age somewhat in the rat. The animals used in these experiments were about 1-2 months old, a time the vast majority of hepatocytes are tetraploid. However, the diploid karyotypes predominate in the focal cells. They do in nodular and hepatocellular carcinomas as has been shown by another group. It raises the very interesting point that Dr. Hennings also brought up. If one argues that Dr. Knudson's two-hit hypothesis of neoplasia is correct, and that for a single cell the hits probably are in the same allele, then it's very likely that the target cell that will ultimately develop into a neoplasm probably is diploid. A tetraploid cell such as in the liver would be a very unsuccessful target for developing into cancer because of four doses of each gene.

Duesberg: So it is true, then, that the tumor cells are all diploid, or near diploid? Is that true?

Pitot: I'm sorry, what was that?

Duesberg: That the malignant tumor cells that come out of these liver cells are diploid?

Pitot: The malignant cells are virtually all aneuploid. As I showed in this case, the focal cells, which occur during initiation and promotion with one of the protocols requiring the least toxicity are essentially all diploid.

Duesberg: When you say aneuploid, do you mean based on a tetraploid complement or on a diploid complement?

Pitot: Well, you recall that euploidy and aneuploidy differ in distinction by the multiple of the haploid number, so a diploid is twice the haploid number, and tetraploid is four times the haploid number, but both are euploid, with structurally normal chromosomes. Aneuploidy indicates a structural change and/or a change in the number of one or more specific chromosome.

Duesberg: Let me rephrase the question. The tumor cells, do they have approximately half as much DNA as the normal cells? Is that what you're saying?

Dr. Pitot: Yes. The well differentiated tumor cells are mostly diploid. This was shown many years ago by Dr. Nowell and others looking at the highly differentiated cells in the Morris hepatocellular carcinomas. However, once these tumors continue to be transplanted and grow, aneuploidy is the rule.

Farber: But, I think we have to point out in this context that when normal liver cells proliferate, there's some evidence that tetraploid goes to tetraploid, but there is also evidence that they go through diploid. It is hard to say at the moment, but it could simply be that the fact that the foci and nodules are diploid is because they are simply proliferating, and perhaps differentiate later on.

Joe W. Grisham: We just heard about the tyranny of words, and we're still being tyrannized by them. I think part of the lack of communication that I've perceived between questioner and answer relates to our imprecise use of words. Henry, I wish you'd say quasi-diploid or pseudo-diploid for those near diploid cells in tumors, instead of diploid. Then maybe we wouldn't be so confused.

Yoakum: I think the question raised about the boundary and about promotion as a justified phenomenon is something that has some validity and still to my mind needs some answering. I'm not sure that clonal expansion justifies everything about the phenomenon. One of the striking differences that we've touched upon but not really focussed on, is that agents that are involved in promotion often affect the organ and perhaps the entire milieu of those foci which will ultimately become tumor foci. Now, that will also relate to the difficulty in accomplishing a model system that requires a synchronous dancing tumor that starts off as a set of foci that go along to become carcinomas. I think that carcinomas rarely happen that way in the clinical arena. I'm not sure that that is a limitation in terms of our being able to study them, but if the concepts of promotion and progression are going to be useful, they should contribute to our development of specific experiments. These experiments should give us specific answers that generate mechanistic pictures and allow us to ask testable and arguable questions. I think that promotion, if it is going to justify itself, needs to be clarified in some way to explain the difference in degree on the cell originating the carcinoma and the effect on the nearby cells.

Boone: Consider an epithelial field, undergoing normal cell turnover, that is exposed to a mutagenic experience. Now for clonal overgrowth of mutated cells to occur the initiating mutation has to produce some degree of block either 1) in a mitosis-suppression pathway (controlled by calcium concentration, for instance) resulting in speeded proliferation of the mutated clonal cells, or 2) in a maturation (differentiation) pathway, with accumulation of normally proliferating mutant stem cells that fail to continue through the usual programmed sequence of non-mitosis, senescence, and disappearance. Either or both blocks could be produced, for example, by a mutation which alters gap junctions in such a way that cell-cell transfer of regulatory small molecules was prevented. "Promotion" is an exogenous stimulation to proliferation, that accelerates tumor development by stimulating the overgrowth of mutated clones. The proliferative stimulation associated with normal cell turnover is correspondingly defined as "endogenous promotion." This idea of continuously occurring "endogenous promotion" of mutated clones due to the proliferative stimuli of normal cell turnover is supported by the general rule, originally enunciated by Bergonie and Tribondeau, that the frequency of tumors in a given type of epithelium is related to the cell turnover time of that epithelium, or, stated another way by Robbins (3), that "pathologic states which induce reparative, hyperplastic, or metaplastic cell proliferations provide settings where, relatively speaking, cancers are more likely to develop." The comment was made that progression could occur to cells in the stage of promotion. Couldn't there be endogenous progression occurring from the very first initiation, supported by turnover proliferation? So that in actual fact promotion is occurring during the stage of progression, according to Foulds hypothesis of neoplastic progression from the very first mutation on through. Is promotion an acceleration of progression?

Lieberman: I don't have an answer to that question. I'm wondering if I couldn't get us to think a little more broadly about progression. It is implicit in the way we all think about progression that like development, it's sequential. That is, trait A must be acquired before trait B, and B before C, and so on. And yet, there's really not a lot of evidence that that need be so all the time. Cells must acquire a great many different traits in

66

order to be able to invade and then metastasize. I wonder if we aren't box-
ing ourselves in a little bit in our conceptual framing of questions, thinking that progression needs to be sequential all the time. I just throw that
out for anyone's thoughts.

Sudilovsky: I would agree with you that progression could occur in different
ways, in various manners in different cells, and the question is to what specific system it applies. Dr. Farber has some comments.

Farber: This is in response to Dr. Boone. The small intestine has the highest turnover of any organ of the body, but the lowest incidence of cancer.
Another good example is psoriasis. In this condition there is enormous
proliferation of the epidermis, but no cancer. So you have two very good
examples in human experience, clearcut, where cell turnover doesn't seem to
relate to cancer.

Duesberg: I would like to ask Dr. Pitot once more about a statement that he
made. He said studying oncogene activation was a good theoretical concept to
understand cancer. I think he added "because of the relationship to viruses". Now, viral oncogenes and proto-oncogenes are genetically fundamentally
different. All viral oncogenes are regulated by viral promoters and often
the coding sequences are changed. I wonder why, considering these differences, is it a good idea to study proto-oncogenes if you want to understand
cancer. You should look at the specifics rather than at the common things.
If you want to understand birds and why they fly, you don't study reptiles.
The reptiles and the birds are related, but you'd rather look at birds if you
want to see what's specific about them. So, I wonder why Dr. Pitot thinks
studying proto-oncogenes is relevant to cancer.

Pitot: We had to start somewhere. Clearly from work that we and others had
done many years ago examining a series of different metabolic characteristics
of many hepatocellular carcinomas, primary and transplanted, one could find
no characteristics ubiquitous to all malignant neoplasms. Perhaps I should
have prefaced my discussion in this area with a distinction between transcriptional activation and mutational activation of proto-oncogenes. The high
level of expression of viral oncogenes in viral induced cancers is closely
related to the transformation event in those particular cells. Similarly one
finds high levels of expression of one or more of a number of proto-oncogenes
in a variety of malignant neoplasms. Granted, there is no single proto-oncogene expression characteristic of all neoplasms and, from our previous experience, this was not unexpected. Still we had to start somewhere and in view
of the enhancements of viral oncogene expression in such cancers, it is reasonable to look at proto-oncogene activation at the beginning. We may be
wrong but only the future will tell us that.

Rubin: I think finding activation of oncogenes is what's called an epistemol-
ogical problem. You find what you look for, and if you look for more you
find more. So when, for instance, I guess Harold Weintraub and Grudien
looked at Rous infected cells for the activation of genes, not just oncogenes, they found that something like a thousand genes were activated. And
if you label them all oncogenes, pretty soon you'll have pretty much the
whole genome of the cell involved, which may be so. But that's a real
problem.

Yoakum: This is a comment to Dr. Farber's answer a while ago about cell
proliferation. I think that the example he mentioned was psoriasis. In that
case the experiment is in progress, and the data will be coming out in the
next 3-5 years. People with psoriasis have been treated with psoralen for
about 10 years which should be a very good initiator (psoralen), and I think
even in only this short period (10 years), the early data suggests that these
people get additional skin cancers. So, that may be an answer to your

statement, Dr. Farber, about the lack of a role for cell proliferation in promotion.

Farber: But I'm not talking about <u>treatment</u> for psoriasis. I'm talking about psoriasis. I've checked, I've talked to dermatologists, I've talked to many people, and in the books, and none has observed increased incidence of any skin neoplasms in people who have psoriasis. Patients have this disease for many years with much cell proliferation. So I'm not talking about psoralen, which is obviously another problem entirely.

Yoakum: This is a case of natural extraproliferation of cells and when you add on an initiating carcinogen you observe increased carcinogenesis in humans.

Farber: But that's another story. Granted, if you take a patient with psoriasis and then you add an initiator, that person may be at risk for developing cancer. But that's irrelevant. The fact is that pure psoriasis by itself doesn't seem to.

Boone: With regard to cancer of the colon, I wanted to mention for the record that Dr. Lipkin at Memorial Sloan Kettering has spent...

Farber: The small intestine, not colon. The small intestine has the highest rate of turnover of epithelial cells; it's enormous.

Boone: Ah. Well, then I can still make this point. In people with familial polyposis, Dr. Lipkin has measured the labeling index of colonic epithelium and showed that that is even more increased, so that's still in favor of the idea.

Farber: But he doesn't know why!

Boone: As far as the small intestine, I think everyone is mystified by the fact that tumors so rarely occur there.

Farber: But nobody knows the role of cell proliferation in the colon. You can argue...

Boone: Anything which accelerates proliferation in the colon would tend to be promotional. I think that's fair.

Farber: that in the places where we know, stomach, liver and in others, it's inhibition of cell proliferation that favors focal proliferation and cancer, not hyperplasia. In the stomach, you get atrophic gastritis which is the common precursor for cancer, not hypertrophic gastritis.

Moderator (Oscar Sudilovsky): Dr. Rubin has a comment and then we will adjourn.

Rubin: I think it is too easy to believe that there is a simple correlation between proliferation and transformation or malignant conversion. We take the case of chemical carcinogenesis in cell culture, the Heidelberger system, and there you don't see transformation until the culture has become confluent and there has been a shutdown in the rate of growth in most of the cells. And actually we see the same thing in spontaneous transformation of NIH 3T3 cells. That is, when the culture is multiplying exponentially, there is the least likelihood of malignant transformation. It is only when they become confluent that they seem to be either induced or selected for malignant transformation. Therefore, I think one cannot make a simple and general equation, like that made by Dr. Moore.

Sudilovsky: We can also confirm Dr. Rubin's observations in regards to
cultured cells from rat liver. The immortalized outbred Sprague Dawley
epithelial liver cell line, K22, obtained by Bernie Weinstein around 1973 or
1974, cultured only under logarithmic conditions does not grow when
transplanted in the nude mice. However, with my associate Dr. Tom K. Hei, we
obtained several years ago their malignant transformation (K22 innocula grew
in NIH nude mice) by culturing them when they became overconfluent. And now,
ladies and gentlemen, this session is really adjourned.

REFERENCES

1.   O. Sudilovsky and T.K. Hei: Aneuploid nuclear DNA content in some
        enzyme-cultured rat liver foci during tumor promotion. Fed. Proc.
        42:7 (1983).

2.   J.H. Wang, L. Hinrichsen, C. Mansilla and O. Sudilovsky: Hepatocarcino-
        genesis: Aneuploid nuclear DNA content in enzyme-altered foci (EAF)
        in rats. Lab. Invest. 58:100A (1988).

3.   S.L. Robbins and R.S. Cotran, "The Pathologic Basis of Disease", 2nd
        Edition, W.B. Saunders Company, Philadelphia, PA, 1979, p. 182.

PROGRESSION IN TERATOCARCINOMAS

G. Barry Pierce and Ralph E. Parchment

Department of Pathology
University of Colorado School of Medicine
Denver, CO

One of the joys of preparing this paper was the necessity of rereviewing the elegant work of Leslie Foulds (1969), Harry Greene (1951), and Jacob Furth (1953), and of recalling many pleasant and stimulating discussions with each of them. Foulds studied solid tumors, primarily adenocarcinomas of the breast, and his definition of progression as the independent and irreversible gain or loss of unit characters with time leading to the autonomous state has served oncologists well. Similarly, Greene's study of adenocarcinoma of the endometrium and breast which demonstrated increased malignancy with time are classics, and Furth's demonstration of conditioned and dependent tumors and loss of dependency as a manifestation of progression has markedly influenced endocrine therapy of cancer. The mechanism of progression appears to involve the selection of cells best able to survive under the conditions. Thus, the irreversible gains and losses of attributes by a tumor with time are the result of extermination of subpopulations of cells with consequent loss of their phenotypic traits, leaving a tumor composed of more rapidly growing subpopulations expressing their particular traits. Thus, the phenotype and behavior of tumors change irreversibly with time.

Like other tumors, teratocarcinomas undergo progression, a process requiring heterogeneous populations of stem cells and a means of selecting for those best able to survive under the conditions. Most oncologists view mutation or genomic instability as the most important means of developing heterogeneity (Nowell, 1982); however, we view them as late mechanisms. Early tumors [for example, adenocarcinomas of the colon (Cox and Pierce, 1982), melanoma (Gray and Pierce, 1964) and testis (Pierce and Dixon, 1959)] are heterogeneous in terms of ability of stem cell components to differentiate, and we believe these are selected for not only by host factors, but also by growth regulating factors secreted by the stem cell components. Synthesis of growth regulating factors by a malignant stem line is a reflection of similar synthetic processes by the corresponding normal cell lineage during its development (Pierce and Speers, 1988). These ideas have been generated from studies of teratocarcinoma.

A teratocarcinoma is defined as a malignant tumor composed of an admixture of benign tissues representing the three germ layers, and embryonal carcinoma cells. These embryonal carcinoma cells are malignant multipotential stem cells capable of differentiating into the differentiated tissues of the tumor (Pierce and Dixon, 1959; Pierce et al. 1960b, Kleinsmith and Pierce, 1964). The latter are usually benign although extraembryonic

Boundaries between Promotion and Progression during Carcinogenesis
Edited by O. Sudilovsky *et al.*, Plenum Press, New York, 1991

71

differentiations of embryonal carcinoma are extremely malignant (choriocarcinoma and yolk sac carcinoma) (Pierce, 1967). In addition to being present in teratocarcinomas, embryonal carcinomas can occur alone without evidence of somatic differentiation. Such tumors when treated with retinoic acid or dimethylsulfoxide express the differentiated phenotype in vitro (Strickland and Mahdavi, 1978) and in vivo (Speer, 1982). It can be concluded that the embryonal carcinoma cells have the potential for differentiation, although they may not express it under particular environmental conditions.

These interesting tumors develop spontaneously from primordial germ cells (Stevens, 1967), and they can be produced experimentally by heterotopic transplantation of mouse embryos at various stages of development (Stevens, 1967; Damjanov et al., 1971). We postulated that embryonal carcinoma cells were the neoplastic equivalent of the inner cell mass cells of the mouse blastocyst because of their ability to differentiate into the three germ layers (Pierce, 1967). This was confirmed in studies in which inner cell masses cultured in vitro underwent spontaneous transformation into embryonal carcinoma cells (Evans and Kaufman, 1981; Martin, 1981).

The environment contributed by host tissues was shown to have an important effect on teratocarcinogenesis as a result of transplanting embryos to heterotopic sites: the highest incidence of tumors was obtained when the embryos were transplanted into the testes, much lower incidences were obtained when the transplants were made into the kidney (Stevens, 1970; Damjanov et al., 1971). Similarly, teratocarcinogenesis by spontaneous transformation of inner cell masses in tissue culture was only successful in the presence of feeder cells (Evans and Kaufman, 1981; Martin, 1981). It is concluded that positive growth regulators were present in the successful situations. In similar vein, Fidler (1973) has demonstrated the selective effects of host tissue on the ability of B16 melanoma cells to colonize in the lung. Dissociated B16 cells were injected I.V. and the resultant lung colonies were dissociated to single cells and reinjected I.V. The process of injection, selection and reinjection was repeated. Eventually a line of B16 cells was obtained that preferentially colonized the lung. It is concluded that the host plays an important role in progression.

When murine teratocarcinomas are serially transplanted subcutaneously over long periods of time they lose the ability to differentiate their somatic tissues and evolve as embryonal carcinomas or yolk sac carcinomas. This is a clear example of progression (Pierce and Dixon, 1959a, 1959b), which can be expedited by ascites conversion. When teratocarcinomas were converted to the ascites many strains produced embryoid bodies which floated in the ascites fluid (Pierce and Dixon, 1959a,b). Some of these embryoid bodies were complex and contained elements representing each of the primary germ layers. Other more primitive ones were composed only of embryonal carcinoma, overlain by a layer of visceral yolk sac. Subcutaneous transplantation of such embryoid bodies always gives rise to solid teratocarcinomas with twelve or more somatic tissues. With serial passage of ascites fluid, however, this multipotency was lost and solid intraperitoneal implants of embryonal carcinoma or parietal yolk sac carcinoma appeared (Pierce and Dixon, 1959b). This transition occurred rapidly with testicular teratocarcinomas, but when an ovarian teratocarcinoma was converted to an ascites, conversion was slow and the tumors lost their differentiated tissues in the reverse order that the embryo acquired them (Pierce et al., 1960a). For example, the potential to make bone marrow, bone, and cartilage was lost relatively early, but the ability to form brain and trophoblastic giant cells was lost late. The intriguing conclusion about this process is that ascites conversion did not select against these tissues per se, rather it selected populations of embryonal carcinoma cells that were unable to express their potential for complex differentiations. Although the environment provided by the host's tissue plays an important role in progression of teratocarcinoma, we believe that

Table 1: Stem Lines of OTT6050 and Their Counterparts
in the Blastocyst

OTT6050 Stem Line	Normal Counterpart
Primary	
-Totipotent	Totipotent blastomere
Secondary	
-Multipotent line Mintz and Illmensee (1975)	ICM, late blastocyst
-ECa 247 (trophectoderm potential)	ICM, early blastocyst, trophectoderm potential
-JC44 (blastocyst potential)	Late morula, early blastocyst
-F-9 (multipotent, endoderm)	Peri-implantation blastocyst
PC13 ?	?

the microenvironment created by secretions of stem lines results in "autoselection of stem lines" and that this mode of selection plays an important
role in progression.

This is best illustrated by experiments using OTT6050, an embryo derived
transplantable teratocarcinoma (Stevens, 1970), from which many sublines have
been isolated by cloning or other manipulations. Five of these sublines and
their normal counterparts are shown in Table 1. F9 does not spontaneously
differentiate (Bernstine et al., 1973), but under experimental conditions it
can be induced to differentiate into endoderm (Strickland and Mahdavi, 1978)
and under other conditions into a variety of tissues (Speers, 1982). It has
been shown to secrete PDGF- and TGFß-like molecules (Rizzino and Bowen-
Pope, 1985; Rizzino, 1985; Rizzino et al., 1983) as well as an autocrine
growth factor (Rizzino and Crowley, 1980; Pierce et al, 1989). ECa 247
(Lehman et al., 1974) does not spontaneously differentiate but preferentially
differentiates into trophectoderm when placed in the blastocyst (Pierce et
al., 1987). It is responsive to a toxic factor made by C44 (Parchment et
al., 1989). C44 makes some differentiated tissues in subcutaneous transplants, but it makes embryoid bodies that closely resemble blastocysts when
converted to the ascites (Monzo et al., 1983; Parchment et al., 1989). Mintz
and Illmensee (1975) used a totipotential subline of OTT6050 that resulted in
fertile chimeric mice after injection into blastocysts. This cell line has
been lost. Finally, PC13 produces embryonal carcinoma derived growth factor
which is stimulatory of growth of embryonal carcinoma cells and other cells
(Heath and Isake, 1984; van Veggel et al., 1987).

One can thus envision the original OTT6050 as a tumor composed of multiple lines of embryonal carcinoma stem cells each with its particular potentials and some that secrete potent growth regulators (Table 2). The toxic
secretions of C44 would certainly eliminate nearby ECa 247 components, while
the secretions of positive growth factors by F-9 and PC13 components would
stimulate other stem lines with appropriate receptors. Finally, if a cell
line made an autocrine growth factor that stimulated only itself to faster
growth, it would eventually overgrow the other non-responsive cells. With
any of these possibilities the result would be autoselection leading to progression. Thus, two selective factors must be operative in progression,

Table 2: Autocrine and Paracrine Secretions of Sublines of OTT6050

Secretion	Embryonal Carcinoma	Blastocyst	Author
PDGF/like	F-9	+	Rizzino & Bowen-Pope, (1985)
TGFß/like	F-9	+	Rizzino (1985) Rizzino et al. (1983)
ECDGF	PC13	?	Heath & Isake (1984) van Veggel et al. (1987)
?	F-9	?	Rizzino & Crowley (1980)
Toxic Factor	JC44	Blastocyst	Pierce et al. (unpublished)

autoselection by the tumor stem lines and host selection.

Since tumors may be composed of multiple stem lines, it is reasonable to presume that they are mosaics. It would be interesting to know the threshhold number of cells (patch size) of each stem line that would permit the functions noted above. Could patch size also influence the ability of a stem line to metastasize as in Fidler's experiments?

To return to teratocarcinoma. A master stem cell with the potential to make all of the other stem lines should be present in OTT6050. What is the normal counterpart of this master stem cell and of its secondary stem cells? As mentioned previously, embryonal carcinomas may be derived from and have the functions of inner cell mass cells. Because tumors are caricatures of the process of tissue renewal (Pierce and Speers, 1988), it follows that the normal equivalent of the totipotent stem cell of OTT6050 must be a totipotent blastomere (Table 1). ECa 247 must be a caricature of cells of early ICM with trophectodermal potential (Pierce et al., 1987). The line of OTT6050 used by Mintz and Illmensee (1975) must caricaturize the cells of late ICM that make embryo, because when injected into blastocysts their offspring colonized all tissues including the germ line. C44 must be a caricature of cells of the morula which have the potential to make blastocysts. Thus, OTT6050 was heterogeneous on the basis of the potential of its primary stem cell for differentiation, which corresponded to the normal totipotent cell that gave rise to it by carcinogenesis. Its secondary stem cells, F-9, ECa 247, etc., which are derived by differentiation, can be selected as cell lines, or as a result of progression can be selected for or against just as were the particular stem lines of a melanoma (Gray and Pierce, 1964). We believe two factors operate in early tumors as the mechanism of progression: the production of growth factors by the stem cell lines (Table 1), and selection of stem lines best able to survive in the environment of the host.

Secretion of growth factors by tumors is now well known, but overlooked is Armin Braun's (1956) demonstration of production of such factors by plant teratomas. Normal adult plant cells could not make these factors, but embryonic plant cells from which the tumor developed could make them. We, thus, concluded that if a tumor makes a growth factor, that factor will have played an important autocrine or paracrine role in the development of the normal cell lineage in the embryo (Pierce and Speers, 1988). This conclusion is supported by the observations that a variety of growth factors have been isolated from lines of embryonal carcinoma, and similar factors have been found in the blastocyst (Table 2).

What are the roles of the factors in normal development? To answer this question, ECa 247 which preferentially differentiates into trophectoderm and P19 which preferentially differentiates into the three germ layers were exposed to blastocele fluid (Pierce et al., 1989). Whereas single P19 cells flourished in this fluid, 44% of ECa 247 cells died when placed in it. The cell responsible for synthesizing the toxic factor is not known, but because tumors are a caricature of the process of tissue renewal (Pierce and Speers, 1988), a search was made for embryonal carcinomas that might make the toxic factor found in blastocele fluid. To this end, C44 when converted to the ascites makes embryoid bodies that closely resemble blastocysts (Monzo, 1983) and these blastocysts often contained dead cells (Parchment et al., 1989). It was reasoned that the fluid of these neoplastic blastocysts might contain the toxic activity. When this fluid was tested, it preferentially killed ECa 247 and had no effect on P19. Purification of the molecular mediator of this activity is nearing completion. Space will not permit discussion of the normal function of this factor in the normal blastocyst, but it appears to be responsible for the programmed cell death that occurs in blastocysts (El-Shershaby and Hinchliffe, 1974; Handyside and Hunter, 1986).

PDGF- and TGF-like molecules are also found in the blastocyst (Table 2). Unlike the toxic factor, the function of PDGF-like molecules or TGFß in the blastocyst is not known, but they may serve as inducers of differentiation (Pierce and Speers, 1988). In this regard basic FGF is an inducer of mesenchyme in amphibian embryos (Slack et al., 1987; Grunz et al., 1988). Thus, it appears that molecules with inductive activity in the embryo are reprogrammed in the adult to subserve other activities.

Our hypothesis that if a tumor makes a factor, that factor will play an important role in the development of the normal cell lineage is not restricted to teratocarcinoma and the blastocyst. We believe it applies to all of the paraneoplastic syndromes of which lung cancers, that produce enough ACTH to cause a Cushing's syndrome, are good examples (Pierce and Speers, 1988). If the hypothesis is correct, fetal lung cells should synthesize and be responsive to ACTH at some point in their development. Synthesis of ACTH has not yet been demonstrated, but the responsiveness of embryonic lung to ACTH is well known (Becker, 1984). Finally, ACTH could serve as a positive growth factor in the progression of lung cancer. Similarly, the need for negative growth factors are demonstrated in studies of progression of a malignant melanoma. A malignant melanoma when converted to the ascites, progressed to a rapidly growing amelanotic melanoma (Gray and Pierce, 1964). The wild type tumor was cloned and slowly growing pigmented melanomas were obtained as were fast growing amelanotic melanomas similar to the progressed tumors. It was concluded that progression depended upon selection of cells best able to survive under the conditions. Because the amelanotic stem line grew four times faster than the melanotic one and the wild type tumors failed to segregate into black and white areas, it was postulated that the slow growing melanotic cells somehow inhibited the fast growing amelanotic ones. The negative growth factors could operate by killing fast growing cells or by slowing their growth. The evidence of factors produced by one population of cells that are toxic to another as reported here for teratocarcinoma reinforces these ideas, but such factors have not been identified in the melanoma system as yet.

There is no reason to believe that mechanisms similar to those of progression should not operate during the latent period in carcinogenesis between initiation of tumor cells and the appearance of tumors as the result of promotion. In this situation one could envision host factors and autocrine and paracrine secretions affecting subpopulations of tumor cells either positively or negatively. Although Foulds (1969), Greene (1951), and Furth (1953) did not consider progression as a mechanism of carcinogenesis, there is no reason why the same principles shouldn't occur during the evolution of a tumor. It should be kept in mind that regulatory factors may also be produced by surrounding normal tissues, and thereby contribute to progression as it occurs when only a few neoplastic cells are present.

Finally, whereas the vast majority of malignant tumors undergo progression, very rarely a primary tumor and its widespread metastases may undergo spontaneous regression and either disappear or become differentiated into benign cells of no danger to the host (Cushing and Wolbach, 1927). In the past, immune mechanisms have been invoked to explain the spontaneous disappearance of tumors (Lewison, 1976). Because the mechanisms of differentiation were and are unknown, scant heed was paid to the idea that the production of toxins and differentiating factors by tumors in the appropriate ratios could in the rare instance account for spontaneous regression. We all know how rare it is to find a rapidly proliferating cell type that retains the ability to differentiate. This is true in the adult but perhaps not in tumors, or in the embryo which the tumor caricaturizes, (Pierce and Speers, 1988) where all cells are rapidly dividing.

ACKNOWLEDGEMENTS

This work was supported in part by a gift from RJR Nabisco Co., and NIH grants #CA 35367 and CA 47369. We wish to thank V. Starbuck for editorial assistance.

REFERENCES

Becker, K. L., 1984, The endocrine lung, in: "The Endocrine Lung in Health & Disease," K. L. Becker and A. F. Gasdar, eds., W. B. Saunders Co., Philadelphia.

Bernstine, E. G., Hooper, M. G., Grandchamp, S. and Ephrussi, B., 1973, Alkaline phosphatase activity in mouse teratoma, Proc. Natl. Acad. Sci., U.S.A., 70:3899-3903.

Braun, A. C., 1956, The activation of two-growth-substance systems accompanying the conversion of normal to tumor cells in crown gall, Cancer Res., 16:53-56.

Cox, W. F., Jr., and Pierce, G. B., 1982, The endodermal origin of the endocrine cells of an adenocarcinoma of the colon of the rat, Cancer, 50:1530-1538.

Cushing, H. and Wolbach, S. B., 1927, The transformation of a malignant paravertebral sympathicoblastoma into a benign ganglioneuroma, Am. J. Path. 3:203-216.

Damjanov, I., Solter, D., Belicza, M. and Skreb, N., 1971, Teratomas obtained through extrauterine growth of seven-day mouse embryos, J. Natl. Cancer Inst., 46:471-475.

El-Shershaby, A.M. and Hinchliffe, J. R., 1974, Cell redundancy in the zona-intact preimplantation mouse blastocyst: a light and electron microscopic study of dead cells and their fate, J. Embryol. Exp. Morph., 31:643-654.

Evans, M. J. and Kaufman, M. H., 1981, Establishment in culture of pluripotential cells from mouse embryos, Nature 292:154-156.

Fidler, I. J., 1973, Selection of successive tumor lines for metastasis, Nature New Biol., 242:148-149.

Foulds, L., 1969, "Neoplastic Development," Vol. 1, Academic Press, 1969.

Furth, J., 1953, Conditioned and autonomous neoplasms. A review, Cancer Res., 13:477-492.

Gray, J. M., and Pierce, G. B., 1964, Relationship between growth rate and differentiation of melanoma in vivo, J. Natl. Cancer Inst., 32:1201-1210.

Greene, H. S. N., 1951, A conception of tumor autonomy based on transplantation studies: A review, Cancer Res., 11:899-903.

Grunz, H., McKeehan, W. L., Knochel, W., Born, J., Tiedemann, H., and Tiedemann, H., 1988, Induction of mesodermal tissues by acidic and basic heparin binding growth factors, Cell Differentiation, 22:132-190.

Handyside, A. H., and Hunter, S., 1986, Cell division and death in the mouse blastocyst before implantation, Roux's Arch. Dev. Biol., 195:519-526.

Heath, J. K., and Isake, C. M., 1984, PC13 embryonal carcinoma-derived growth factors, The EMBO Journal, 3:2957-2962.

Kleinsmith, L. J., and Pierce, G. B., 1964, Multipotentiality of single embryonal carcinoma cells, Cancer Res., 24:1544-1551.

Lehman, J. M., Speers, W. C., Swartzendruber, D. E., and Pierce, G. B., 1974, Neoplastic differentiation: Characteristics of cell lines derived from a murine teratocarcinoma, J. Cell Physiol., 84:13-28.

Lewison, E. F. (ed.), 1976, Conference on Spontaneous Regression of Cancer, Natl. Cancer Inst. Monogr., 44:1-150.

Martin, G. R., 1981, Isolation of a pluripotential cell line from early mouse embryos cultured in medium conditioned by teratocarcinoma stem cells, Proc. Natl. Acad. Sci., U.S.A. 78:7634-7638.

Mintz, B., and Illmensee, K., 1975, Normal genetically mosaic mice produced from malignant teratocarcinoma cells, Proc. Natl. Acad. Sci. U.S.A., 72:3585-3589.

Monzo, M., Andres, X., and Ruano-Gil, D., 1983, Etude morphologique d'une population homogene de cellules de terato-carcinome, Bulletin De L'Association Des Anatomistes, 67:91-98.

Nowell, P. C., 1982, Genetic instability in cancer cells: Relationship to tumor cell heterogeneity, in: "Tumor Cell Heterogeneity", A. H. Owens, D. S. Coffey, and S. B. Baylin, eds., Academic Press, New York.

Parchment, R. E., Damjanov, A., Damjanov, I., Gramzinski, R. A., and Pierce, G. B., 1989, Neoplastic embryoid bodies of embryonal carcinoma as a source of blastocele fluid, in preparation.

Pierce, G. B., 1967, Teratocarcinoma: Model for a developmental concept of cancer, in: "Current Topics in Developmental Biology," Vol. 2, A. A. Moscona and A. Monroy, eds., Academic Press, New York.

Pierce, G. B., Arechaga, J., Jones, A., Lewellyn, A., and Wells, R. S., 1987, The fate of embryonal-carcinoma cells in mouse blastocysts, Differentiation, 33:247-253.

Pierce, G. B., and Dixon, F. J., Jr., 1959a, Testicular teratomas. I. The demonstration of teratogenesis by metamorphosis of multipotential cells, Cancer, 12:573-583.

Pierce, G. B., and Dixon, F.J., Jr., 1959b, Testicular teratomas. II. Teratocarcinoma as an ascitic tumor, Cancer, 12:584-589.

Pierce, G. B., Dixon, F.J., Jr., and Verney, E. L., 1960a, An ovarian teratocarcinoma as an ascitic tumor, Cancer Res. 20:106-111.

Pierce, G. B., Dixon, F. J., Jr., and Verney, E. L., 1960b, Teratocarcinogenic and tissue-forming potentials of the cell types comprising neoplastic embryoid bodies, Lab. Invest., 9:583-602.

Pierce, G. B., Lewellyn, A., and Parchment, R. E., 1989, Detection of different regulatory mechanisms for subpopulations of embryonal carcinoma and inner cell mass cells in the mouse blastocyst, in preparation.

Pierce, G. B., and Speers, W. C., 1988, Tumors as caricatures of the process of tissue renewal: Prospects for therapy by directing differentiation, Cancer Res., 48:1996-2004.

Rizzino, A., 1985, Early mouse embryos produce and release factors with transforming growth factor activity, In vitro Cell. Dev. Biol., 21:531-536.

Rizzino, A., and Bowen-Pope, D. F., 1985, Production of PDGF-like growth factors by embryonal carcinoma cells and binding of PDGF to their endoderm-like differentiated cells, Dev. Biol., 110:15-22.

Rizzino, A., and Crowley, C., 1980, Growth and differentiation of embryonal carcinoma cell line F9 in defined media, Proc. Natl. Acad. Sci., U.S.A., 77:457-461.

Rizzino, A., Orme, L. S., and DeLarco, J. E., 1983, Embryonal carcinoma cell growth and differentiation production of and response to molecules with transforming growth factor activity, Experimental Cell Res. 143:143-152.

Slack, J. M. W., Darlington, B. G., Heath, J. K., and Godsave, S. F., 1987, Mesoderm induction in early Xenopus embryos by heparin-binding growth factors, Nature, 326:197-200.

Speers, W. C., 1982, Conversion of malignant murine embryonal carcinoma to benign teratomas by chemical induction of differentiation in vivo, Cancer Res., 42:1843-1847.

Stevens, L. C., 1967, Origin of testicular teratomas from primordial germ cells in mice, J. Natl. Cancer Inst., 38:549-552.

Stevens, L. C., 1970, The development of transplantable teratocarcinomas from intratesticular grafts of pre- and post-implantation mouse embryos, Dev. Biol. 21:364-382.

Strickland, S., and Mahdavi, V., 1978, The induction of differentiation in teratocarcinoma stem cells by retinoic acid, Cell, 15:393-403.

van Veggel, J. H., van Oostwaard, T. M. J., de Laat, S. W., and van Zoelen, E. J. J., 1987, PC13 embryonal carcinoma cells produce a heparin-binding growth factor, Experimental Cell Res., 169:280-286.

DISCUSSION

Michael Lieberman: This morning, Henry mentioned that progression, in the traditional view of those working with chemical carcinogenesis, was largely irreversible. Yet many of the experiments, the Mintz experiment for one, would suggest that progression is not irreversible at all, under certain circumstances. I'm wondering if you can talk a little about Brinster's experiments in terms of ploidy. If you use a diploid or a near diploid line to begin with versus an aneuploid line, when you get the chimeras, what is the ploidy of the chimeric cells?

G. Barry Pierce: Down's syndrome people are aneuploid: this means that some aneuploid cells are capable of going through embryonic development and becoming well differentiated. In the case of Down's syndrome the individual is well differentiated but not particularly functional. Apparently the outcome depends upon the chromosomes involved. In the Brinster experiment, it has been shown that some aneuploid cells are regulated in terms of becoming benign and apparently normal functional cells in the milieu of the embryo-derived cells (1,2,3. The cancer derived cells are considered normal even though they are derived from aneuploid precursors because they respond to homeostatic regulation (4). This is the best criterion of normalcy that we have. Chimera formation with embryonal carcinoma cells under the best of circumstances is a rare event (2), and it is impossible to say how adversely aneuploidy affects the process. It is clear from studies in which inner cell mass cells are transformed in vitro to become embryonic stem cells (or in vitro derived embryonal carcinoma cells) that these transformed cells are little altered from the normal. They are capable of forming fertile chimeras in a much higher proportion of cases than has been achieved with spontaneously arising or egg transplant-derived embryonal carcinomas (5). This would suggest that aneuploidy is a decrement in chimera formation.

There is no reversal of progression in embryonic regulation of cancer. The undifferentiated tumors that have been employed in chimera formation have all undergone progression as Foulds defined it. Like the tumors employed by Foulds, Furth, and Greene, our tumors remain undifferentiated in serial transplantation and do not regress. What was not known to these investigators, who so accurately described progression, was that embryonal carcinoma cells, and presumably other tumors as well, retain the potential for differentiation even though it is not expressed in the usual transplantation circumstances. Thus, when the tumors are placed in the appropriate environment they express the potential for differentiation, become benign and possibly even normal in terms of their regulation. In the case of chimera formation the multipotent embryonal carcinomas become benign, unipotent somatic cells. They are no longer malignant and cannot be construed as an element of progression. As for the ploidy of the cancer-derived chimeric cells, there is little information. I presume that aneuploid cells remain aneuploid, but, again, there is little information.

Jose Russo: I enjoyed your presentation, it is a very nice piece of work. My concern is the following: Is the presence of a putative growth factor or factors in the cells an indication that they are functional?

Pierce: John Heath at Oxford and Angie Rizzino at Nebraska have done the most work in this regard. Rizzino usually refers to the factors as PDGF-like molecules or TGF-like molecules and he has shown that they are functional in terms of their reactions in tests for them (6). What their function might be in the blastocyst is not known because mesenchyme has not developed at the time the factors appear. So it is likely that they have autocrine or paracrine functions in the blastocyst, but what those functions are we do not know at this time.

Harry Rubin: How aneuploid can the tumors get and still make a chimera?

Pierce: I can't answer that specifically either, Harry, but we have a tumor that has two modal peaks at 57 and 74 chromosomes. This tumor is regulated in terms of tumor and colony forming ability by the blastocyst, and makes trophectodermal chimeras, but not chimeric mice. I wondered which of the chromosomal modes would be the most efficient in regulation of tumor and colony formation so I cloned the tumors, but by the time I had enough cells to karyotype, the tumors-lines had chromosomes numbers of 57 and 74 (unpublished). I now hesitate to accept the view that the neoplastic karyotype is unstable. I think there have to be stringent regulatory mechanisms that make a cell in tissue culture, with either 57 or 74 chromosomes, produce offspring with 57 and 74 chromosomes and not some other number.

Ivan Damjanov: I could just add that I don't think that the ploidy makes any difference with regard to the capacity of the tumor to differentiate. People have looked for chromosomal abnormalities and couldn't find much. We have a human teratocarcinoma stem line that has 53 chromosomes and several diploids and differentiates nicely.

Pierce: I agree. Spears induced aneuploid embryonal carcinoma cells to differentiate with chemicals (7). His cells had Robertsonian translocations and these translocations were found in the differentiated cells of the experiment (Speers, unpublished). You cannot say that the differentiated cells were normal or that they might have been produced in a chimera, but they were benign when assayed in vitro. Little is known about the nature of the benign cells in chimeras. Are they similar to SV_{40} transgenic cells, most of which behave quite normally when integrated into an animal, but when put in culture transform almost immediately? Are the cancer derived "normal" cells initiated cells that only require promotion for the expression of the neoplastic state? Are they truly normal cells? Little information is available.

Damjanov: Talking about progression and promotion, is there a possibility that there might be some progression in your system since you were talking about primary and secondary stem cells? Is there something that you could really pin it down to and say these are primary and secondary stem cells? Is there something that we could call progression in this system?

Pierce: Oh yes. I think we see classical tumor progression, because if you select teratocarcinomas by repeated intraperitoneal transplantation, the tumors simplify their morphologic patterns, increase their growth rate and never revert to the wild type. This is classical progression. As for secondary stem lines in a teratocarcinoma, these are embryonal carcinoma cells that for reasons unknown are no longer totipotent. In embryologic terms they are restricted in their differentiation.

In the normal embryo there is a totipotent stem in the 8 or 16 cell embryo. If these totipotent cells undergo carcinogenesis the resulting embryonal car- cinoma should be totipotent and in my terminology a primary stem cell. If a chimera were made with this totipotent cell its progeny would be able to popu- late the germ line, which seems to be the best test of totipotency that we have at that time. Inner cell mass derived embryonal carcinomas are totipo- tent according to this definition. As these totipotent tumors are manipulat- ed in vitro and in vivo the cells lose part of their potency and become secon- dary stem cells. This might be similar to determination in the embryo. As an example ECa 247 has a potential largely restricted to trophectoderm. I view it as a secondary stem cell line in comparison to the totipotent OTT6050 primary stem cell line that Mintz and Illmensee employed in their chimera ex- periments. Because ECa 247 is more restricted in relation ship to the totipo- tency of OTT6050, this raises the question that you are concerned with: Is ECa 247 a restricted or secondary stem line because of progression or is it a restricted secondary stem line because of determination and differentiation? Remember that the scientists who first described progression were unaware that cancer cells could differentiate and did not include the effect of dif- ferentiation in their deductions. I believe ECa 247 arose via induction and selection of the induced cells. I cannot prove the idea and cannot rule out that ECa 247 did not develop exclusively by the unknown mechanisms of progression.

Charles Boone: Around 10% of Down's syndrome patients get leukemia.

Pierce: Yes, I'm aware of that Chuck, and I don't know why. Down's syndrome individuals do not get squamous cell carcinomas or any other kinds of tumor. It is an interesting phenomenon, possibly there are people in the room that can explain why these people should be so susceptible to leukemia. I can't.

Peter Duesberg: Barry, did I understand correctly that you have chimeric mice that are aneuploid, that have completely normal tissue and have 57 chromosomes in a cell?

Pierce: All of my chimeras made with ECa 247 (57 and 74 chromosomes) are chimeric in the extraembryonic tissues only. I have never produced a viable chimera with ECa 247cells, but I must admit that I have not tried too many times. Papaioannou produced chimeras from aneuploid embryonal carcinoma cells. She was impressed with the strain specificity of embryonal carcinoma cells able to make chimeras. For example, she had a line C145 that never made chimeras, but in a few cases the extraembryonic endoderm of these ani- mals were chimeric. I would refer you to her studies which are the most complete. P19, an embryo transplant derived embryonal carcinoma, was euploid and seldom gave rise to a living chimera. Analysis of embryos at mid-gesta- tion indicated that 60% of blastocysts injected with P19 cells were chimeric at mid-gestation. Clearly chimera formation is a complex process which can be aborted as a result of deficiencies of cells anywhere along the developmental pathway.

Duesberg: So, is there any evidence in your lab or in the literature that there is a normal mouse tissue somewhere, with 57 chromosomes?

Pierce: The answer is yes, chimeric mice have been produced by inclusion of aneuploid embryonal carcinoma cells into blastocysts. The resulting animal may be chimeric in only a few of its tissues. But chimeric animals have been produced using aneuploid cells.

Duesberg: But none of yours are chimeric?

Pierce: I have produced chimeric mice with P19 cells but I have never produced chimeric mice with ECa 247 cells for the reasons given above. These cells preferentially localize in trophectoderm and I am using them as a neoplastic model of early inner cell mass with trophectodermal potential, to study how pretrophectodermal cells are regulated in the blastocyst. For a control for these studies, I used P19 cells which form a high percentage of midgestation chimeras. This allows, then, comparisons to be made between regulation of pretrophectodermal cells and embryonic cells. The former give rise to extraembryonic tissues and the latter to embryonic ones. To date we have found a toxic factor in blastocele fluid that specifically kills ECa 247 cells, but not P19 cells. We believe that this is the mechanism that ensures that trophectoderm or trophoblast, its derivative, is eliminated from the embryo and therefore cannot appear as an inclusion in the newborn animal.

REFERENCES

1. Papaioannou, V.E., Gardner, R.L., McBurney, M.W., Babinet, C. and Evans, M.J., Participation of cultured teratocarcinoma cells in mouse embryogenesis. J. Embryol. Exp. Morph., 44:93-104, 1978.
2. Papaioannou, V.E., Evans, E.P., Gardner, R.L. and Graham, C.F., Growth and differentiation of an embryonal carcinoma cell line (C145b). J. Exp. Morph. 54:277-295, 1979.
3. Illmensee, K., Reversion of malignancy and normalized differentiation of teratocarcinoma cells in chimeric mice. In, Genetic Mosaics and Chimeras in Mammals. L.B. Russell, Ed., pp. 3-25, Plenum, New York, 1978.
4. Pierce, G.B., and Speers, W.C., Tumors as caricatures of the process of tissue renewal: prospects for therapy by directing differentiation. Cancer Res. 48:1996-2004, 1988.
5. Evans, M. J., and Kaufman, M. H., Establishment in culture of pluripotent cells from mouse embryos. Nature 292:154-156, 1981.
6. Rizzino, A., and Bowen-Pope, D.F., Production of PDGF-like growth factors by embryonal carcinoma cells and binding of PDGF to their endoderm-like differentiated cells. Dev. Biol. 110:15-22, 1985.
7. Speers, W.C., Conversion of malignant murine embryonal carcinoma to benign teratomas by chemical induction of differentiation in vivo. Cancer Res. 42:1843-1847, 1982.

DEVELOPMENTAL POTENTIAL OF MURINE

PLURIPOTENTIAL STEM CELLS

Allan Bradley

Institute for Molecular Genetics
Baylor College of Medicine
One Baylor Plaza
Houston, TX   77030

PLURIPOTENT STEM CELL LINES

The early embryo contains a group of pluripotential inner cell mass (ICM) cells which give rise to cells of the differentiated tissues of the adult, including the germ line[1-2]. As the ICM cells proliferate, groups of these cells become committed to specific developmental pathways. Pluripotential cells appear to persist in the embryonic portion of the embryo up until 7.5 days[3]. These undetermined pluripotent stem cells have been established as permanent tissue culture cell lines either directly from the embryo (embryonic stem (ES) cells)[4,5] or indirectly from teratocarcinoma tumors (embryonal carcinoma (EC) cells) (See Fig. 1).

Origin of EC Cells

Teratoma tumors were first discovered as spontaneous tumors in the testes of strain 129 mice. These tumors arise from actively proliferating germ cells and may be recognized histologically within the seminiferous tubules of the 15 day mouse fetus[6]. At this time the cells resemble the ectodermal cells of normal day 5 or day 6 embryos. Transplanted germinal ridges from strain 129 mice also give rise to tumors at similar frequencies. Germ cells were demonstrated to be the initiator cells in such transplants because ridges derived from Steel homozygous donors (germ cell deficient) do not give rise to teratomas[6].

Teratoma tumors also arise spontaneously in the ovaries of strain LT mice[7]. In this instance these tumors arise from oocytes that have initiated development spontaneously (i.e., parthenogenetically). These postmeiotic oocytes[8] develop as fairly normal preimplantation embryos[6] but subsequent growth in this ectopic site leads to abortive embryogenesis.

The observation of the formation of teratomas of an embryonic origin in LT mice led to the artificial induction of teratomas by transplanting early embryos (1-7.5 days) to ectopic sites in histocompatible hosts[9-13]. Teratomas formed under these conditions closely resemble those which develop spontaneously in LT mice.

Many testicular-, ovarian-, and embryo-derived teratomas are benign. In these tumors all of the cells differentiate and lose their ability to proli-

ferate. Some tumors will proliferate after transplantation between histocompatible hosts. These tumors are known as teratocarcinomas. The transplanta-
bility of such tumors resides in individual stem cells within them, so-called
embryonal carcinoma (EC) cells. The diverse array of differentiated cells
associated with teratocarcinomas has also been demonstrated to be a property
of the embryonal carcinoma cells: single EC cells transferred in vivo have
reformed teratocarcinoma tumors and following proliferation have given rise
to the entire array of differentiated cell types.

Three main factors determine whether a grafted embryo will give rise to
a teratoma or a teratocarcinoma:

1. The developmental stage of the embryo[3,15]
2. The portion of the grafted embryo[6,17]
3. The genetic background of the embryo and host[18-20]

The efficiency of teratocarcinoma formation appears to increase with the age
of the transplanted embryo up to 7.5 days development, which presumably
reflects the increasing number of pluripotent stem cells in the transplanted
embryo. After 7.5 days efficiency decreases abruptly and embryos give rise
to teratomas, reflecting the loss of pluripotent cells from these later stage
embryos[3,15]. The transfer of dissected egg cylinders has demonstrated that
teratocarcinoma formation is a property restricted to the embryonic ectoderm
cells[16,17].

Embryo grafting using a variety of mouse strains has also shown that
teratocarcinoma formation is effected by genetic factors. The genotype of
the host is crucial for the proliferation rather than differentiation of EC
cells[19]. Some strains are permissive (e.g., C3H and BALB/c) with up to 70%
of embryos forming teratocarcinomas[20], while other strains (e.g. AKR and

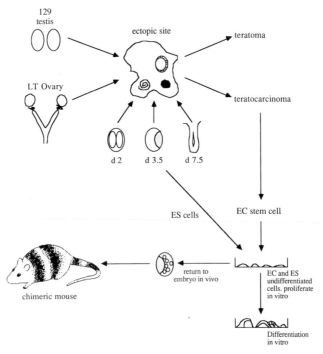

Fig. 1. Origin and differentiation of EC and ES
        cells.

C57BL) are non-permissive. Additionally, there is maternally transmitted permissiveness in both host and graft[20].

EC cells can be propagated indefinitely in a tumor and can also be isolated and subsequently propagated in vitro[21]. Normally the frequency of obtaining EC cultures from primary tumors is very low[22,23], but this may be increased in subsequent generations by increase in EC cell components of the tumor.

Origin of ES Cells

Embryos may be removed prior to implantation and grown successfully in vitro beyond the normal implantation time in the uterus. In vitro the embryo attaches to the surface of the petri dish and the trophectoderm cells spread ultimately exposing the ICM cells to the culture environment[24]. The ICM will continue to proliferate and if left undisturbed will form a variety of differentiated cell types. However, the undisturbed growth of an embryo in vitro is not a strong selection for growth of pluripotent stem cells. Presumably these cells exist but ultimately the developmental differentiation signals are strong enough to stimulate complete differentiation in preference to continued proliferation of the stem cells within the egg-cylinder structure.

It is possible to disturb the developing embryo 4-5 days after implantation in vitro and disassociate the cells into smaller groups. This deprives the cells of the normal differentiation signals and groups of embryonic stem cells, detected by virtue of their morphological similarity with EC cells, can be seen to proliferate as primary colonies. These colonies can be subcultured and expanded into cell lines[4,5]. The isolation of ES cells does not appear to show strain restrictions[24].

STEM CELLS IN VITRO

EC and ES cells have many advantages over embryonic cells in experimental situations[26,28]. Stem cells facilitate the study of cells equivalent to those isolated from the early post implantation embryos which would otherwise be obscured by the maternal environment. In addition, cellular homogeneity and the availability of large numbers of cells facilitate biochemical analysis.

Differentiation of EC Cells In Vitro

When EC cells differentiate in vitro, the first differentiated cells to appear are epithelioid in type, and are recognized as primary endoderm. These cells closely resemble the parietal or visceral endoderm of the early post implantation mouse embryo[26,29,30] (See Fig. 2). Primary endoderm may appear spontaneously in various cultures, although typically EC cells can be induced to differentiate routinely by denying the cells a feeder layer and culturing them as aggregates on a non-adhesive surface[29,31] (See Fig. 2). The aggregates of cells which typically form under these conditions have an inner core of ectodermal cells and an outer rind of endoderm cells. These structures resemble the egg cylinder of the early post-implantation embryo.

Simple embryoid body-like structures are also seen when immuno-surgically isolated ICMs are cultured for a short time in vitro[32], and are also a common feature of intraperitoneally passaged tumors.

Those EC cell lines which differentiate well, will further differentiate from simple embryoid bodies to become "cystic" embryoid bodies. These are composed of an endodermal outer shell and a mesodermal inner core. Quite complex but disorganized development has been observed in cystic embryoid bodies. Although the basic process of endoderm formation appears to be very similar, the morphogenesis of mesoderm would appear to be dissimilar from that which occurs in the normal 6-7 day embryo[33]. If cystic or simple embryoid bodies are allowed to re-attach to tissue culture dishes, further cellular differentiation will occur over a number of weeks. While the degree of differentiation can be quite considerable, it is often of a disorganized nature.

One of the dissimilarities between the teratocarcinoma model system and the normal embryo is the appearance of parietal endoderm, or occasionally a mixture of parietal and visceral endoderm, as the primary differentiated cell type which delaminates on the surface of the EC cell aggregate[33]. This contrasts with the true embryonic situation where visceral endoderm, and not parietal endoderm, is found in contact with the surface of the egg cylinder.

This observation would appear to contradict the true embryonic situation where it appears that the local environment is important for the segregation of the endoderm into two distinct populations. It is thought that this might be affected by contact with different underlying cell types[34-36].

F9 EC cells have been induced to differentiate in vitro by treatment with retinoic acid[38] and primary endoderm cells so formed could be induced to differentiate further in a visceral or parietal direction (determined by the appearance of basement membrane structures and alpha-fetoprotein[38-40]). Thus the further extension of the in vitro model system to a slightly later and more subtle set of interactions which are comparable to events seen in normal embryonic development, demonstrates that influences resulting from cell-cell interactions occurring in vivo can be mimicked in vitro. Other EC cell lines have been stimulated to differentiate with retinoic acid[41]. P19 EC cells are particularly interesting since they can be manipulated to develop into either a neuronal or muscle direction in vitro[42-44].

Homology between EC cells and true embryonic stem cells has been further investigated using various antisera which react against cell surface antigens on EC cells. A number of antisera recognize similar antigenic sites on EC cells, germ cells and early embryos[45]. The monoclonal reagent M1/22.25 has been particularly informative in this regard[46]. It reacts with ICM cells, but does not label the ICM-derived embryonic ectoderm of the 6.5 day embryo, although it does react with the determined embryonic endoderm. The ectoderm only shows a positive response until late on day 5[47]. By this stage the inner cell mass has delaminated to form an outer layer of endoderm and additionally contains a rudimentary proamniotic cavity (See Fig. 2).

By cell surface criteria this is the most advanced stage at which cells homologous to EC cells appear to be present in the developing embryo, its in vitro equivalent being the simple embryoid body. Primordial germ cells are not easily distinguishable until later in development when they are located in the germinal ridges. At this time they are M1/22.25 positive. This data suggests that the ectodermal cells before 5.5 days are the embryonic homologs of EC cells.

However the cell surface phenotype may not be the best indicator of the determinative state of a particular cell type, because this only views a single protein moiety on the surface of a cell. 2-D gel electrophoretic comparisons of protein synthesis are probably a better method of analysis because the complete spectrum of proteins produced by a cell type may be compared. Such comparisons of 3.5 day ICM cells and EC cells, however, do not support the monoclonal results[48-50]. Various other comparisons between

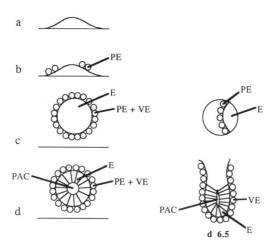

Fig. 2. Differentiation of stem cells in vivo and in vitro.
   a. Stem cell colony attached to substrate in vitro.
   b. Formation of endodermal cells on surface of the
      stem cell colony.
   c. Subculture of attached colonies from (b) on a
      non-adhesive surface results in formation of
      simple embryoid bodies with an outer rind of
      endoderm cells. Formation of endoderm resembles
      that observed with the 4 day blastocyst.
   d. Simple embryoid bodies progress in culture. The
      inner core of ectodermal cells become columnar
      and a proamniotic cavity forms. This resembles
      events observed in the 6.5 d embryo in vivo.

PE = Primary endoderm (b) or Parietal endoderm (c) & (d)
E   = Ectoderm
VE = Visceral endoderm
PAC = Proamniotic cavity

embryonic and EC cell proteins have been made. It appears that EC cells aggregated into small clumps and labeled during the first three hours of suspension culture and the cells of the 5 day embryo bear the closest similarity[51].

Some of the most compelling data which supports the model teratocarcinoma system is the ability of EC cells to differentiate normally following reintroduction into the blastocoele of a blastocyst[54-56]. However, it is also in this environment that differences between tumor-derived EC cells and embryo-derived ES cells become most apparent as reflected by the fate of the injected cells within both the chimaeric embryo and adult animal.

STEM CELLS IN VIVO

The pluripotency of both ES and EC cells may be unequivocally demonstrated by their ability to form chimaeric associations when combined with normal mouse embryos, ultimately participating in the formation of a wide variety of tissues in chimaeric individuals[52-58].

Normal Differentiation

EC Cells. Many differences are evident in the differentiative abilities of EC cell lines in vivo and this is reflected in their differing abilities to form chimaeras (See Table 1). Some EC cell lines fail to show either normal or neoplastic growth in the embryonic environment (for example in the R5/3, F1/9, and 1009 cell lines[59]). The reason for this is not known, but either might be due to the death of the introduced cells, or to their rapid differentiation into a non-proliferative state. Other lines, for example, PSA4 TG12, apparently only contribute to extraembryonic tissues[60]. Lines such as C17, C86, and P19 produce chimaeras with abnormal development or tumors[54,55,61].

The range of differentiative abilities of EC cell lines in chimaeras is correlated with poor differentiation in vitro. Those cell lines which require artificial induction of differentiation clearly do not exhibit normal differentiation in an embryonic environment[61,62].

The possession of a normal karyotype, although a useful characteristic of an EC cell line, does not necessarily ensure normal differentiation in a chimaera. Conversely, cell lines which show extensive somatic chimaerism do not always possess a normal karyotype[56,60,63].

If EC cells are to be involved in the production of functional gametes, it is essential that they should possess a normal chromosome complement, otherwise the failure of chromosome segregation at meiosis will result in the production of aneuploid gametes. Functional gametes have been reported from two lines maintained as in vivo tumors, OTT6050 and LT2484-395[53,64,65] and in the in vitro line METT-1[57,66]. Only 6 individuals which have shown this characteristic have been reported, out of many chimaeras produced from these and other lines[59].

The reasons for the almost total exclusion of EC cells from the embryonic cellular pool which gives rise to the gametes are not clear; it might reflect a bias or limitation in the differentiative ability of a particular cell line. An analysis of a number of chimaeric individuals constructed using the EC cell line OTT6050 (ascites-derived) or LT72482 (solid tumor-derived), has indicated sporadic distribution of tumor cell-derived tissue; EC-derived cells appearing in only one or two tissues (out of 12 examined), in the majority of chimaeric individuals analyzed[65]. However, from a consideration of the spectrum of colonization of somatic tissues by those cell lines, which have successfully produced germ line chimaeras, it appears that

Table 1. Derivation and Chimaera Forming Efficiencies of a Number of
EC Cell Lines Examined Before and After Birth

		Midterm			Term		
	Strain	Number Analysed	Number Normal	Number Abnormal	Number Analysed	Number Normal	Number Abnormal
OTT6050	129	179	8	6	279	32[a]	2
OTT5568	129	272	63	68	168	28	1
METT-7	129	-	-	-	324	46[b]	1
LT72484	LT	30	2	10	74	8[c]	3
C17	C3H	-	-	-	77	5	8
C86	C3H	13	3	0	71	0	6
C145b	C3H	404	5	0	201	0	0
P19	C3H	422	36	73	62	0	10
P10	C3H	9	3	0	58	31	0
F1/9	F1	71	0	0	18	0	0
1009	129	15	0	0	137	0	0
		1415	120	157	1469	150	31
			(8.4%)	(11.1%)	(10.2%)	(2.1%)	

Note: Cell lines grouped under original tumor for clarity.

[a]Two germ cell line chimaeras reported.
[b]Two germ cell line chimaeras reported.
[c]One germ cell line chimaera reported.

References: 52-57, 59-61, 63-67

exclusion of EC cells in the formation of any particular tissue is not a general feature of EC cells (although some EC cell lines have shown restricted patterns of differentiation within many individuals examined[55,60]).

The three germ line chimaeras reported prior to 1981[53,64,67] had been derived from EC cell lines maintained entirely in vivo by serial passage of tumor material in syngeneic hosts, a method of culture presumed to be less selective than that incurred during maintenance in vitro. However, the METT-1 line has given rise to two germ line chimaeras[68] and has been maintained under in vitro culture conditions prior to blastocyst injection[69]. When the EC-chimaeras are compared with ICM-chimaeras it is clear that the tissue distributions of EC chimaeras are very patchy, thus, the EC cells appear to remain somewhat clumped during embryogenesis. ICM chimaeras do, however, show considerable mixing with host cells, indicating that the introduced cells do not remain clumped, even when introduced into the embryo as a single clumped inner cell mass. The differences in cellular mixing presumably result in the limited distributions of EC-derived cells in chimaeras.

There are indications that the derivation of the cell lines might also be a significant factor in chimaera formation efficiency. To date, EC-chimaeras have only been produced from embryo-derived cell lines, whether of spontaneous parthenogenetic derivation[63,70] or from fertilized embryos which had been used to induce tumors[59]. No chimaeras have so far been reported using EC cell lines derived from spontaneous testicular germ cell tumors. It is possible that premeiotic germ cell-derived EC cell lines in testes are not capable of participating in normal embryogenesis, and this might be a reflection on the different derivation of the cell types involved. That is, germ cell lines are derived from neoplastically transformed germ cells, while embryo-derived lines are isolated from definitive embryonic pluripotential cells which have undergone induced neoplasia in an ectopic site.

ES Cells. A large study of the differentiation potential of ES cells in vivo has been carried out on 12 independently derived ES cell lines (See

Fig. 3. Chimaeric mouse generated by injecting an ES cell line from a pigmented mouse into an albino host embryo. The fine grained intermingling of pigmented and Albino hairs is clearly visible.

Table 2(a). Chimaeras Constructed Following Injection of
1-3 ES Cells/Embryo

Cell Line	Number Injected	Number Born	% Born	Number Chimaeric	% Chimaeric
B2B2	66	40	60.1	14	35.0
CP2	123	94	76.4	18	19.1
CP3	156	111	71.1	43	38.7
CP4-2[+]	29	8	27.6	5	62.5
A13	160	109	68.1	36	33.0
X1	79	52	73.5	4	7.7
CL6	161	103	64.0	21	20.4
Rm5.3	63	48	76.2	8	16.7
CZ4+	8	6	75.0	2	33.3
	845	571	67.6	151	26.4

Table 2a and 2b)[71,72]. Without exception all have proliferated and integrated into the host embryo normally. The chimaeras obtained had many similarities to the chimaeras formed following injection of ICM cells into host blastocysts (See Fig. 3). Thus, the proportion of chimaeras generated was very high and did not involve prenatal loss; the level of contribution to the chimaeras as assayed by GPI isozymes and coat color was very high (over 99% in some cases) and the mixing of host and ES derived cells could be detected in almost every other tissue analyzed. Chimaeras generated with ES cells have mainly utilized the injection of karyotypically male (XY) cell lines into embryos of undetermined sex. Chimaeras generated in such a manner have demonstrated a sex distortion effect (See Table 2b)[58] caused by the dominant effect of the XY cells pushing development in the direction of a male. In cases where a female embryo has become a male chimaera, the germ line of this individual will be composed entirely of sperm derived from the introduced ES cells. Appropriate combinations of XY stem cells injected into XY embryos will also transmit through the germ line, albeit at a lower frequency. Typically about 40% of the XX embryos which are chimaeric are converted to males and of these about 33% will transmit ES derived sperm[58].

Sterile mosaics are also an outcome from such an experiment and these have a range of bilaterally distinct morphological abnormalities affecting the reproductive system. About 25% of the XY-XY combinations will also transmit the ES derived cells through the germ line.

ABNORMAL DIFFERENTIATION

One striking difference between EC and ES chimaeras is that abnormal differentiation is only observed with the former cells. The extent of abnormal development that has occurred in EC chimaeras is difficult to estimate, as very few studies have terminated in the experiment at mid-gestation. However, when this has been done it has revealed large numbers of abnormal embryos[61,63] which would be unlikely to come to term. The malignant EC phenotype is also not invariably suppressed following introduction into the embryo. In some EC-chimaeras the resulting tumors may also be a contributing factor to extensive prenatal loss.

One method of assessing prenatal losses where mid-gestation analysis has not been performed is to consider the proportion of injected embryos which

Table 2(b). Chimaeras Constructed Following Injection of 12-15 ES Cells/Embryo

Cell Line	Number Injected	Number Born	% Born	Number Chimaeric	% Chimaeric	Chimaeras Female	Male[a]
CP1	246	161	65.7	70	43.4	27	40
CC1.2	321	205	64.0	127	62.8	29	97
CC1.1	83	61	73.5	37	60.6	13	24
	650	427	65.7	234	54.8	69	161

[a]33% of fertile males are germ cell chimaeras.

survive to term. Summarizing the data from EC cell lines which have exhibited above average rates of chimaerism, the survival to birth was 35.9% (Lines OTT6050, LT2484, METT-1, P19, P19S18, P10[20]). This figure compares with an average for ES cell lines of 67% (Table 2a and b)[58]. It is also clear that if development is assessed midterm (Table 1), 20% of EC conceptuses are chimaeras, however at birth this has declined to 10%. This loss is presumably accounted for by the 11% of conceptuses which are visibly abnormal at the time of analysis.

It appears that extensive participation of many EC cell lines to an embryo may jeopardize the development of that conceptus. It is entirely possible that high levels of contribution to specific organs, particularly when EC cell descendants remain clumped, rather than undergoing cell mixing, may result in abortive development of that organ. This would lead to the selection for animals with much reduced levels of EC derived contributions when examined at birth. Indeed, where midterm examinations have been performed, malformed embryos tend to exhibit the highest level of EC derived participation[61,64]. However, this is not an absolute rule, which again reflects the heterogeneity of EC cells. Cell lines such as P10 have produced chimaeras with high levels of contribution and no abnormalities[73].

Many of the large number of EC-embryo chimaeras have developed tumors both before and after birth[59]. Tumors have mostly been teratocarcinomas though tumors of specific differentiated cells have also been observed. In most instances it is not possible to distinguish between the possibilities that a single cell did not differentiate and subsequently formed a tumor, or that it did differentiate and subsequently reverted, especially as most chimaeras with tumors do exhibit normal differentiated tissues. It is also not clear whether tumor bearing chimaeras received irreversibly tumorigenic cells from an indistinguishable population to the EC cell line. Experiments with one particular cell line P19S18 which forms both normal and abnormal tissues, have shown that injecting a single cell will produce chimaeras with both normal and abnormal tissues. Thus in this cell line neoplastic or differentiated phenotype is the property of single clones of cells. The abnormal pattern of growth may reflect an inability of the embryo to fully control the growth and differentiation of this particular cell line, perhaps its division rate exceeds that of the embryo cells, and ultimately the quantity of EC-derived tissue exceeds that which the embryo could regulate. Indeed experimental data from aggregation chimaeras has indicated that the ratio of embryo to EC-derived cells in a chimaeric embryo is crucially important to the successful development of the chimaera.

CONCLUSIONS

What might be the cause of the differences between ES and EC cells? This is a difficult question to answer because of the spectrum of cell types encompassed by the label "EC." In many senses these cells are very similar and the variety of outcomes from a chimaera construction experiment is merely a sensitive test for a subtle change in cell phenotype. Indeed some EC cell lines, such as P10, behave in a manner which is very similar to ES cells, while at the other end of the spectrum some EC cell lines such as C145b and C86 can not make successful normal contributions to a live born animal. Abnormal development may be attributed to either a change in cell surface phenotype which prevents cell mixing with the embryo or a cell cycle time that exceeds that of the cells that will eventually contribute to the embryo. It is tempting to assign these changes to the selection that takes place during the induction of the tumor. In this instance, there is clearly selection in the tumor for cells that continue to proliferate rather than differentiate in a tumor environment. In most instances this is unlikely to represent a genetic change since the frequency of teratocarcinoma generation can be as high as 80%. Thus it appears to be relatively routine for undifferentiated embryo cells to continue to proliferate when the normal relationship with the embryo is disrupted. In vitro the derivation of an ES cell line does not require selection for cells that do not respond to the normal developmental signals, rather the physical isolation of cell-cell contacts is ensured by the disruption of the embryo into small clumps of 2-3 cells. Consequently the stem cells are deprived of the normal differentiation signals which would operate during growth in an ectopic site in vivo. Ultimately these procedures result in an outcome that differs when measured by the regulation of the cell once it is returned to its normal environment in vivo.

The nature of the changes in the phenotype of a pluripotential embryonic cell to an embryonal carcinoma cell is currently not defined. Clearly it is highly variable and occurs with high efficiency (up to 80%), which tends to discount the possibility of a genetic change such as oncogene activation being responsible. This leads to the conclusion that the change must be of an epigenetic nature and the reversibility of this change must depend on other factors, such as the progression of these cells in vitro, and result in the accumulation of secondary events which can lead to a permanent alteration in cell phenotype. This will affect their extent of interaction once returned to the normal embryonic environment and determine the absolute level of contribution, which may be potentially detrimental to a developing embryo. Embryonic stem cells can be transmitted through the germ line with high efficiency. This has led to the development of these cells as a system which can be manipulated to change the genome of the mouse to study functions of genes by performing the real reverse genetics[74-76] either in a random[74], random and selected[76] and ultimately in a directed manner[77].

ACKNOWLEDGEMENTS

I would like to thank my colleagues Phil Soriano, Ann Davis, and Paul Hasty for their comments on this manuscript. I would also like to acknowledge the current support of my laboratory from the Searle Scholars Program.

REFERENCES

1. R. L. Gardner and M. F. Lyon, X-chromosome inactivation studied by injection of a single cell into the mouse blastocyst, Nature 231:385-386 (1971).
2. R. L. Gardner and J. Rossant, Investigation of the fate of 4.5 day postcoitum mouse inner cell mass cells by blastocyst injection, J. Embryol. Exp. Morph. 52:141-52 (1979).

3. I. Damjanov, D. Solter, and N. Skreb, Teratocarcinogenesis as related to the age of embryos grated under the kidney capsule, Wilhelm Roux Arch. 167:288-90 (1971).

4. M. J. Evans and M. H. Kaufman, Establishment in culture of pluripotential cells from mouse embryos, Nature 292:154-5 (1981).

5. G. R. Martin, Isolation of a pluripotential cell line from early mouse embryos cultured in medium conditioned by teratocarcinoma stem cells, Proc. Natl. Acad. Sci. USA 78:7634-8 (1981).

6. L. C. Stevens, Testicular ovarian and embryo derived teratomas, Cancer Surveys 2:75-91 (1983).

7. L. C. Stevens and D. S. Varnum, The development of teratomas from parthenogenetically activated mouse eggs, Dev. Biol. 37:369-80 (1974).

8. J. J. Eppig, L. P. Kozak, E. M. Eicher, and L. C. Stevens, Ovarian teratomas in mice are derived from oocytes that have completed the first meiotic division, Nature 269:517-18 (1977).

9. L. C. Stevens, The development of teratomas from intratesticular grafts of tubal mouse eggs, J. Embryol. Exp. Morph. 20:329-41 (1968).

10. L. C. Stevens, The development of transplantable teratocarcinomas from intratesticular grafts of pre- and post-implantation mouse embryos Dev. Biol. 21:364-82 (1970).

11. D. Solter, N. Skreb, and I. Damjanov, Extrauterine growth of mouse eggs results in malignant teratoma, Nature 227:503-504 (1970).

12. I. Damjanov, D. Solter, M. Belicza, and N. Skreb, Teratomas obtained through the extrauterine growth of seven day old mouse embryos, J. Natl. Cancer Inst. 46:471-482 (1971).

13. D. Solter, I. Damjanov, and H. Koprowski, Embryo derived teratomas, a model system in developmental and tumor biology, in: "The Early Development of Mammals," M. Balls and A. E. Wild, eds., Cambridge University Press, Cambridge (1975).

14. L. J. Kleinsmith and G. B. Pierce, Multipotency of single embryonal carcinoma cells, Cancer Res. 24:1544-52 (1964).

15. D. Solter, M. Dominis, and I. Damjanov, Embryo derived teratocarcinoma II: Teratocarcinogenesis depends on the type of embryonic graft, Int. J. Cancer 25:341-9 (1980).

16. D. Solter and I. Damjanov, Explantation of extra-embryonic parts of 7 day old mouse egg cylinder, Experientia 29:701-5 (1973).

17. S. Diwan and L. C. Stevens, The development of teratomas from endoderm of mouse egg cylinders, J. Natl. Cancer Inst. 57:937-42 (1976).

18. D. Solter, M. Dominis, and I. Damjanov, Embryo derived teratocarcinoma I: The role of strain and gender in the control of teratocarcinogenesis, Int. J. Cancer 24:770-2 (1979).

19. D. Solter, M. Dominis, and I. Damjanov, Embryo derived teratocarcinoma III. Development of tumors from teratocarcinoma permissive and nonpermissive embryos transplanted to F_1 hybrids, Int. J. Cancer 28:479-85 (1981).

20. I. Damjanov and D. Solter, Maternally transmitted factors modify development and malignancy of teratomas in mice, Nature 296:95-7 (1982).

21. J. Jami and E. Ritz, Multipotentiality of single cells of transplantable teratocarcinomas derived from mouse embryo grafts, J. Natl. Cancer Inst. 52:1547-1552 (1974).

22. B. Mintz and C. Cromiller, METT-1: a karyotypically normal in vitro line of developmentally totipotent mouse teratocarcinoma cells, Somatic Cell Genetics 7:489-505 (1981).

23. M. W. McBurney and B. J. Rogers, Isolation of male embryonal carcinoma cell lines and their chromosome replication patterns, Dev. Biol. 89:503-8 (1982).

24. E. J. Robertson, M. H. Kaufman, A. Bradley, and M. J. Evans, Isolation properties and karyotype analysis of pluripotent (EK) cell lines from normal and parthenogenetic embryos, in: "Cold Spring Harbor Conference on Cell Proliferation," Vol. 10, L. M. Silver, G. R.

Martin, and S. Strickland, eds., Cold Spring Harbor Laboratory, Cold Spring Harbor, NY (1983).

25. L. M. Silver, G. R. Martin, and S. Strickland, eds., "Cold Spring Harbor Conference on Cell Proliferation," Cold Spring Harbor Laboratory, Cold Spring Harbor, NY (1983).

26. C. F. Graham, Teratocarcinoma stem cells and normal mouse embryogenesis, in: "Concepts in Mammalian Embryogenesis," M. I. Sherman, ed., MIT Press, Cambridge (1977).

27. G. R. Martin, Teratocarcinomas and mammalian embryogenesis, Science 209:678-76 (1980).

28. M. Evans, Origin of mouse embryonal carcinoma cells and the possibility of their direct isolation into tissue culture, J. Reprod. Fertil. 62:625-31 (1981).

29. G. R. Martin and M. J. Evans, Differentiation of clonal lines of terato-carcinoma cells: formation of embryoid bodies in vitro, Proc. Natl. Acad. Sci. USA 72:1441-1445 (1975).

30. B. L. M. Hogan, D. P. Barlow, and R. Tilly, F9 Teratocarcinoma cells as a model for the differentiation of parietal and visceral endoderm in the mouse embryo, Cancer Surveys 2:115-40 (1983).

31. M. J. Evans and G. R. Martin, The differentiation of clonal teratocarci-noma cell cultures in vitro, in: M. I. Sherman and D. Solter, eds., "Teratomas and Differentiation," Academic Press, New York (1975).

32. D. Solter and B. B. Knowles, Immunosurgery of the mouse blastocyst, Proc. Natl. Acad. Sci. USA 72:5099-5102 (1976).

33. G. R. Martin, L. M. Wiley, and I. Damjanov, The development of cystic embryoid bodies in vitro from clonal teratocarcinoma stem cells, Dev. Biol. 61:230-44 (1977).

34. M. Dziadek, Modulation of alphafoetoprotein synthesis in the early post implantation embryo, J. Embryol. Exp. Morph. 46:135-146 (1978).

35. B. L. M. Hogan and R. Tilly, Cell interactions and endoderm differentia-tion in cultured mouse embryos, J. Embryol. Exp. Morph. 62:379-394 (1981).

36. R. L. Gardner, Investigation of the cell lineage and differentiation in the extraembryonic endoderm of the mouse embryo, J. Embryol. Exp. Morph. 68:175-198 (1982).

37. S. Strickland and V. Mahdavi, The induction of differentiation in tera-tocarcinoma stem cells with retinoic acid, Cell 15:393-403 (1978).

38. E. L. Kuff and J. W. Fewell, Induction of neural-like cells and acetyl choline esterase activity in cultures of F9 teratocarcinoma treated with dibutyryl cyclic adenosine mono-phosphate, Develop. Biol. 77:103-115 (1980).

39. S. Strickland, K. K. Smith, and K. R. Marotti, Hormonal induction and differentiation in teratocarcinoma stem cells: Generation of parie-tal endoderm by retinoic acid and dibutyryl cAMP, Cell 21:347-355 (1980).

40. B. L. M. Hogan, D. P. Barlow, and R. Tilly, F9 cells as a model for the differentiation of parietal and visceral endoderm in the mouse embryo, Cancer Surveys 2:115-140 (1983).

41. A. M. Jetten, M. E. R. Jetten, and M. I. Sherman, Stimulation and dif-ferentiation by several murine embryonal carcinoma cell lines by retinoic acid, Exp. Cell Res. 124:381-391, (1979).

42. E. M. V. Jones-Villeneuve, M. W. McBurney, K. A. Rogers, and V. I. Kalnius, Retinoic acid induces embryonal cells to differentiate into neurons and glial cells, J. Cell Biol. 94:253-262 (1982).

43. E. M. V. Jones-Villeneuve, M. A. Rudnick, F. Harris, and M. W. McBurney, Retinoic acid-induced neuronal differentiation of embryonal carcinoma cells, Mol. Cell Biol. 3:2271-2279 (1983).

44. M. W. McBurney, E. M. V. Jones-Villeneuve, M. K. S. Edwards, and P. J. Anderson, Control of muscle and neuronal differentiation in a cul-tured embryonal carcinoma cell line, Nature 299:165-167 (1982).

45. F. Jacob, Mouse teratocarcinomas and embryonic antigens, _Immunol_. _Rev_. 33:3-32 (1977).

46. P. L. Stern, K. Willison, E. Lennox, G. Galfie, L. Milstein, D. Secker, A. Zeigler, and T. Springer, Monoclonal antibodies as probes for differentiation and tumour associated antigens: A Forssman specificity on teratocarcinoma stem cells, _Cell_ 14:775-783 (1978).

47. M. G. Stinnakre, M. J. Evans, K. R. Willison, and P. L. Stern, Expression of Forssman antigen in the post implantation mouse embryo, _J_. _Embryol_. _Exp_. _Morph_. 61:117-131 (1981).

48. M. J. Dewey, R. Filler, and B. Mintz, Protein patterns of developmentally totipotent mouse teratocarcinoma cells and normal early embryo cells, _Develop_. _Biol_. 65:171-182 (1978).

49. R. H. Lovell-Badge and M. J. Evans, Changes in protein synthesis during differentiation of embryonal carcinoma cells and a comparison with embryo cells, _J_. _Embryol_. _Exp_. _Morph_. 59:187-206 (1980).

50. C. Failly-Crepin and G. R. Martin, Protein synthesis and differentiation in a clonal line of teratocarcinoma and in pre-implantation mouse embryos, _Cell_. _Diff_. 8:61-73 (1979).

51. M. J. Evans, R. H. Lovell-Badge, P. L. Stern, and M. G. Stinnakre, Cell lineage in the mouse embryo: Forssman antigen distributors and patterns of protein synthesis, _in_: "Cell Lineage Stem Cells and Cell Determination," Inserm Symposium 10, N. LeDouarin, ed., Elsevier, North Holland (1979).

52. R. L. Brinster, The effect of cells transferred into the mouse blastocyst on subsequent development, _J_. _Expt_. _Med_. 140:1049-1056 (1974).

53. B. Mintz and K. Illmensee, Normal genetically mosaic mice produced from malignant teratocarcinoma stem cells, _Proc_. _Natl_. _Acad_. _Sci_. _USA_ 72:3585-3589 (1975).

54. V. E. Papaioannou, M. W. McBurney, R. L. Gardner, and M. J. Evans, Fate of teratocarcinoma cells injected into early mouse embryos, _Nature_ 258:7073 (1975).

55. V. E. Papaioannou, R. L. Gardner, M. W. McBurney, C. Babinet, and M. J. Evans, Participation of cultured teratocarcinoma cells in mouse embryogenesis, _J_. _Embryol_. _Exp_. _Morph_. 44:93-104 (1978).

56. M. J. Dewey, D. W. Martin, Jr., G. R. Martin, and B. Mintz, Mosaic mice with teratocarcinoma derived mutant cells deficient in hypoxanthine phosphoribosyltransferase, _Proc_. _Natl_. _Acad_. _Sci_. _USA_ 74:5564-5568 (1977).

57. T. A. Stewart and B. Mintz, Successive generations of mice produced from an established culture line of euploid teratocarcinoma cells, _Proc_. _Natl_. _Acad_. _Sci_. _USA_ 78:6314-6318 (1981).

58. A. Bradley, M. J. Evans, M. H. Kaufman, and E. J. Robertson, The formation of functional germ line chimaeras from embryo-derived teratocarcinoma cell lines, _Nature_ 309:255-256 (1984).

59. V. E. Papaioannou and J. Rossant, Effects of embryonic environment on proliferation and differentiation of embryonal carcinoma cells, _Cancer Surveys_ 2:165-183 (1983).

60. C. L. Stewart, Formation of viable chimaeras by aggregation between teratocarcinomas and preimplantation mouse embryos, _J_. _Embryol_. _Exp_. _Morph_. 67:167-179 (1982).

61. J. Rossant and M. W. McBurney, The developmental potential of an euploid male teratocarcinoma cell line after blastocyst injection, _J_. _Embryol_. _Exp_. _Morph_. 70:99-112 (1982).

62. M. W. McBurney and B. J. Rogers, Isolation of male embryonal carcinoma cells and their chromosome replication patterns, _Develop_. _Biol_. 89:503-508 (1982).

63. J. T. Fujii and G. R. Martin, Developmental potential of teratocarcinoma stem cells in utero following aggregation of cleavage stage mouse embryos, _J_. _Embryol_. _Exp_. _Morph_. 74:79-90 (1983).

64. K. Illmensee and B. Mintz, Totipotency and normal differentiation of

single teratocarcinoma cells cloned by injection into blastocysts, Proc. Natl. Acad. Sci. USA 73:549-553 (1976).

65. K. Illmensee, Reversion of malignancy and normalized differentiation of teratocarcinoma cells in chimaeric mice, in: "Genetic Mosaics and Chimaeras in Mammals," L. B. Russell, ed., Plenum Press, New York (1978).

66. T. A. Stewart and B. Mintz, Recurrent germline transmission of the teratocarcinoma genome from the METT-1 culture line to progeny in vivo, J. Exp. Zool. 224;465-471 (1982).

67. B. Mintz and C. Cronmiller, Normal blood cells of anemic genotype in teratocarcinoma derived mosaic mice, Proc. Natl. Acad. Sci. USA 75:6247-6251 (1978).

68. B. Mintz and C. Cronmiller, METT-1, a karyotypically normal in vitro line of developmentally totipotent teratocarcinoma cells, Som. Cell Genet. 7:489-505 (1981).

69. R. L. Gardner and V. E. Papaioannou, Differentiation in the trophectoderm and inner cell mass, in: "The Early Development of Mammals," M. Balls and A. E. Wild, eds., Cambridge University Press, Cambridge (1975).

70. C. Cronmiller and B. Mintz, Karyotypic normalcy and quasi-normalcy of developmentally totipotent mouse teratocarcinoma cells, Develop. Biol. 67:465-477 (1978).

71. A. Bradley and E. J. Robertson, Embryo derived stem cells: A tool for elucidating the developmental genetics of the mouse, Current Topics in Dev. Biol. 20:357-371 (1986).

72. E. J. Robertson and A. Bradley, Production of permanent cell lines from early embryos and their use in studying developmental problems, in: "Experimental Approaches to Mammalian Embryonic Development," J. Rossant and R. A. Pederson, eds., Cambridge University Press, Cambridge (1986).

73. J. Rossant and M. W. McBurney, Diploid teratocarcinoma cell lines differ in their ability to differentiate normally after blastocyst injection, in: "Cold Spring Harbor Conf. on Cell Proliferation," Vol. 10, L. M. Silver, G. R. Martin, and S. Strickland, eds., Cold Spring Harbor Laboratory, Cold Spring Harbor, NY, (1983).

74. E. J. Robertson, A. Bradley, M. Kuehn, and M. J. Evans, Germ Line transmission of genes introduced into cultured pluripotential cells by a retroviral vector, Nature 323:445-448 (1986).

75. M. R. Kuehn, A. Bradley, E. J. Robertson, and M. J. Evans, A potential animal model for Lesch-Nyhan Syndrome through introduction of HPRT mutations into mice, Nature 326:295-298 (1987).

76. M. L. Hooper, K. Hardy, A. Handyside, S. Hunter, and M. Monk, HPRT-deficient (Lesch-Nyhan) mouse embryos derived from germ line colonization by cultured cells, Nature 326:292-295 (1987).

77. S. L. Mansour, K. R. Thomas, and M. R. Capecchi, Disruption of the proto-oncogene int-2 in mouse embryo-derived stem cells: A general strategy for targeting mutations to non selectable genes, Nature 336:348-352 (1988).

DISCUSSION

Moderator (Ivan Damjanov): Let me lead off with some of the questions that come immediately to mind. The difference between embryonal carcinoma cells that were produced from embryos and the ICM derived ES cells is that in the first case you start with the normal embryo, put it into the living mouse and then isolate the cells from a well-formed tumor. In the case of ES cells you take ICM, put it into culture and propagate the embryonic cells ad infinitum. What would happen if you took ES cells that were never in a mouse and injected them into the mouse, produced tumors and then transplanted these

tumors several times? If the ES cells are passaged through the living mice
in a tumorigenic mode and then explanted in vitro, do they still retain their
developmental potential? Are the cells "passaged" in vivo different from ES
cells propagated in vitro only? Or do these cells resemble more closely the
EC cells produced initially in vivo by transplanting the embryos?

Allan Bradley: We haven't done the experiment so I can't give you a definitive answer, but I can speculate. What I suspect is happening is that it is
the initial event that is important; in other words, how long is it before
you have a substantial contribution of EC cells in a tumor? And I suspect
that once you have a cell line and you passage it either in vitro or in vivo,
as you are suggesting, that probably there is not going to be too much
change. I think there is a lot of data to suggest that karyotypic changes
occur with time although most likely these are minimal., I would say that
perhaps you would expect the stem cells, specifically the EC cells, to behave
still normally despite of being derived from a tumor mainly because that's
the way they were initially made. In other words, growth as a tumor is not
the primary selection.

George Michalopolous: I would like to ask you if there is anything known
about the behavior of these ES cells in tissue culture. Do they grow as an
immortalized cell line? Do they grow as a tumor cell line or are they
contact inhibited? What is the nature of their response to specific growth
factors?

Bradley: I don't know if their response to growth factors has actually been
tested. They are not contact inhibited. They grow very much as I showed
you, as a clump of cells, and they will continue to do that. If you don't
segregate and passage them regularly the clump of cells will begin to
differentiate and at that stage you will invariably select for cells that
can't differentiate in culture. That's really the only information I have.
As far as growth factors are concerned, I don't know.

Michalopolous: Do they depend on serum for their growth?

Bradley: Yes. Obviously there are factors that allow them to grow. There
is a factor that I think has been identified which inhibits their
differentiation in tissue culture, but at this stage the absolute
requirements for growth factors are not understood. They do require feeder
cells, but this can be negated to some extent, and you can select for ES
cells that do not require feeder cells. So whether it is a surface problem
or whether it is a growth factor, again we don't know at this stage.

Damjanov: Along the same lines, when you establish embryo derived stem cells
in culture, is it possible to alter them by exposing them to carcinogens,
radiation or any other external influence that would visibly change either
their morphology or growth pattern, growth requirements, or differentiation
capacity?

Bradley: That is the whole problem with cell lines. They are subject to
change, and if you abuse them then you can select all sorts of things out in
tissue culture. But also, if you treat them well, in other words you feed
them regularly and segregate them regularly and do the things they like, give
them lots of serum, then you can do quite diverse experimental manipulations,
particularly with regard to changing the genome. For instance, we've been
able to infect cells with a mutagenic agent, in this case a retrovirus, and
select out clones of interest which have insertions of particular genes.
These virus infected ES cells can even recolonize the germ lines. So you can
do quite extensive tissue culture manipulations and still go back through the
germ line. In other words, you are not really changing the fate of these
cells by doing that.

Unidentified speaker: How karyotypically stable are your cultured lines?

Bradley: I think they are pretty stable. We've taken lines up to about 25 passage generations and still gone through the germ line. We haven't always extensively tested this, and I suspect there always are going to be cell lines that will lose or gain chromosomes as history passes. But I don't know the frequency with which it occurs at this stage.

Unidentified speaker: So you don't get the usual murine heteroploid transformation after about 5 subculture passages?

Bradley: No. These karyotypes are probably taken around about passage generation 10, so you are talking about a lot of cell doublings.

SUMMARY BY MODERATOR

Moderator (Ivan Damjanov): The tumor models described by the two speakers
are important for the understanding of cell to cell interactions and for how
this interaction can modify the development or reversal of malignancy. In
the case of ES cells produced in vitro from explanted embryos the moment of
explantation could be considered as initiation. It is arguable whether there
is any promotion or progression. It is, however, unquestionable that the
embryo derived ES cells have all the features of "immortalized" neoplastic
cells, although these characteristics are reversible. I think that is the
beauty of this system. Obviously, one cannot monitor all the characteristics
of the EC or ES cells, but the message obtained even in a limited
experimental system is clear: the genetic determinants of malignancy can be
overpowered by the epigenetic control mechanisms. These epigenetic factors
may be within the embryo or the culture medium or in the in vivo environment
traditionally called the tumor.

TUMOR PROGRESSION IN TRANSGENIC MICE

CONTAINING THE BOVINE PAPILLOMAVIRUS GENOME

Peter M. Howley

Laboratory of Tumor Virus Biology
National Cancer Institute
Bethesda, MD 20892

Transgenic mice are an interesting system to study a variety of biological phenomena. It has been a particularly interesting way of studying oncogenesis. It has provided a tractable system to study and to dissect the multiple steps involved in carcinogenesis, essentially by fixing one of these steps[1]. In transgenic mice, DNA is introduced into the germ line and that gene is stably inherited to the offspring of the mice. Transgenic mice have been developed using a 1.69 tandemly reiterated copy of the bovine papillomavirus type 1 (BPV-1) genome[2]. BPV-1 encodes two different viral oncogenes, and in the transgenic line of mice that has been developed, these two viral oncogenes are under the control of the transcriptional regulatory sequences of the virus to ensure proper tissue specific expression of the viral oncogenes.

BPV-1 induces cutaneous fibropapillomas in cattle. These lesions are fibropapillomas and histologically consist of two tissue types: proliferating squamous epithelial cells and proliferating dermal fibroblasts. In the transgenic BPV 1.69 mice, it is the dermal fibroblasts which express the specific viral proteins. The bovine papilloma virus genome, as all papilloma viruses, contains about 8,000 base pairs of genetic information. Papilloma viruses are double stranded DNA viruses and BPV-1 has been the prototype virus for our understanding of the molecular biology of this group of viruses. A map of BPV-1 is depicted in Figure 1.

BPV-1 contains a series of genes or open reading frames (ORFs) which are indicated by these arcs on the outside. The virus contains a region where there are no significantly large ORFs which is about 1,000 bases long and contains the origin of DNA replication, a series of transcriptional promoters, and transcriptional enhancers which are involved in the tissue specific expression of the virus. In addition, there are eight designated ORFs which are located in the "early" portion of the genome which are involved in the DNA replication and in the transcriptional regulation of the virus. The L1 and L2 ORFs encode the major and minor capsid proteins.

There are two oncogenes encoded by BPV-1, E5 and E6. The E5 onco-protein is a 44 amino acid protein. It is the smallest onco-protein known to date. It is strikingly hydrophobic, and it is found in cellular membranes in transformed cells[5]. The E6 onco-protein is a cysteine rich protein which is predicted to contain cysteine fingers. It is found in the nucleus as well as in membrane fractions[6]. Exactly how E5 and E6 function as onco-proteins is not yet understood. The transgene which Lacy et al.

Boundaries between Promotion and Progression during Carcinogenesis
Edited by O. Sudilovsky *et al.*, Plenum Press, New York, 1991

originally injected to generate the BPV 1.69 line of mice consisted of 1.69
copies of the viral genome[2]. This reiterated copy of the viral genome
contains a duplicated segment of the transforming region and one copy of the
late region. Douglas Hanahan specifically engineered more than one copy of
the viral genome in generating this transgene to provide the opportunity for
homologous recombination to occur, allowing the excision of viral genome from
the host chromosome. This was thought to be important because BPV-1 is not
normally integrated within transformed cells[7].

Work from Hanahan's laboratory[2], demonstrated that mice containing
this transgene develop abnormal skin with protuberant tumors heritably at
about 8 to 9 months of age. There are two pathological changes that are
characteristic in these mice: The first is abnormal skin, and the second is
that of protuberant fibroblastic tumors. The transgene is integrated as a
tandem repeat of approximately 5 copies within the germ line of this strain
of mice[2]. The DNA is found in an extrachromosomal form in all of the
abnormal skin tissues and in the skin tumors. In non-affected skin and the
internal organs, in the germ line tissues from these mice, the viral DNA is
exclusively integrated[2].

In a collaborative study with Doug Hanahan's laboratory, we were interested in defining at a molecular level the various pathological changes
observed. Did the pathology correlate with the expression of the viral
genome and was there any evidence of other non-viral genetic events that were
associated with them? Examples of the pathological changes observed are

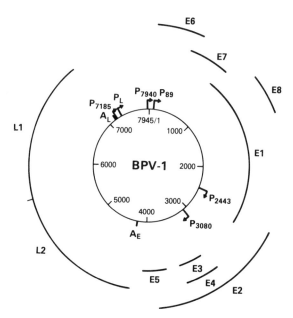

Fig. 1. A schematic representation of the
BPV-1 circular double-stranded DNA
genome. Mapped promoters are
indicated by the arrowheads[3].
The arcs represent the ORGs of
this virus. A detailed discussion
of the molecular biology and genetics of BPV-1 has recently been
published[4].

shown in Figure 2. This is an example of the normal skin as seen in one of these mice at 14 weeks of age, with no abnormalities. The sections labeled fibromatosis and aggressive fibromatosis are from areas of abnormal skin. There is a loss of the appendages with proliferation of dermal fibroblasts. There is a spectrum of changes from a mild form called "fibromatosis" to a dense and diffuse form which we call "aggressive fibromatosis". This is not yet a protuberant tumor, but there is a dense proliferation of dermal fibroblasts which can extend down into the subcutaneous fat. A certain percent of these animals will develop protuberant tumors which are dermal fibrosarcomas, with a high mitotic index and occasionally abnormal mitoses.

In order to characterize the molecular biology of these different lesions, cell lines from these different pathologic entities were established[8]. The cells from the fibromatoses could grow to a monolayer but they were contact inhibited. Those from the aggressive fibromatoses were no longer contact inhibited, in fact they were anchorage independent, and were tumorigenic when injected into nude mice -even though they were not tumorigenic in the original mouse. Cell cultures of the tumors were not contact inhibited and formed dome shaped foci[8]. Cells from the tumors were also anchorage independent and tumorigenic when assayed in nude mice[8]. Growth curves were carried out with the various cells[8]. Essentially, the dermal fibroblasts grew very poorly in tissue culture. The cells from the non-aggressive fibromatosis grew slightly better than normal cells but did

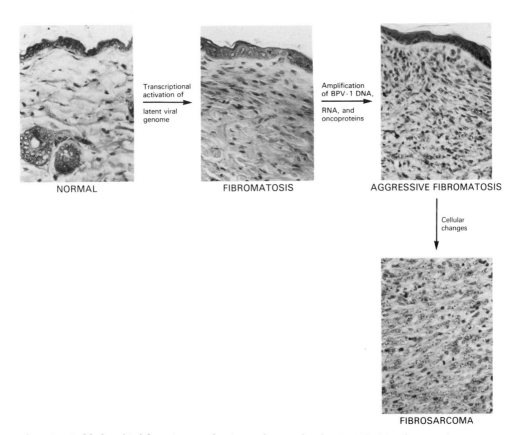

NORMAL   FIBROMATOSIS   AGGRESSIVE FIBROMATOSIS

Transcriptional activation of latent viral genome

Amplification of BPV-1 DNA, RNA, and oncoproteins

Cellular changes

FIBROSARCOMA

Fig. 2. Cell-heritable stages in tumorigenesis in BPV 1.69 mice. Shown are specific stages in the development of fibrosarcomas in these mice. This figure is reproduced from an article by Sippola-Thiele et al.[8].

not grow nearly as well as the cells from either of the tumors or from the aggressive fibromatoses[8].

The tumor tissues, the abnormal skin tissues, and the derived cell cultures from these mice were examined for evidence of viral gene expression[8]. RNA analysis from the normal skin of a newborn mouse, from the normal skin of a 14 week old mouse, or from a 16 week old mouse revealed no evidence of any BPV-1 transcription. Thus in normal tissues in which the DNA is integrated, the viral genome is transcriptionally quiet. In contrast in the fibromatoses, mild or aggressive, there is BPV-1 specific mRNA indicating that activation of transcription is associated with the proliferation of the dermal fibroblasts, even though it is a benign proliferation[8]. In the fibromatoses, the level of transcription correlated at least to some degree with the level of aggression of the fibroblastic proliferation. The fibrosarcomas are also transcriptionally active[8]. Studies with cell cultures exactly paralleled those with the tissues from the mice. Cultures of dermal fibroblast from a newborn and from a 14 week old transgenic revealed no evidence of viral expression[8]. The fibromatoses and the fibrosarcomas are transcriptionally active for the BPV-1 genome.

Analysis of the E5 onco-protein within these cells parallels that of viral specific transcription. The E5 protein can be demonstrated in the fibrosarcomas and in the aggressive fibromatoses but not in cultures of normal dermal fibroblasts[8]. There was not a consistent difference in the levels of E5 between the aggressive fibromatoses and the tumors, suggesting that some additional genetic change must be involved in tumor progression.

To summarize the points that can be made from the viral transcriptional and from the E5 protein analyses, tumorigenesis of dermal fibroblasts in the BPV 1.69 mice involves distinct proliferative stages. Cell cultures derived either from normal skin, from benign proliferative fibromatoses, or from malignant fibrosarcomas each evidence distinguishable, heritable cell characteristics. Expression of the BPV 1 genome is restricted to the proliferative lesions. The proliferative capacity of the benign fibromatoses appears to directly correlate with the levels of viral gene expression and E5 onco-protein. Progression to malignancy, however, is not accompanied by further increase in the transgene activity, strongly implicating cellular genetic changes in these later stages of tumorigenesis.

To address this, a karyotypic analysis of the cell lines and primary tissues from these mice was carried out[9]. The analysis was done by Dr. Valerie Lindgren, who performed a karyotypic analysis of the various cell lines and tissues from these mice. The karyotype of the dermal fibroblasts from newborn mice was diploid and normal. The karyotype from the mild fibromatosis cell lines were also essentially normal. Although four out of the five examined were diploid, and in one of them three was an abnormality consisting of an extra chromosome 15[9]. In the aggressive fibromatoses, however, none of the cells were diploid. Indeed they were all aneuploid, but without any specific chromosomal abnormalities[9]. Analysis of the tumor lines and of the tumors revealed two specific chromosomal abnormalities. The first was either a trisomy of chromosome 8 or a portion involving chromosome 8[9]. The second was a monosomy of chromosome 14. To determine whether these changes were specific to the tumor progression in these cells or a consequence of the location of the transgene within these mice, its location was determined by in situ hybridization[9]. The BPV-1 genome was found to be integrated on band C of chromosome 15[9]. Summarizing the above: 1) Normal cells from the BPV 1.69 mice have a normal diploid karyotype. 2) Integration of the transgene is on chromosome 15. 3) Benign proliferative fibromatoses are not characterized by specific chromosomal abnormalities. 4) Malignant fibrosarcomas show consistent abnormalities of one or both of two chromosomes: chromosome 8 (where there is a trisomy or duplication of a portion of

the chromosome) and chromosome 14 (where there is a monosomy or translocation). The finding of a loss of a portion of chromosome 14 implicates the
possibility that there may be a suppressor gene involved in progression. I
should point out, interestingly, that the mouse retinoblastoma gene homologue
has been mapped to chromosome 14[10]. With regards to the chromosome 8,
amplification of a portion of this chromosome has been noted before with
tumor progression in mice[11]. In this study, Frost noted such change in
metastatic mouse tumors. The finding of abnormalities involving one of two
chromosomes, raises the possibility that there may be a suppressor gene on
chromosome 14 and another alteration on chromosome 8 which may be somehow
regulating the suppressor gene on chromosome 14.

In summary, these studies, with reference again to Figure 2, indicate
that there are distinct stages of tumor progression that are occurring in the
BPV 1.69 transgenic mice. Progression from the latent state in normal cells
to the mild fibromatosis implies transcriptional activation of a latent viral
genome. Going from a mild fibromatosis to aggressive fibromatosis involves
amplification of the viral DNA, an increase in the level of expression of the
viral RNA, and the viral onco-proteins. Finally, progression to malignancies
entails specific cellular genetic changes which in this model affect chromosomes 8 and/or 14.

REFERENCES

1. D. Hanahan, Oncogenesis in transgenic mice, in: "Oncogenes and Growth
    Control," T. Graf and P. Kahn eds., Springer-Verlag KG, Berlin
    (1986), pp. 349-363.
2. M. Lacey, S. Alpert, and D. Hanahan, The bovine papillomavirus virus
    genome elicits skin tumors in transgenic mice, Nature 322:609-612
    (1986).
3. C. C. Baker and P. M. Howley, Differential promoter utilization by the
    bovine papillomavirus in transformed cells and in productively in-
    fected wart tissues, EMBO J. 6:1027-1035 (1987).
4. P. F. Lambert, C. C. Baker, and P. M. Howley, The genetics of bovine
    papillomavirus type 1, Ann. Rev. Genetics 22:235-258 (1988).
5. R. Schlegel, M. Wade-Glass, M. S. Rabson, and Y. C. Yang, The E5
    transforming gene of bovine papillomavirus encodes a small
    hydrophobic polypeptide. Science 233:464-467, (1986).
6. E. J. Androphy, J. T. Schiller, and D. R. Lowy, Identification of the
    protein encoded by the E6 transforming gene of bovine papillomavirus,
    Science 230:442-445 (1985).
7. P. M. Howley and R. Schlegel, Papillomavirus transformation, in: "The
    Papovaviridae 2: the papillomaviruses," N. P. Salzman and P. M.
    Howley eds., Plenum Publishing Corp., New York (1987), pp. 144-166.
8. M. Sippola-Thiele, D. Hanahan, and P. M. Howley, Cell-heritable stages
    of tumor progression in transgenic mice harboring the bovine papillo-
    mavirus type 1 genome, Mol. Cell. Biol. 9:925-934 (1989).
9. V. Lindgren, M. Sippola-Thiele, J. Skowronski, E. Wetzel, P. M. Howley,
    and D. Hanahan, Specific chromosomal abnormalities characterize
    fibrosarcomas of bovine papillomavirus-1 transgenic mice, Proc. Natl.
    Acad. Sci. USA, in press (1989).
10. J. C. Stone, J. L. Crosby, C. A. Kozak, A. R. Shievella, R. Bernards,
    and J. H. Nadeau, The murine retinoblastoma homologue maps to chromo-
    some 14 near Es-10, Genomics, in press (1989).
11. P. Frost, R. S. Kerbel, B. Hunt, S. Man, and S. Pathak, Selection of
    metastatic variance with identifiable karyotypic changes from a
    nonmetastatic murine tumor after treatment with 2'-Deoxy-5-azacyti-
    dine or hydrourea: implications for the mechanism of tumor progres-
    sion, Cancer Res. 47:2690-2695 (1987).

DISCUSSION

Michael Lieberman: Peter, you didn't mention E6. Is anything happening with it?

Peter M. Howley: Immunoprecipitations with antisera to E6 reveal the protein in the aggressive fibromatoses and in the tumors. Unfortunately the immunologic reagents for #6 that are available (6), are probably not sensitive enough to detect the protein in the mid fibromatosis. Thus the observations with E6, parallel those with E5.

George Michalopolous: I think it is very interesting that the enhanced proliferation of the fibroblasts appears at a certain point of age in the development of the mouse. Would you care to speculate as to what the mechanisms might be? Have you considered taking some of the fibroblasts at the stage and transplanting them back into the younger animals to see if their proliferation is suppressed?

Howley: We haven't done the experiment that you suggest, and I think it is a very good experiment. I should point out that the tumors that develop generally occur in areas of abnormal skin, often in areas of injury. In fact, it was first noted in doing this that the tumors often occurred at the end of the mouse tails which had been clipped for DNA analysis (2). Thus wounding may play a role, perhaps by providing a continued stimulus in the development of fibromatosis or possibly also independently involved in the progression of the tumors.

Unidentified Speaker: What happens if you promote the mice?

Howley: Some experiments have been done with TPA, with no positive results to date. That actually was the question that I was asking about, about the alternative tumors promoters. I think those are important questions to go back to and examine. On the other hand, knowing that beta TGF is able to induce fibroplasia, we've examined its effects in these mice, again with no positive results to date.

George Yoakum: I found your data on gene expression very interesting, and I have a two-part question. What do you know about any variation in the structure of this transgene in the cell lines you've isolated and in the tumors you observe in situ in the mouse? Could you speculate on how you might approach the problems with these secondary genetic changes that may be occurring?

Howley: Let me answer the second part first. Given the location of the retinoblastoma gene on chromosome 14 in the mice (10), we're certainly looking at the mouse rb expression within the benign and the malignant tissues. In a less directed analysis, one could anticipate making differential cDNA libraries between the various stages, in the hope of identifying RNA's that are differentially expressed.

Yoakum: I was asking about Southern analysis of your cell lines to see what your tandem duplication structure is like in each of them.

Howley: We've done some experiments to examine the DNA in the cell lines and the data is still preliminary, but we see no evidence that would indicate that the DNA is excising through homologous recombination. Rather, it appears to be coming out through onion skinning replication. There appears a direct correlation between the levels of extrachromosomal viral DNA and the level of viral gene expression. Extra-chromosomal DNA is noted only in abnormal skin and tumors, correlating well with evidence for viral gene expression.

Peter C. Nowell: This is a nice story. I didn't understand your last suggestion concerning a suppressor of a suppressor gene. Are you suggesting that the extra dose on chromosome 8 might be carrying something that is suppressing a suppressor on chromosome 14?

Howley: There are two possible scenarios. One could imagine that the gene on chromosome 14 is a tumor suppressor gene, which is under the negative transcriptional control of a gene on chromosome 8. As such, it would be the balance of these two factors that are critical in maintaining the cell norm. Given the recent exciting data of the interactions of specific viral onco-proteins with the retinoblastoma protein, one could also imagine that there may be cellular proteins which can associate with the products of cellular suppressor genes. As such, it may be then be the balance of these factors with the product of tumor suppressor genes that may eventually result in whether the cell is under normal proliferative control or not. Thus the interaction between the two proteins could either be at the transcriptional regulatory level or at the level of a protein-protein interaction.

Henry C. Pitot: In view of the morphology of these lesions, has anyone looked for human papilloma virus in fibromatoses in the human?

Howley: Not to my knowledge. One candidate may be fibrosarcoma protuberans which looks very similar to the fibroblastic tumors that are associated with bovine papilloma virus. Most of the human papilloma virus that have been characterized so far, however, are associated with purely epithelial proliferative lesions.

Peter Duesberg: I was wondering about what is the evidence that these genes are actually oncogenes? You call them oncogenes and onco-proteins, but apparently these genes are present in a virus that induces primarily benign fibromas in the cow. In the transgenic mouse they are in every single cell, and you observe a month later a clonal tumor in some cases, which is 1 out of 10^{12} mouse cells. I wonder what the evidence is on which you base the name oncogenes or onco-proteins in such a situation.

Howley: One of the points that you may have missed is that the abnormal pathology or proliferation actually correlated with the expression of this gene, not merely with the presence of it. When the gene quiescent, one does not see the abnormal proliferation. The reason for calling it an onco-protein really has nothing to do with the studies presented here but rather with the extensive genetic analysis that my laboratory, that of Doug Lowy at the NIH, and that of Dan DeMaio at Yale have done over the past decade. BPV is able to transform NIH 3T3 cells and a variety of other rodent cells in tissue culture, and it is the genetic dissection of the viral genes that are involved in this transformation that leads us to designate E5 and E6 as the viral oncogenes.

Duesberg: Is it still kosher to call the morphological changes induced in a 3T3 cell an oncogene? I thought that that is no longer acceptable. If normal cultured diploid cell in the mouse doesn't do it, do you think it's legitimate to call it an oncogene just because the 3T3 cell is transformed by it?

Howley: BPV transforms a variety of established rodent cell lines. It also transforms a variety of primary cells. In fact, Bill Jarrett's laboratory in Scotland uses a diploid line of bovine conjunctiva cells to score the transforming potential of these genes.

Duesberg: Isn't that what counts is what happens in your system? I mean, in the whole mouse every cell contains them and the cells are normal.

Howley: However, Peter, it is only in the cells in which E5 and E6 are expressed that one notes the abnormal proliferation. So, to reiterate the major point of my presentation, it is the expression, not the presence of the viral genome that correlates with the abnormal pathology. I agree that a gene that is not expressed is probably not going to be important.

Duesberg: But isn't it expressed in the fibromas which you called benign as well as in the cow, which are diploid?

Howley: Yes, I would agree, and that was one of the points I was trying to make, Peter, that the protein itself is associated with the benign proliferation of the cells, and that it requires other cellular genetic events that are important for malignancy.

Duesberg: Under these conditions, I don't understand why it is helpful to call it an oncogene.

Howley: Because, as I've just said, it is able to malignantly transform a variety of cells in culture.

Michael Lieberman: I was going to ask a similar question. It has got to do with the fact that you only see fibromatoses and fibrosarcomas in these animals, and yet one can see a spectrum of other lesions under other circumstances. In a variety of cells in culture, are the E5 and E6 promoters expressed in non-fibroblastic lines as well, and does that have anything to do with the rather limited repertoire in these transgenes?

Howley: The viral transgene that was put into these mice is under the homologous viral transcriptional regulatory elements, and expression is only seen in the dermal fibroblasts. There is no expression that is seen in the internal organs. Within the transcriptional control region of this virus, there are viral enhancers with specifically target to these cells.

SV40 T ANTIGEN TRANSGENIC MICE: CYTOTOXIC T LYMPHOCYTES

AS A SELECTIVE FORCE IN TUMOR PROGRESSION*

Barbara B. Knowles[1], Susan Faas[1], Antonio Juretic[1],
Niles Fox[+1], Roseanne Crooke[1], Douglas Hanahan[2],
Davor Solter[1], Lorraine Jewett[1]

[1] The Wistar Institute
3601 Spruce Street
Philadelphia, PA 19104

[+]Present address: Lilly Research Laboratories
Indianapolis, IN 46285

[2]Department of Biochemistry
University of California Medical School
San Francisco, CA

INTRODUCTION

The specific immune response to some antigenic determinants expressed on
the initiated cell can be a contributing factor to the prevention of tumor
appearance. Indeed, one aspect of tumor progression may be the evolution and
selection of tumor cell variants capable of avoiding the immune response of
the host. The oncogenic virus-induced tumors have provided the most direct
evidence for these points; viral gene products elicit specific immune responses, and the normal cell-virus infected cell-virally transformed cell
praxis provides an experimental system to test these concepts. However, in
vivo tumorigenicity testing of cells transformed by viruses in vitro does not
provide proof for these hypotheses since the characteristics of these cells
following growth in vitro cannot reflect those of the analogous tumor cells
arising in the selective environment of the intact organism. The development
of the simian virus (SV40) tumor (T) antigen (ag) transgenic mouse[1] has
provided a model system in which the specific contribution, if any, of host
immunity to the control of these endogenous tumors can be evaluated. Moreover, because expression of the viral transforming gene can be targeted to
different tissues, by linking the sequences encoding the SV40 T/t antigens to
those controlling transcription of other genes in specific cell types, a wide
range of tumor types are available for comparison[2-5]. Expression of sufficient levels of the viral oncogene potentiates tumor formation but their
appearance is controlled by subsequent events which may vary depending on the
cell type and on the developmental time in which SV40 Tag is expressed.

*This work was supported by grants from the U.S.P.H.S. CA 10815, CA 21124, CA
18470, CA 25875, and DK 39808

Our previous investigation of SV40-infected adult mice had indicated that the genetically determined ability to mount a cytotoxic T lymphocyte (CTL) response to SV40 Tag determined whether or not virus-induced tumors appeared in aging animals[6]. The differing ability of various mouse strains to mount an SV40 Tag-specific CTL response maps to the major histocompatibi-

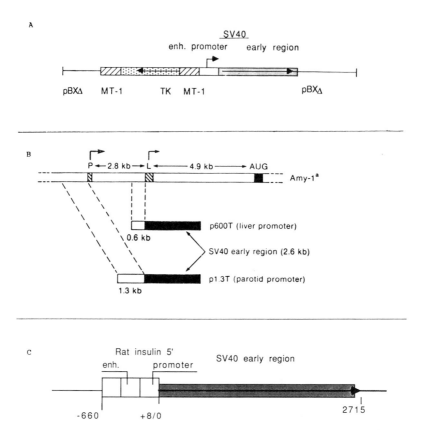

Fig. 1. Gene constructs used to prepare SV40 Tag transgenic mice. **A**. The SV-MK gene contains the SV40 early region gene under the control of its own promoter and enhancer and, in the opposite orientation, the herpes simplex virus thymidine kinase (TK) gene under the control of the metallothionein (MT-1) gene promoter; plasmid sequences (pBX△) were linearized at the Pvu 1 site in the vector[1]. **B**. The p600T and p1.3T transgenes contain the indicated sequences from the α-amylase gene and the SV40 early region coding sequences and polyadenylation sites. The hybrid gene was excised and purified from the vector sequences prior to injection into the male pronucleus[5] (and Fox et al., unpublished). The parotid (P) and liver (L) promoter and their relative position within the intact amylase gene are indicated. These gene constructs were prepared by Dr. U. Schibler, University of Geneva. **C**. The RIP Tag gene contains the indicated sequences from the 5' end of the rat insulin II gene and the SV40 early region gene coding sequences and polyadenylation site; plasmid sequences were linearized within the vector[3].

Table 1. Characteristics of SV40 Transgenic Mice Surveyed

SV40 Tag transgenic line	Promoter/ enhancer	Tumors	CTL response to SV40 Tag*	Developmental time of expression
419[1]	SV40	none	+	adult
427[1]	SV40	choroid plexus	-	neonatal
194**	amylase "parotid"	none	+	designed adult
334[5]	amylase "liver"	hibernoma	-	designed neonatal
RIP Tag 2[3]	rat insulin	insulinoma	+/-	fetal
RIP Tag 4[3]	rat insulin	insulinoma	+	adult

*Following SV40 inoculation
**Fox et al., in preparation
+/-Variable and low
[1,3,5] See References

lity complex (MHC) class I genes[7], whereas mice of each MHC haplotype tested all mount a high titer antibody response to SV40 Tag[8]. Correlative experiments with SV40-transformed cells in mice containing SV40 Tag-specific CTL revealed that expression of the MHC class I genes (H-2 K/D) was required for tumor cell rejection[8,9], a finding which reflects the obligatory dual recognition of specific antigen and MHC gene products by the T cell receptor[10]. Our initial approach with SV40 Tag transgenic mice was to investigate their ability to mount a CTL response after SV40 challenge. We then determined whether deregulation of H-2 K/D gene expression accompanied tumorigenicity of the differentiated cells in which SV40 Tag was expressed. We suggest that the evolution of immune responsiveness or tolerance in the various SV40 Tag lineages depends on the tissue and cell type targeted and the developmental time of transgene expression. Tolerance to SV40 Tag may arise as a result of its expression during development of the T cell repertoire; in such mice immune surveillance does not contribute to tumor growth control. In responsive animals SV40 Tag-specific CTL may be activated either immediately, following onset of transgene expression, or later, after the tissue reorganization that results from hyperplastic growth; the time of CTL activation appears to depend on the cell type that initially synthesizes SV40 Tag. SV40 Tag-specific CTLs may thus encounter cells shortly after they express the viral oncogene and before they develop their full tumorigenic potential, or after they have developed a multiplicity of changes that characterize the transformed cell. In the latter case, SV40 Tag-specific CTL can be thought of as one of the selective forces operating at, or forming, the boundary between tumor promotion and progression.

We will discuss our investigation of several different SV40 transgenic lineages prepared by injection of various gene constructs that contain the SV40 early region coding sequences and polyadenylation sites (Fig. 1 and Table 1).

Both the 427 and 419 SV40 Tag transgenic lineages are derived from mice injected with the SV-MK plasmid[1] in which transcription of the SV40 early region gene products is under the control of the viral enhancer and promoter (Fig. 1A). In mice of the 427-line and in those of other SV40 Tag lineages that contain the intact viral early region gene, expression of SV40 Tag is readily detected in the choroid plexus, thymus, and kidney[1,11]. In fact, each mouse of the 427 lineage succumbs to choroid plexus tumors by seven months of age, by which time one-third of them have also developed thymic stromal hyperplasia[1,12]. Expression of SV40 Tag in the 419-line mouse has been more difficult to document since neither hyperplasia nor tumors appear in any tissue of these mice. Nonetheless, SV40 Tag has been detected in the kidney and in the skin of adult 419-line mice[13,14].

To determine whether these mice were capable of generating an SV40 Tag immune response they were injected with 2×10^2 infectious units of SV40. SV40 Tag is not a capsid antigen but rather a viral gene product which is synthesized by SV40 infected cells. SV40 does not replicate in mouse cells,

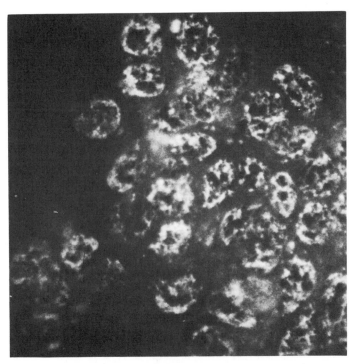

Fig. 2. A permanent cell line derived from the kidney of a 419-line mouse. Cells were grown on coverslips and fixed in -20°C acetone. SV40 Tag was localized to the nuclei by reaction with the 503-31 monoclonal antibody specific for SV40 Tag[25] and then with fluorescence isothiocyanate (FITC) tagged-goat immunoglobu-lin (Ig) G anti-mouse IgG heavy and light chain specific.

Table 2. Frequency of CTL Precursors in Normal B6, 419-
and 427-Line Transgenic Mice

Mice	Frequency (1/N)	P
B6-SV40 immunized	18,914 (12,614-28,362)	‾0.717
B6- not immunized	0*	--
419-SV40 immunized	23,780 (17,008-33,247)	0.538
427-SV40 immunized	0*	--

*Not calculable.

Table 3. Expression of MHC Class I Antigens on Cell Lines Derived from
Hyperplastic Tissues of SV40 Transgenic Mice

Cell line	Probable cell type of origin	Antibody		
		P3X63Ag8 Control	Y3 No treatment	Y3 Interferon-γ
427.1	subcapsular epithelium-thymus	6*	37	96
1308.1	cortical reticular epithelium-thymus	3	87	92
1307-6.1.1	interdigitating cell-thymus	6	68	81
1307-6.1.7	medullary epithelium-thymus	3	90	84
IgSV195	ß-islet-pancreas	7	12	72

*% reactive cells

Cells, pretreated or not for 4 days with 500 units recombinant interferon-γ (Genentech), were trypsinized, resuspended in cell culture
supernatant from the P3X63Ag8[24] myeloma cell line, which secretes an
IgG that does not react with cultured mouse cells, or with that from the
Y3[25] hybridoma which secretes an antibody reactive with determinants
expressed on the MHC class I H-2K^b and K^a molecules. Antibody reactivity was assessed, after incubation with FITC-tagged goat IgG anti-
mouse IgG, by flow microfluorimetry. Interferon-γ treatment did not
change the % positive cells incubated with control antibody (P3X63Ag8).

the coat proteins are not made, but mouse cells can be abortively infected and these cells, like SV40 transformed cells, synthesize SV40 Tag which can then immunize the host. In contrast to normal mice of the same strain combination, 427-line transgenic mice are specifically tolerant to SV40 T antigen; no CTL are detected in assays of their lymph node or spleen cells after primary or secondary in vivo immunizations with the virus, or after secondary in vitro restimulation with syngeneic SV40-transformed cells[12]. The frequency of SV40 Tag-specific precursor T cells capable of differentiating into CTL is below the level of detection by limiting dilution analysis[15] (Table 2). In contrast, some 419-line mice spontaneously make antibody to SV40 Tag[12] and after immunization, active SV40 Tag-specific CTL are elicited. The frequency of SV40 Tag-specific CTL precursors, 1/23,780 lymphocytes from SV40-immunized 419-line mice[15] (Table 2), is similar to that found in SV40-immunized normal C57B1/6 mice (419-line mice are of C57B1/6 origin). Profound specific immune tolerance, such as that exhibited by 427-line mice to SV40 Tag, is thought to result from interactions which occur during development and differentiation of T cells in the thymus. Rearrangement of the ß- and α-chains of the T cell receptor and initial cell surface display of the mature receptor occur during T cell maturation in the thymus[17,18]. T cells bearing receptors that recognize self-molecules encountered in the thymus are not found among the population of T cells that exit the thymus and migrate to the periphery[19-21]. This deletion of potentially autoreactive clones is considered the basis of self tolerance. Thus, if the SV40 Tag transgene is expressed in the thymus during development of the T cell repertoire, those T cells capable of recognizing SV40 Tag would not survive negative selection in the thymus.

This may indeed be the case in 427-line mice. We investigated the cellular basis of the thymic hyperplasia that characterizes the 427-mouse line and localized the abnormality to the thymic stroma; areas of neoplasia were found in both the cortical and medullary regions of the thymus[22]. Portions of these hyperplastic thymuses were transferred under the kidney capsule of nude mice and cells from the resultant tumors were cultured in vitro, or cells from a portion of a hyperplastic thymus were placed directly in vitro. Single cell clones of the resulting cellular outgrowths were characterized on the basis of morphology, ultrastructure, and reaction with monoclonal antibodies as SV40-transformed epithelial cells from: the subcapsular cortex (427.1, thymic nurse cells); the inner cortex (1308.1, thymic reticular cells); and the medullary region (1307-6.1.7). A clonal cell line with properties of the bone marrow-derived thymic interdigitating cell (1307-6.1.1) was also obtained[22]. Each of these cell lines synthesize SV40 Tag and each expresses MHC class I antigens (Table 3) and can be induced to express MHC class II antigen after treatment with recombinant interferon-γ[22]. Therefore, representatives of each cell type with which the differentiating prothymocyte could interact can express not only SV40 Tag but also the MHC gene products, which are necessary for T cell receptor interactions. Expression of SV40 Tag during embryogenesis is inferred from the SV40 Tag specific tolerance observed in young adult 427-line animals; however, the earliest developmental time at which SV40 Tag is expressed is unknown. Presumably, late fetal expression would suffice to ensure elimination of T cell clones capable of responding to SV40 Tag. Although we have not made a systematic study of the fetal thymus, SV40 Tag expression was detected in the 427-line neonatal thymus (Albee Messing, Univ. of Wisconsin, personal communication).

The evidence that SV40 Tag is expressed in 419-line mice is both direct [SV40 Tag can be immunoprecipitated from adult kidneys[13]], and indirect [some 419-line mice contain antibody to SV40 Tag[12]; engraftment of tissues from 419-line mice immunizes normal mice to an SV40 Tag-specific response[14]; and SV40 Tag-positive permanent cell lines are readily derived from explants of 419-line kidneys (Fig. 2) and other tissues]. Therefore, it seems likely that the different sites of integration of the transgene in 419- and 427-line

Table 4. Tumor Appearance and Immune Status of Mice of the SV40 Tag Transgenic Lineages

Transgenic	Site of Tag expression	Antibody to SV40 Tag*	Tumor	Immunized	Tumors following immunization
427[1]	brain, thymus, kidney	−	choroid plexus	tolerant	N.D.
419[1]	kidney, skin	+[12]	none	responsive	N.D.
334[5]	liver, adipose tissue	−	hibernoma	tolerant	N.D.
194*	liver, adipose tissue	+	none	responsive	N.D.
177-5[27]	exocrine pancreas	N.D.	exocrine pancreas 5 mos.	tolerant	exocrine pancreas 5 mos.
RIP Tag 2[3]	pancreatic β-islet	+/−[28]	insulinoma 3-4 mos.	low response	insulinoma 3-4 mos.
RIP Tag 4[3]	pancreas β-islet	+[28]	insulinoma 7-10 mos.	responsive	insulinoma 13-16 mos.

*Presence of circulating antibody reactive with SV40 Tag in mice not exogenously immunized
**Fox, et al., in preparation.
[1,3,5,12,27,28]See References

mice influence SV40 Tag expression. If expression of this cell-bound antigen occurs only in mature mice and/or extrathymically, CTL precursors with the capacity to recognize SV40 Tag may survive thymic selection because they have not encountered this antigen during their maturation. Such immunocytes in the periphery would then be capable of eliminating SV40 Tag-expressing cells when they first appear in the animal. Whether CTLs generated in this manner are necessary and sufficient to eliminate SV40 Tag-expressing cells when they first appear in the intact organism is unresolved. Two experimental approaches, one genetic (out-and-back crossing 419-line mice to mouse strains which mount but weak CTL response), and one surgical (transplanting 427-line thymuses into neonatal 419-line mice), are being pursued to determine whether such mice, in which the CTL response is attenuated, develop tumors.

To summarize, 427-line mice develop SV40 Tag-induced tumors and are tolerant to this viral oncogene product, whereas 419-line mice which bear the same construct are capable of an immune response and do not develop SV40 Tag-induced tumors. Permanent cell lines derived from each of these lineages express SV40 Tag and the MHC class I gene products required for CTL recognition.

THE ROLE OF CELL TYPE AND OF DEVELOPMENTAL TIME
OF EXPRESSION IN TOLEROGENESIS

Several questions were raised by this initial study in the 419 and 427 lineages which we have begun to investigate in other transgenic lineages. The first concerns the cell type which expresses SV40 Tag and the second, the developmental time of expression of SV40 Tag. How do these impact on immune recognition? A related issue to be considered is whether or not the immune response plays a role in the control or selection of emergent endogenous tumors. Although mice of the SV40 Tag-tolerant 427 lineage express SV40 Tag early in life and in the thymus, immune tolerance operationally extends to cell-bound molecules expressed on many cell types other than those found in the thymus. To determine whether mice in which SV40 Tag was expressed on other cell types were SV40 Tag tolerant or responsive, we investigated the ability of a series of SV40-Tag transgenic mice, prepared by us or by others, to generate an SV40 Tag-specific CTL response following immunization (Table 4). Furthermore, to determine the role of the developmental time of expression of SV40 Tag in tolerogenesis, two gene constructs were designed to target SV40 Tag expression in transgenic mice at different developmental stages.

Activation of expression of the α-amylase gene (Amy-1[a]) is determined by two different untranslated regions in the mouse genome[26]. Transcription is initiated from the early "liver" (L) promoter in the neonate, and transcription from the "parotid" (P) promoter is activated at two weeks of age. Adult levels of α-amylase are not reached until three weeks of age, due to late activation of transcription from the P promoter. Transgenic lineages of mice bearing the SV40 Tag coding sequences under control of either the early (p600T) or late (p1.3T) promoter were prepared (5 and Fox et al., manuscript in preparation). Mice from lineages bearing the p600 T transgene succumb with SV40 Tag-induced tumors of the brown fat (hibernomas), and some also exhibit hepatocellular carcinomas[5] (Table 1). On the other hand, no tumors appear in mice of the p1.3T lineage, although our preliminary evidence suggests that there is expression of SV40 Tag within fat depots. The ability of mice of the 334 (p600 T, early) and 194 (p1.3T, late) lineages to mount an SV40 Tag CTL response following SV40 injection was assessed. Mice of the 334 lineage are tolerant to SV40 Tag while those of the 194 line are SV40 Tag-responsive (Crooke et al., manuscript in preparation). Taken at face value, these results once again suggest a correlation between the time of expression of SV40 Tag and specific immune tolerance (early expression) or responsiveness (late expression). Moreover, tumors appear in the tolerant and not in the immuno-responsive lineages. The Amy-1ᵃ gene is now known to be expressed in parotid, liver, pancreas and in depots of both brown and white adipose tissue[26]. Expression of the α-amylase SV40 Tag transgene is even further restricted since no expression in the pancreas is seen[5]. This distribution of SV40 Tag expression leads us to suggest that expression in thymic stromal cells per se is not a prerequisite for induction of tolerance. In this respect the transgene product is equivalent to the other cell-bound gene products within other specialized cells in the normal mouse; low level expression by resident thymic stromal cells, or by individual bone marrow-derived cells, which traffic through the thymus during repertoire formation, may be possible sources of the tolerogen. Experiments to visualize SV40 Tag expression at the single cell level in the thymus of 334-line mice have not been performed.

THE EFFECT OF IMMUNIZATION ON TUMOR APPEARANCE

    419- and 194-line mice are immunologically responsive to SV40 Tag, but they don't develop SV40-induced tumors and although the 427- and 334-line mice bear SV40 Tag-positive tumors, they are SV40 Tag tolerant. To find a workable experimental system in which we could determine whether immunization could affect the appearance of endogenously programmed SV40 Tag-induced tumors we expanded our studies to other transgenic lineages. We tested the ability of mice from an elastase-1-SV40 Tag transgenic lineage, 177-5, which succumb to tumors of the exocrine pancreas[27], to generate SV40 Tag-specific CTLs following inoculation of SV40 Tag. These mice, which express SV40 Tag early in life, are unable to generate an SV40 Tag-specific CTL response; like 427- and 354-line mice, they are tolerant to SV40 Tag. A group of eight mice from the 177-5 lineage were injected at 5 weeks of age with SV40 and their age and cause of death were compared to that of their untreated littermates. Each mouse in the immunized and control groups died at approximately 5 months of age and bore one or more pancreatic tumors (Knowles, Brinster and Jewett, unpublished). As expected, a single inoculum of SV40 did not inhibit tumor appearance in these SV40 Tag-tolerant transgenic mice.

    Our investigation of an SV40 Tag transgenic lineage which develops tumors of the endocrine pancreas produced results indicating that inoculation with SV40 could control tumor appearance. Several of the RIP Tag lineages[3] (Fig. 1c) had previously been shown to synthesize SV40 Tag-specific antibody spontaneously, a response which was stimulated by immunization with purified SV40 Tag[28]. Two of the RIP Tag lineages, 2 and 4, were chosen for further study. Mice of the RIP Tag 2 lineage express high levels of SV40 Tag within their ß-islet cells during the last week of embryogenesis and throughout their short adult life. In contrast, RIP Tag 4 mice express SV40 Tag focally, in some but not all ß-islets, expression is initiated at some point within the first 2½ months of their lives and they die in 8 to 10 months bearing insulinomas[28]. Rare mice of the RIP Tag 2 lineage spontaneously produce low titer antibody to SV40 Tag. Spontaneous production of such antibody is more frequent in RIP Tag 4 mice[28]. Immunization with purified SV40 Tag resulted in production of low titer antibody in only some of the RIP Tag 2 mice whereas each immunized RIP Tag 4 mouse produced high titer SV40 Tag-specific antibody[28]. Therefore, unlike the transgenic lineages previously discussed (427, 354, 177-5), the RIP Tag 2 and 4 mice revealed no correlation between tumor formation and SV40 Tag tolerance.

    When these mice were inoculated with infectious SV40, those from the RIP Tag 4 lineage mounted a strong SV40 Tag-specific CTL response whereas the CTL response in those of the RIP Tag 2 lineage was sporadic and weak (Table 1). The RIP Tag 4 mice thus provide tumor-bearing SV40 Tag-immunoresponsive mice, in which the effect of SV40 inoculation on tumor development could be

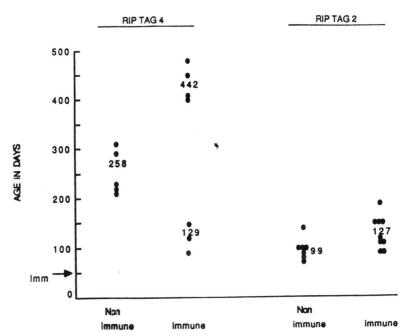

Fig. 3. Age at death of RIP Tag 4 and RIP Tag 2 mice
after immunization with SV40. Mice were bred at
Cold Spring Harbor Laboratories and either immunized at the Wistar Institute (RIP Tag 4) or at
Cold Spring Harbor Labs (RIP Tag 2) at 50 days of
age. Littermates formed the unimmunized control
groups. The mean age of the animals at death is
indicated; ● represents the age at death of individual animals.

assessed. Mice of both lineages were inoculated with 2x10[7] infectious
units of SV40 at 50 days of age, a time chosen to approximate the time of
SV40 Tag expression in the RIP Tag 4 mice. At this time the ß-islet
hyperplasia in the RIP Tag 2 mice is markedly advanced. The mean age at
death of unimmunized mice of the RIP Tag 4 lineage was determined to be 258
days. Immunized RIP Tag 4 mice fell into two groups: those that died within
79 days of immunization (129 days of age) and those that died at well over a
year of age (day 442). The cause of death of the mice that died prior to the
unimmunized mice was not determined; however, blood glucose levels of each
animal that survived to 200 days and beyond was monitored weekly. In
addition, each mouse was necropsied. Each of these mice, whether immunized
or not, was hypoglycemic and bore an insulinoma at death. In contrast to the
obvious difference in life span of SV40-inoculated and control RIP Tag 4
mice, the life span of SV40-inoculated and control RIP Tag 2 mice was
similar; each mouse exhibited decreasing blood glucose levels with age and
each bore a large insulinoma at necropsy (B. B. Knowles, L. Jewett, and D.
Hanahan, manuscript in preparation). These results, considered together with
the observation of lymphocyte infiltrates in the hyperplastic islets of SV40
Tag-immunized RIP Tag 3 mice (which like those of the RIP Tag 4 lineage
express SV40 Tag only as adults[28]) uphold the hypothesis that the immune
response can effectively control promoted cells expressing the endogenous
viral oncogene.

Although these data support the involvement of immune recognition of
SV40 Tag expressing cells in tumor growth control, some questions remain. If
an active CTL response can be generated in RIP Tag 4 mice, why do these mice
fail to reject their SV40 Tag expressing ß-cells without exogenous immunization? Why do immunized RIP Tag 4 mice eventually succumb with insulinomas?
Why does immunization fail to significantly curtail the time of tumor appearance in RIP Tag 2 mice? An answer to each of these questions may rest in
MHC antigen expression on the pancreatic ß-cell surface. In normal mice,
ß-cells probably express low levels of MHC class I and no MHC class II antigen[29]. Aberrant expression of MHC class II molecules on ß-cells has been
proposed to expose islet cell-sequestered self-antigens to the scrutiny of
mature T cells setting up the possibility of an autoimmune response[30].
This proposal infers that islet cell-bound antigens are neither secreted into
the general circulation nor processed and presented by phagocytes, so that
self-tolerance to them cannot be established. MHC class II-restricted helper
T cells may or may not be required for immune recognition of islet cell-bound
antigens but MHC class I-restricted CTL are probably essential for their
immune destruction. If the H-2 K/D antigens are normally expressed at very
low levels, the probability of intracellular interaction with SV40 Tag, and
extracellular recognition by CTLs decreases. Therefore escape from either
initial sensitization of, or destruction by, CTLs becomes more likely.

We have analyzed MHC antigen expression on permanent pancreatic ß-cell
lines established from insulinomas of SV40 Tag transgenic mice[31,32] and
find that MHC class I antigen expression is barely detectable. As in other
differentiated cell types H-2 K/D expression can be upregulated by pretreatment with interferon-γ [31] (Table 3). These SV40 Tag positive ß-cells are
not lysed by SV40 Tag-specific CTLs in in vitro assays unless H-2 K/D expression is increased by previous exposure to interferon-γ [31]. We therefore
suggest that inoculation of these mice with SV40 guarantees that a population
of primed SV40 Tag-specific CTL is present in RIP Tag 4 mice when SV40 Tag is
first expressed in the pancreatic ß-cells, before they have acquired the
characteristics of transformed cells. If such SV40 Tag-positive ß-islet
cells express sufficient H-2 K/D antigens, they can be recognized by these
activated CTL and eliminated. Localized interferon-γ secretion by the activated CTL can positively affect H-2 K/D expression on proximal cells and contribute to the surveillance cycle. The eventual appearance of SV40-induced
tumors in SV40-inoculated RIP Tag 4 mice may be attributed to the decline of

SV40-Tag specific CTL in the aging mouse, and/or expansion of variant ß-cells capable of eluding immune surveillance because the level of MHC class I antigen expressed is insufficient for T cell recognition. In RIP Tag 4 mice, that are not SV40-inoculated, the spontaneous generation of an SV40 Tag-specific CTL response may not occur until sufficient numbers of hyperplastic cells are destroyed and SV40 Tag antigen is processed and presented by other host cells. If sufficient changes in the hyperplastic cell occur before an effective CTL response is established, cells that have developed the capacity to escape the immune response may be present in the population. Regarded in this light the CTL response may be _part_ of the boundary between tumor promotion and progression, eliminating the majority of tumor cells and selecting variants able to elude an efficient CTL response.

We suggest that if SV40 Tag is expressed on appropriate cell types, during development of the T cell repertoire it is subjected to the same immune scrutiny as that of other endogenous molecules. The outcome of such scrutiny is SV40 Tag-specific immune tolerance. Of the three profoundly tolerant transgenic lineages surveyed only 427-line mice are known to express SV40 Tag on thymic stromal cells. Adult mice of the 334 and 177-5 lineages express SV40 Tag in specialized cells in other tissues. As is the case with endogenous cell-bound molecules we are unable to state how tolerance to SV40 Tag is induced in mice of these latter two lineages. SV40 Tag may be transiently expressed either by thymic stromal cells themselves or by bone marrow-derived cell types that transit the thymus during T cell repertoire formation. Mice of the RIP Tag 2 lineage appear to be partially tolerant to SV40 Tag, a state that might reflect the cell type bearing SV40 Tag and its ability to effectively present antigen. In SV40 Tag tolerant mice, when a given cell expresses a sufficient level of SV40 Tag, it can continue to proliferate and even perhaps forming a hyperplastic nodule. Further genetic change of such immortalized cells can then lead to the development of a frank tumor which can invade other tissues and metastasize. Tumors in tolerant mice can result in death of the animal. This may occur early or late in the progression of the tumor depending on the tissue affected. 427-line mice developing brain tumors die without evidence of metastasis, however each hibernoma-bearing mouse in the 354-lineage develops metastatic lesions before death[5]. Immunization of SV40 Tag-tolerant mice cannot result in an effective CTL response and does not affect tumor appearance. If SV40 Tag is not expressed until after formation of the T cell repertoire, or if it is expressed on cell types not normally subject to immune surveillance, then the possibility of developing an SV40 Tag-specific immune response alters the probability of tumor appearance. In mice such as those of the 419 and 194 lineages, we conjecture that SV40 Tag is expressed on a cell type which is readily accessible to immune surveillance; animals become immunized when SV40 Tag is expressed and SV40 Tag-positive cells which appear are promptly destroyed. SV40-induced tumors do not appear in these lineages. SV40 Tag expression on some cell types, such as those of the pancreatic ß-islet, may escape initial immune surveillance either because of inadequate presentation in association with the cell's MHC antigens or because SV40 Tag-MHC antigen display on this specific cell type does not stimulate any immune response. In the absence of immune surveillance SV40 Tag-induced hyperplasia can occur followed by tissue remodeling and cell destruction. The resultant SV40 Tag antigen presentation by phagocytic cells could result in an effective specific immune response, but this would occur after these hyperplastic cells have accumulated the multiple genetic changes associated with transformation. Those cells which have developed mechanisms allowing them to escape the surveillance of SV40-specific primed CTLs, such as defective regulation of expression of their MHC class I antigens, can form progressive tumors. Exogenous immunization of these mice at the time when SV40 Tag expression is low, and before the cells have developed the multiple changes which characterize progressive tumors, can curtail tumor appearance.

Our studies within the various transgenic lineages suggest that the endogenous viral oncogene product is effectively recognized by host lymphocytes. Recognition in a specific time and/or tissue leads to specific immune tolerance or responsiveness states, which correlate with tumor appearance or the tumor-free phenotype, respectively. Functional tolerance is found when the viral gene product is exclusively expressed on tissues not normally subject to immune surveillance. In this case specific immunization overrides tolerance and some measure of tumor growth control is seen.

REFERENCES

1. R. L. Brinster, C. Howe, A. Messing, T. Van Dyke, A. Levine, and R. Palmiter, Transgenic mice harboring SV40 T-antigen genes develop characteristic brain tumors, Cell 37:376-387 (1984).
2. A. Messing, H. Y. Chen, R. D. Palmiter, and R. L. Brinster, Peripheral neuropathies, hepatocellular carcinomas and islet cell adenomas in transgenic mice, Nature 316:461-463 (1985).
3. D. Hanahan, Heritable formation of pancreatic ß-cell tumors in transgenic mice expressing recombinant insulin/simian virus 40 oncogenes, Nature 315:115-122 (1985).
4. L. J. Field, Atrial natriuretic factor-SV40 T antigen transgenes produce tumors and cardiac arrhythmias in mice, Science 239:1029-1033 (1988).
5. N. Fox, R. Crooke, L.-H. Hwang, U. Schibler, B. B. Knowles, and D. Solter, Expression of an α-amylase SV40 T antigen hybrid gene in transgenic mice results in metastatic tumors of brown adipose tissue, Science (1989) (In Press).
6. J. Abramczuk, S. Pan, G. Maul, and B. B. Knowles, Tumor induction by simian virus 40 in the mouse is controlled by long term persistence of the viral genome and the immune response of the host, J. Virol. 49:540-548 (1984).
7. K. Pfizenmaier, S. Pan, and B. B. Knowles, Preferential H-2 association in cytotoxic T-cell responses to SV40 tumor-associated specific antigens, J. Immunol. 124:1888-1891 (1980).
8. S. Pan, J. Abramczuk, and B. B. Knowles, Immune control of 40-induced tumors in mice, Int. J. Canc. 39:722-728 (1987).
9. L. R. Gooding, Characterization of a progressive tumor from C3H-fibroblasts transformed in vitro with SV40 virus. Immunoresistance in vivo correlates with phenotypic loss of H-2K^k, J. Immunol. 129:1306-1312 (1982).
10. P. C. Doherty, B. B. Knowles, and P. J. Wettstein, Immunological surveillance of tumors in the context of major histocompatibility complex restriction of T cell function, Ad. Canc. Res. 42:1-65 (1984).
11. R. D. Palmiter, C. Y. Howe, A. Messing, and R. L. Brinster, SV40 enhancer and large T antigen are instrumental in development of choroid plexus tumors in transgenic mice, Nature 316:457-460 (1985).
12. S. J. Faas, S. Pan, C. A. Pinkert, R. L. Brinster, and B. B. Knowles, Simian virus 40 (SV40)-transgenic mice that develop tumors are specifically tolerant to SV40 T antigen, J. Exp. Med. 165:417-427 (1987).
13. T. VanDyke, C. Finlay, and A. J. Levine, A comparison of several lines of transgenic mice containing the SV40 early genes, CSH Symp. Quant. Biol. 50:671-678 (1985).
14. P. J. Wettstein, L. Jewett, S. Faas, R. L. Brinster, and B. B. Knowles, SV40 T antigen is a histocompatibility antigen of SV40-transgenic mice, Immunogenetics 27:436-441 (1988).
15. A. Juretic, and B. B. Knowles, Frequency of SV40-specific cytotoxic T lymphocyte precursors in two SV40 T antigen transgenic mouse lines (1989) (Manuscript submitted).

16. S. Fazekas de St. Groth, The evaluation of limiting dilution assays, J. Immunol. Methods 49:R11-R23 (1982).

17. H. R. Snodgrass, R. Kisielow, M. Kieter, M. Steinmetz, and H. von Boehmer, Ontogeny of the T-cell antigen receptor within the thymus, Nature 313:592-594 (1985).

18. D. H. Raulet, R. D. Gorman, H. Saito, and S. Tonegawa, Developmental regulation of T-cell receptor gene expression, Nature 314:103-106 (1985).

19. P. Kisielow, H. S. Teh, H. Bluthmann, and H. von Boehmer, Tolerance in T-cell-receptor transgenic mice involves deletion of nonmature $CD4^+8^+$ thymocytes, Nature 333:742-746 (1988).

20. J. W. Kappler, U. Staerz, J. White, and P. C. Marrack, Self-tolerance eliminates T cells specific for MLs-modified products of the major histocompatibility complex, Nature 332:35-38 (1988).

21. H. R. MacDonald, R. Schneider, R. K. Lees, R. C. Howe, H. Acha-Orbea, H. Festenstein, R. M. Zinkernagel, and H. Hentgartner, T-cell receptor V-beta use predicts reactivity and tolerance to MLs a-encoded antigens, Nature 332:40-44 (1988).

22. S.J. Faas and B. B. Knowles, Establishment and characterization of thymic cortical medullary and IDC-like cell lines that support stem cell proliferation, manuscript submitted.

23. S. Pan and B. B. Knowles, Monoclonal antibody to SV40 T-antigen blocks lysis of cloned cytotoxic T-cell line specific for SV40 TASA, Virology 125:1-6 (1983).

24. G. Kohler and C. Milstein, Continuous cultures of fused cells secreting antibody of predefined specificity, Nature 256:495-497 (1977).

25. B. Jones and C. A. Janeway, Cooperative interaction of B lymphocytes with antigen specific helper T lymphocytes is MHC restricted, Nature 292:547-549 (1981).

26. U. Schibler, P. H. Shaw, F. Sierra, O. Hagenbuchle, P. K. Wellauer, et al., Structural arrangement and tissue-specific expression of the two murine alpha-amylase loci Amy-1 and Amy-2, in: "Oxford Surveys on Eucaryotic Genes," N. MacLean, ed., 3:210-234, Oxford University Press, Oxford (1986).

27. D. M. Ornitz, R. E. Hammer, A. Messing, R. D. Palmiter, and R. L. Brinster, Pancreatic neoplasia induced by SV40 T-antigen expression in acinar cells of transgenic mice, Science 238:188-193 (1987).

28. T. E. Adams, S. Alpert, and D. Hanahan, Non-tolerance and auto antibodies to a transgenic self antigen expressed in pancreatic ß-cells, Nature 325:223-228 (1987).

29. S. Baekkeskov, T. Kanatsuna, L. Kereskog, D. A. Nielsen, P. A. Peterson, A. H. Rubenstein, D. F. Steiner, and A. Lernmark, Expression of major histocompatibility antigens on pancreatic islet cells, Proc. Natl. Acad. Sci. USA 78:6456-6460 (1981).

30. G. G. Bottazzo, R. Pujol-Borrell, T. Hanafusa, and M. Feldman, Role of aberrant HLA-DR expression and antigen presentation in induction of endocrine autoimmunity, Lancet ii:1115-1119 (1983).

31. A. Gilligan, L. Jewett, D. Simon, I. Damjanov, F. M. Matchinsky, H. Weik, C. Pinkert, and B. B. Knowles, Functional pancreatic beta-cell line from an SV40 T antigen transgenic mouse, Diabetes (1989) (In Press).

32. S. Efrat, S. Linde, H. Kofod, D. Spector, M. Delannoy, S. Grant, D. Hanahan, and S. Baekkeskov, ß-cell lines derived from transgenic mice expressing hybrid insulin-oncogenes. Proc. Natl. Acad. Sci. USA 85:9037-9041 (1988).

DISCUSSION

George Michalopolous: Do you see the immune response operating at the boundary between promotion and progression? Would you care to put it on one side or the other of the boundary?

Barbara B. Knowles: One part we could put on one side and the other part we could put on the other side. If you immunologically eliminate cells that start expressing the viral oncogene, just about the time that they start expressing it, then you've eliminated promoted cells right away. If promoted cells that lose control of expression of the major histocompatibility complex genes arise, they could escape the selection and form progressive tumors. I would suggest that the immune response is a selective force operating at, or actively forming part of the boundary between promotion and progression.

George Yoakum: I notice that the origin of replication was present in the constructs that you used in making your transgenic mice. I wonder if you have looked at Southern's to see if there is reintegration of those, as a consequence of their presence in any relevant sequence in the mouse cells that are growing out in these cases.

Knowles: The 427- and 419-line mice are Ralph Brinster's mice and I have not looked at that question. Those are the only mice in which the SV40 origin of replication is there. And I don't think that they've looked at it, either.

Yoakum: To follow up that line, I'm wondering (since large T is almost always attributed to being expressed in the nucleus and not at the cell surface) if there is any agreement in epitopes between the large T and small T that might account for this different observation. Also, would you speculate as to the relevance of your findings in terms of the humans that wound up being inoculated with SV40? When these humans achieve an age where their immune system is going to lose some of these more sophisticated functions, what do you think may happen?

Knowles: It would be very difficult for me to speculate on what's going to happen with the SV40 inoculated people. We presume that they would be capable of immunologically eliminating any cells that are harboring SV40 and expressing SV40 T-antigen, but we don't know that that's the case. Getting back to your first question about cell surface T: epitopes of SV40 large T-antigen are expressed on the cell surface. This has been found in several labs (mine, Janet Buttel's, S. Tevethia's, and Linda Gooding's) and they are the target for CTLs. This is confirmed also from the peptide work that Tevethia has been doing recently. I don't think that any of the mapped epitopes that are expressed on the cell surface are in the portion shared by T and t.

Harry Rubin: Would you comment on the general significance of immune surveillance in view of the fact that T cell deficient nude mice do not develop tumors in any higher incidence that immunologically competent mice.

Knowles: That is also a difficult question that has been around for quite some time. I don't have a very good answer for it.

José Russo: As a commentary to Dr. Yoakum's question, about eight years ago some people reported that they developed a permanent cell line of human breast cells and they called this cell line HB100. When it was studied seriously, they found it to be SV40 T-antigen positive. That woman received a Salk vaccination contaminated with SV40. It is in order to comment on this because the genome of this virus may immortalize or even transform human cells.

DIFFERENTIAL GENE EXPRESSION DURING TUMOR PROMOTION

AND PROGRESSION IN THE MOUSE SKIN MODEL

G. Tim Bowden*, Lawrence E. Ostrowski*, Keith Bonham*
and Peter Krieg[+]

*Radiation Oncology Department
University of Arizona Medical School
Tucson, Arizona 85724

[+]Institute for Virus Research
German Cancer Research Center
Heidelberg, West Germany

INTRODUCTION

The boundary between promotion and progression in experimental carcinogenesis can be operationally defined as long as stable intermediate stages of tumor formation can be identified. Once operational definitions have been made, investigators can and should pursue questions of molecular mechanisms to explain phenotypic changes that occur during promotion and progression. This paper deals with the identification and characterization of molecular markers (i.e., differentially expressed cellular genes) that identify different stages of mouse skin tumor formation. These marker genes whose steady state levels of messenger are elevated at specific stages in skin tumor formation can serve to define the stages of promotion and progression. There is also the possibility that overexpression of one or a number of these genes actually plays a functional role in tumor formation.

The process of mouse skin tumor formation can be subdivided into the operational stages of initiation, promotion and progression[1]. Initiation appears to involve the induction of permanent genetic alterations that are essentially irreversible. Promotion has been postulated to involve the clonal expansion of initiated cells leading to the appearance of a benign tumor, a papilloma. The two-stage protocol involving initiation by a mutagenic agent followed by promotion with a classical phorbol ester tumor promoter gives rise primarily to benign papillomas, only a few of which progress to malignant, squamous cell carcinomas (SCCs). More recently, the laboratories of Slaga[2], as well as Hennings and Yuspa[3] and Bowden[4], have shown that some initiators, as well as some weakly carcinogenic chemical and physical agents enhance the conversion of benign papillomas to squamous cell carcinomas. Another step in the progression of malignant skin tumors is the acquisition of metastatic potential. Recently Slaga's laboratory[5] has demonstrated the induction by repeated carcinogen treatment of SCCs with a high probability of metastasizing. The majority of SCCs induced with a two-stage protocol show a low probability of metastasis.

Using a two-stage model of mouse skin carcinogenesis and utilizing the

Boundaries between Promotion and Progression during Carcinogenesis
Edited by O. Sudilovsky *et al.*, Plenum Press, New York, 1991

technique of differential screening of cDNA libraries, we recently isolated a number of sequences (mal 1-6 and transin) which were overexpressed at different stages of tumor development[6,7]. We have investigated the expression pattern of these genes to determine if their expression might be correlated with promotion and progression and used as criteria to define the boundary between these two stages.

RESULTS

Expression of Mal Sequences in Benign and Malignant Mouse Skin Tumors

To investigate a potential role of the mal sequences in the process of carcinogenesis, we asked whether there was a correlation between the state of tumor development and the level of expression of the isolated mal sequences. Tumors were induced in the back skin of NMRI mice by initiation with 7,12-

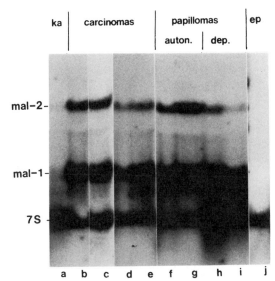

Fig. 1. Expression of mal 1 and mal 2 in different mouse skin tumors. 10 µg of total RNA from normal epidermis or tumors were applied to each lane, size fractionated in 1.4% agarose/2.2M formaldehyde gels, blotted onto cellulose nitrate paper, and hybridized to a mixed probe of nick-translated plasmid DNA pmal 1, pmal 2, and pA6[6,9]. Tumor induction, preparation of tumor and epidermal tissues, and RNA isolation and hybridization were performed as described elsewhere[6] a: RNA from a keratoacanthoma. b: RNA from a squamous cell carcinoma induced by MNNG and TPA. c-e: induced by DMBA and TPA. f,g: RNA from tumor-promoter, independent papillomas. h,i: from tumor-promoter-dependent papillomas. j: RNA from normal epidermis.

dimethylbenz[a]anthracene (DMBA) or with N-methyl-N'nitro-N-nitrosoguanidine (MNNG) and promotion with 12-0-tetradecanoylphorbol-13-acetate (TPA).

Northern blots were performed with RNA isolated from normal mouse epidermis, benign papillomas, a benign keratocanthoma, and malignant squamous cell carcinomas using the mal cDNA clones as probes. As a control in all northern blots the RNA bound to the filters was hybridized either simultaneously or in a second hybridization cycle with a probe specific for the 7S cytoplasmic RNA, present in the same abundance in normal epidermis as well as in tumors[8].

With one exception (mal 4), the transcripts corresponding to the mal cDNA clones were already overexpressed in the papilloma stage of multistep carcinogenesis, and we did not observe further enhanced expression of these sequences during the progression from the benign papilloma to the malignant tumor. There were no detectable differences in the expression of mal 1-related sequences in tumor-promoter-dependent papillomas (Fig. 1h, i), which had been isolated immediately after their appearance in the animals, compared to tumor-promoter-independent (so-called autonomous) papillomas (Fig. 1f, g), which were taken 12 weeks after the end of the TPA treatment.

In contrast, mal 2 expression was slightly enhanced in tumor-promoter-independent papillomas compared to tumor-promoter-dependent tumors (Fig. 1f-i). A keratoacanthoma, another benign tumor sometimes arising during mouse skin carcinogenesis, showed only a very weak expression of mal 1 and 2

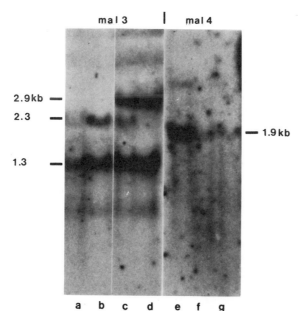

Fig. 2. Expression of mal 3 and mal 4 in different mouse skin tumor. Northern blots were performed as described in the legend to Fig. 1 and elsewhere[6]. a-d: Hybridized with pmal 3. e-g: Hybridized with pmal 4. a,g: RNA from normal epidermis. b,f: RNA from papillomas. c-e: RNA from squamous cell carcinomas.

(Fig. 1a). The expression of these mal sequences was only slightly, if at all, enhanced in the keratoacanthoma compared to normal epidermis (Fig. 1j). In all malignant squamous cell carcinomas we observed a high expression level of mal 1 and mal 2-related sequences. There were only slight differences in the expression levels in several individual carcinomas induced by the two-step protocol with either DMBA/TPA (Fig. 1c-e) or with MNNG/TPA (Fig. 1b).

In contrast, there was a change in the patterns of mal 3-related transcripts during tumor development in the mouse skin. In normal epidermis, small amounts of three transcripts of 1.3, 2.3, and 2.9 kilobases (kb) were detectable (Fig. 2a). In benign tumors, a keratoacanthoma and six different papillomas tested so far, only the 1.3- and the 2.3 kb transcripts were over-expressed (sometimes to a greater extent than in carcinomas[9]), whereas the largest 2.9 kb transcript was not detectable (Fig. 2b). In malignant squa-mous cell carcinomas, this largest transcript related to mal 3 was always overexpressed, whereas the 2.3 kb transcript in many of the carcinomas dis-appeared (Fig. 2d), detectable only in a few carcinomas in lower amounts (Fig. 2c). In all benign tumors tested so far, we only detected an over-expression of the 1.3 kb and 2.3 kb transcripts. Thus, overexpression of the 2.9 kb transcript may be related to the malignant state.

In a similar way, overexpression of mal 4-related sequences appeared to be specific for the malignant state. In northern blot analysis of mal 4, we detected only a slight overexpression of a 1.9 kb transcript in benign papil-lomas compared to normal epidermis (Fig. 2f, g). In malignant squamous cell carcinomas, however, this mal 4 transcript was present in high abundance (Fig. 2e). As estimated by densitometry of the autoradiograms, the factor of elevated transcription levels in carcinomas compared to papillomas and normal epidermis was greater than 10.

Molecular Characterization of the Mal 1, 3 and 4 cDNAs

Since the initially isolated cDNA inserts for the mal sequence were less than full length, we attempted to obtain full length cDNA inserts by screening a gt10 cDNA library made from SCC poly A⁺ RNA and a gt10 cDNA library made from the poly A⁺ RNA isolated from a malignantly transformed epidermal cell line, PDV. Screening with the pmal 1 insert yielded a gt10 cDNA with approximately 1 kb insert which was close to the size of the pri-mary transcript 1.1 kb. The mal 1 cDNA insert was subcloned into a Blue-script vector and the ExoIII, Mung bean nuclease nested deletion sequencing strategy along with Sanger dideoxy technique was used to sequence the insert. Sequencing of this mal 1 insert has indicated that mal 1 possesses a pre-dicted 105 amino acid domain which exhibits 65% homology with the amino acid sequence of a mouse lipid-binding protein (data not shown). This suggested that the gene product of mal 1 may be involved in lipid metabolism or may be a membrane-associated protein. Dideoxy sequencing of a pmal 3 cDNA insert has revealed that mal 3 cDNA is one and a half transcriptional units of mouse ubiquitin plus the 3' untranslated region (data not shown). Sequencing of a full length cDNA for the mal 4 gene has revealed that the mal 4 message is identical to mouse ß-actin (data not shown).

Expression Pattern of the Mouse Transin Gene
in Late Stages of Skin Tumor Progression

In previously published work, we reported that steady state levels of transin RNA transcripts, that code for a secreted proteinase, were over-expressed in invasive, non-metastatic SCCs, in comparison to benign papillo-mas and normal epidermis[7]. We demonstrated that transin RNA encodes a pro-tein with a predicted molecular weight of 53,000 daltons and that the transin protein produced in vivo was a secreted protein migrating as a doublet on a sodium dodecyl sulfate (SDS)-polyacrylamide gel with apparent molecular

130

weights of 58,000 to 60,000 daltons[7]. In addition, we demonstrated that
the protein product of transin was associated with proteolytic activity for
the protein substrate, casein.

Rat transin displayed 48% similarity to human type-1 collagenase[10] and
75% similarity to rabbit stromelysin[11], a secreted protease that degrades
fibronectin and proteoglycans in the basement membrane. These data suggest
that transin is a member of a family of matrix-degrading metalloproteinases.

The expression of proteolytic enzymes by tumors and malignantly transformed cells suggests that matrix-degrading proteinases are required for
invasion and metastasis[12]. The inappropriate expression of transin may,
therefore, be causally involved in one or more steps in tumor progression
from a benign papilloma to a malignant, invasive, and eventually metastatic
SCC. The goal of the research reported here was to investigate the expression of transin in the very late steps of progression from a non-metastatic
SCC to a metastatic SCC.

We first cloned a mouse homolog (TR11A) of the rat transin cDNA from a
gt10 cDNA library made from a SCC[13]. This mouse cDNA clone was used in the
subsequent studies. Next, total RNA isolated from benign papillomas and SCCs
induced by MNNG initiation and TPA promotion or repeated MNNG treatments were
run on a northern gel, blotted, and probed with a nick-translated mouse
transin cDNA probe. This blot was rehybridized with a nick-translated probe
specific to 7S RNA to control for the amount of RNA loaded and transferred to
the nitrocellulose filter (Fig. 3, and data not shown). Three benign papilloma RNAs were investigated on this northern. One of these showed transin
transcripts, perhaps identifying this as a papilloma with an increased probability of conversion. The levels of transin transcripts seen in the SCCs
induced by MNNG initiation and TPA promotion (MNNG/TPA) were more variable
but were consistent with our previous results[7]. With one exception (SCC 55
in Fig. 3) the level of transin transcripts was always higher in SCCs than in
normal epidermis. A very strong 1.9 kb hybridizing transcript was seen in
the RNA from three SCCs induced by repeated MNNG treatment. The level of
transin transcripts induced by repeated MNNG treatment was always higher
than the somewhat variable levels seen in the MNNG/TPA-induced SCCs. Additional tumor samples, some of which were partially degraded, were analyzed by
dot blot analysis (Fig. 4). For comparison, some of the samples in Fig. 3
were also included on this filter. All five of the RNAs from tumors induced
by repeated MNNG showed strong autoradiographic signals at the three dilu-
tions of RNA studied. One of the two RNAs from tumors induced by MNNG/TPA
showed strong autoradiographic signals indicative of high levels of transin
transcripts. Of the four RNA samples from benign papillomas (induced by
MNNG/TPA), one sample showed a strong autoradiographic signal. Therefore, by
both northern and dot blot analysis the levels of transin transcripts were
consistently higher in SCCs induced by repeated MNNG treatment than those
found in either benign papillomas or SCCs induced by MNNG initiation and TPA
promotion.

Since our probing of RNA transcripts involved the use of a cDNA that
contained both amino and carboxy terminal sequences and because at least two
rat transin genes have been detected that differ significantly in their carboxy terminus[14], it was not clear whether we were detecting the mouse transin (or TR11A-related) transcripts or a related transcript with a similarly
conserved amino terminal region. To distinguish between these possibilities,
we subcloned a 0.2 kb HindI fragment from the carboxy terminus of the mouse
transin TR11A clone that did not contain the conserved amino terminus sequence and used this clone as a probe in northern analysis of various tumor
RNAs. Our results (data not shown) using this subclone were the same as
those seen with the large cDNA insert. Therefore, the majority of the hy-

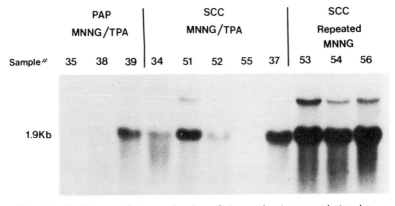

Fig. 3. Northern blot analysis of transin transcripts in various tumors induced by two-stage and complete carcinogenesis protocols. Ten micrograms of total tumor RNA were separated on a 1.4% agarose formaldehyde gel, blotted onto nitrocellulose paper, and probed with a ^{32}P-labeled nick-translated mouse transin cDNA. The blot was hybridized with a ^{32}P-labeled nick-translated probe for 7S ribosomal RNA to verify that the amounts of total RNA loaded were equivalent. The tumor induction protocols included: 1) MNNG initiation followed by TPA promotion, and 2) repeated MNNG treatments of the skin.

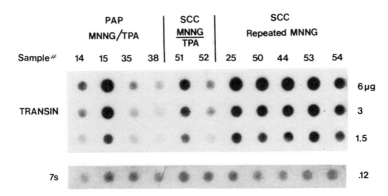

Fig. 4. Dot blot analysis of transin transcripts in various tumors induced by two-stage and complete carcinogenesis protocols. Total RNA (6, 3, 1.5 or 0.12 µg) from various tumors (papillomas (PAP) and SCC) were dotted onto nitrocellulose paper, fixed by baking and then probed with ^{32}P-nick-translated mouse transin or 7S ribosomal cDNA probes. The tumor induction protocols included: 1) MNNG initiation followed by TPA promotion, and 2) repeated MNNG treatments of the skin.

bridizing transcripts seen in the mouse tumor appeared to be homologous to the rat transin-1 gene.

Because the overexpression of the transin gene during late stages of mouse skin tumor progression may be due to amplification or rearrangements of the transin gene, we analyzed the genomic arrangement of the transin gene. High molecular weight genomic DNAs from various mouse tissues including normal epidermis, papillomas, and SCCs induced by different protocols were cut with the restriction enzyme PstI, run on a Southern gel, transferred onto Zeta probe paper, and probed with a nick-translated mouse transin cDNA probe (TR11A). The major hybridizing fragments seen in PstI-restricted normal epidermal DNA were 12, 6.7, 4.1, 3.2, 1.55, and 1.2 kb (Fig. 5). The variations in band intensity, especially the higher molecular weight bands, were apparently due to differences in the loading and transfer of the restricted DNA fragments, as determined by ethidium bromide staining of the gel before and after transfer (data not shown). Therefore, we have no evidence for either rearrangement or amplification of the mouse transin gene in any of the tumor DNAs studied. We also performed Southern analysis on metastatic and non-metastatic SCCs induced by repeated B[a]P treatment and non-metastatic SCCs induced by DMBA as an initiator and TPA as a promoter (data not shown). Analysis of 21 of these tumor samples showed no evidence for either rearrangement or amplification of the transin gene.

Expression of Tumor Associated Genes During the Process of Tumor Promotion

An important unanswered question related to the specificity of tumor promoters is whether the growth stimulatory effect is the only event involved

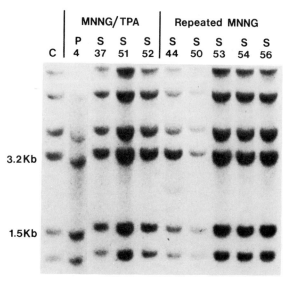

Fig. 5. Southern analysis of DNAs from tumors induced by various carcinogenesis protocols. Ten micrograms of high molecular weight genomic DNA were digested with the restriction enzyme PstI, separated on a 0.8% agarose gel, blotted onto nitrocellulose paper, and probed with a ^{32}P- nick-translated mouse transin cDNA probe. C, control epidermis; S, squamous cell carcinomas; P, papilloma.

in promotion or whether the tumor promoters are able to upregulate the expression of tumor-associated genes whose overexpression may be involved in a functional way in tumor progression. To approach this important question, we have used tumor promoters and a non-promoting, hyperplastic agent to investigate the steady-state levels of expression of tumor-associated, upregulated sequences (mal 1, 2, 4 and transin) in adult mouse epidermis[15].

To study effects of tumor promoters in vivo on the transcription of specific cellular genes, the tumor promoters TPA and 12-0-retinylphorbol-13-acetate (RPA) were applied to adult mouse skin. Total RNA of the epidermis at various times after tumor promoter treatment was probed on northern blots to analyze the steady-state levels of tumor-associated gene transcripts. As a control in all northern blots, the RNA bound to the filters was hybridized

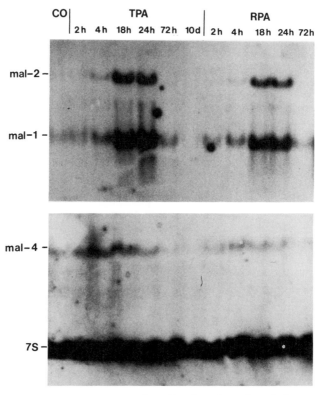

Fig. 6. Time-courses for the levels of mal 1,
2 and 4 transcripts in the epidermis
of mice treated with TPA or RPA. Mice
were treated with 10 nmol of TPA or RPA
(in 0.1 ml of acetone) and groups of
five mice were killed at various times
after treatment. Epidermis was isolated
from the treated back skin and total
RNA was extracted. RNA was separated
on northern gel, blotted and probed
with ^{32}P nick-translated mal 1, 2
and 4 and 7S probes. An autoradiogram
of the hybridized and washed blot is
shown in this Figure. A, time-course
after 10 nmol TPA; B, time-course after
10 nmol RPA.

either simultaneously or in a second hybridization cycle with a probe specific for the 7S cytoplasmic RNA, present in the same abundance in all epidermal
tissues[8]. A single application of 10 nmol of TPA on the uninitiated back
skin of NMRI mice stimulated a transient expression of mal related sequences
(Fig. 6A) and transin (Fig. 7A). The different mal sequences and transin
showed different kinetics of stimulated expression. The expression of mal 1
and mal 2 was enhanced within 4 h after TPA treatment and reached a maximum
~ 24 h after treatment (Fig. 6A). The expression of these sequences returned to near control levels 72 h after treatment. Mal 4 expression was enhanced earlier and for a shorter time (Fig. 6A). Enhanced expression was observed within 2 h, reached a maximum between 4 and 18 h and decreased within 24 h
after TPA treatment. Enhanced expression of transin was seen within 4 h
after TPA treatment, reached a maximum at 18 h and returned to control level
by 24 h (Fig. 7A). The additional smeared hybridization signal in these
northerns was due to partial degradation which commonly was observed in RNA
preparations of TPA-treated epidermis despite the addition of aurintricar-
boxylic acetate as an additional RNase inhibitor. Multiple treatments with
TPA did not further enhance or prolong the enhanced expression of the mal
sequences or transin compared with that seen with a single dose of TPA (data
not shown).

To determine if the response of the mal sequences and transin is related
to tumor-promoting activity, we tested the incomplete second-stage tumor promoter RPA. RPA promotes tumors in NMRI mouse skin following tumor initiation
and a brief treatment with TPA[16]. Treatment of the skin with 10 nmol RPA
resulted in nearly the same responses, except that the level of stimulated
mal expression appeared to be lower (Figs. 6B and 7B).

To investigate whether the tumor-promoter-enhanced expression of the mal
sequences was related to the enhanced proliferative response in the epidermis, mouse skin was treated with a non-tumor-promoting hyperplastic agent
ethyl phenylpropiolate (EPP) using a dose (40 μmol) that induced a hyperplastic response similar to that seen with 10 nmol of either TPA or RPA, but
that is not capable of promoting tumors following tumor initiation[17] (and

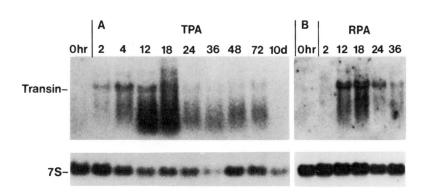

Fig. 7. Time-courses for the levels of transin transcripts
in the epidermis of mice treated with TPA or RPA.
Mice were treated with 10 nmol of TPA or RPA (in
0.1 ml acetone) and groups of five mice were killed
at various times after treatment. Epidermis was
isolated from the treated and non-treated back skin
and total RNA was extracted. RNA was separated on
northern gel, blotted and probed with a ^{32}P nick-
translated mouse cDNA probe for transin and 7S
probe. A, time-course after 10 nmol TPA; B, time-
course after 10 nmol RPA.

135

data not shown). After EPP treatment, there was no enhanced expression of either the mal 4 or transin sequences (Fig. 8). The stronger signal in lane 3 (4 h) of the northern blots hybridized with mal 4 probes was due to the larger amount of RNA loaded onto the gel, as demonstrated by control hybridization with the 7S RNA-specific probe (data not shown). In contrast, expression of mal 1 and mal 2 was slightly enhanced by EPP. The extent of enhancement of these transcripts over control (~ 3 times), however, was less than

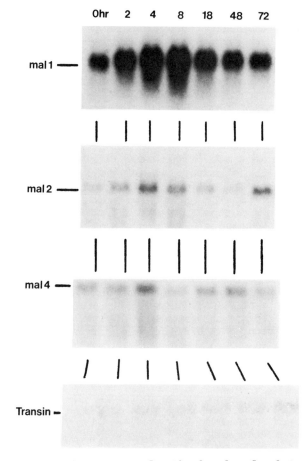

Fig. 8. Time-courses for the levels of mal 1, 2 and 4 and transin transcripts in the epidermis of mice treated with EPP. Mice were treated with 40 μmol of EPP (in 0.1 ml acetone) and groups of five mice were killed at various times after treatment. Epidermis was isolated from the treated and non-treated back skin and total RNA was extracted. RNA was separated on northern gel, blotted and probed with ^{32}P nick-translated mouse transin cDNA probe, mal 1, 2 and 4 as indicated. All lanes had roughly equivalent amounts of RNA as indicated by hybridization with a mouse 7S RNA probe (data not shown).

that seen with TPA (10-20 times). In addition, the time-course for the enhanced expession was different with EPP (Fig. 8) (maximum level between 4 and
8 h) compared with that observed with TPA (Fig. 6 and 7).

DISCUSSION

We have previously described the isolation of sequences activated during
tumor development, using the well-defined in vivo system of multi-stage carcinogenesis in the mouse skin[6,7]. Molecular characterization of these mal
sequences showed that they were not homologous to 16 known retroviral oncogenes. Preliminary DNA sequencing of the mal cDNAs have indicated that mal 1
possesses a domain which exhibits 65% DNA homology with the cDNA sequence of
a mouse lipid-binding protein. This suggested that the gene product of mal 1
may be involved in lipid metabolism or may be a membrane-associated protein.
Sequence analysis of mal 3 and 4 revealed them to be ubiquitin and ß-actin
respectively. The transin gene codes for a secreted proteinase[7] and has a
strong sequence homology with a known metalloproteinase, stromelysin[11]. It
has recently been established that there is a family of transin-like genes
coding for proteinases[14]. These transin genes have in common an N terminus
which codes for proteinase activity while the C termini are different and may
code for substrate specificity or an inhibitor binding site for the secreted
proteinase.

We have reported that the mal genes are overexpressed at specific stages
of mouse skin tumor formation and this differential gene expression could
serve to define the boundaries between tumor promotion and progression.
Overexpression of mal 4 (ß-actin) may be a marker for the malignant state,
because this sequence did not appear to be overexpressed in benign tumors
but transcriptional activation occurred during the progression from benign
papilloma to malignant squamous cell carcinoma. In contrast to the mal 4
expression pattern, overexpression of mal 1 and mal 2 related transcripts was
specific for neoplasia in that they were activated in both papillomas and
squamous cell carcinomas. A change in the transcript pattern of the mal 3
(ubiquitin) related sequences was observed during the progression of a benign
to a malignant tumor. It is not known whether the multiple transcripts were
different precursor molecules reflecting a change in RNA processing or whether they were different mature mRNA species. Southern analysis of the
genomic organization detected at least four copies of the mal 3 related sequences[15]. Nevertheless, in all benign tumors tested only two transcripts
(1.3 and 2.3 kb) were present and overexpressed and the 2.9 kb transcript was
only overexpressed in malignant tumors. Therefore, the overexpression of the
2.9 kb transcript related to mal 3 may be used as a genetic marker to distinguish between the benign and malignant state of tumor progression in mouse
skin.

Besides studying the expression pattern of the mal sequences during
mouse skin tumor progression, we have also studied the expression of the
transin gene during late stages of tumor progression. We have shown that the
steady state levels of transin transcripts in DMBA, TPA induced SCCs were
considerably higher than in benign papillomas[7]. In these initial studies
we used a rat transin-1 probe. Subsequently, we cloned a mouse cDNA which
turned out to have considerable sequence similarity with the rat transin 1
cDNA as well as with rabbit stromelysin. The high degree of similarity
between either rat or mouse transins and rabbit stromelysin suggested a potential role for the overexpression of the transin gene in invasion and
metastasis. It has been shown that rabbit stromelysin degrades many of the
protein components of extracellular matrices including proteoglycans, fibronectin, laminin, soluble type IV collagen and elastin[18]. All of these
proteins are components of the basement membrane which must be degraded by
invasive as well as metastatic tumor cells. The metastatic phenotype results

most likely from a complex series of steps involving the expression of multiple gene products. It is highly likely that the increased metastatic potential of some tumors is due to the increased expression of certain necessary but not sufficient gene products which confer a selective advantage on these cells. We have studied the expression pattern of this gene in SCCs with different potentials for metastasis. Our consistent finding that the levels of transin transcripts in SCCs induced by repeated MNNG were always higher than levels seen in SCCs with a lower probability of metastasis suggests that the higher levels of transin product enhance the probability that the tumor cells penetrate the continuous basement membranes, which separate tissue compartments and surround blood vessels and muscle. A similar observation was made by Garbisa et al[19]. These workers found that transfection of cultured cells with an activated ras oncogene induced metastatic tumors and these metastatic tumors secreted higher levels of type IV collagenolytic protease than non-metastatic tumors. Garbisa et al. suggested that their data supported a biochemical linkage between expression of type IV collagenase activity and the metastatic phenotype. A consideration of the carcinogenic protocol and latency period to induce metastatic tumors in the skin indicated that acquisition of this phenotype is a late event which requires large cumulative doses of a genotoxic complete carcinogen such as MNNG or B(a)P. Despite prolonged genotoxic damage, no tumors demonstrated either amplification or rearrangement of the transin gene. A comparison of genomic alterations to the measured levels of transin transcripts indicated that other mechanisms for enhanced steady state levels of the transin gene must exist.

Molecular hybridization techniques and cDNA cloning techniques have been used to define RNA sequences that change in abundance after phorbol ester tumor promoter treatment[20]. Angel et al[21] have found the upregulation of a number of sequences in human fibroblasts in response to not only the phorbol ester TPA, but also to UV light and mitomycin C. One of the genes transiently induced by these agents in human fibroblasts was the metallothionein gene[21]. This gene was constitutively overexpressed in transformed fibroblasts. We have also observed following promoter treatment of skin the transient expression of sequences constitutively overexpressed in tumors.

The mal 1 sequence, preferentially expressed in differentiated cells of the resting adult epidermis[15], and the mal 2 sequence, which is not expressed at detectable levels in either cell type, showed a transient enhanced expression pattern in adult epidermis after TPA or RPA treatment. These genes were induced in both basal and differentiated cells. Expression of mal 1 and 2 was induced not only by TPA and RPA but also by the non-tumor-promoting hyperplastic agent EPP, although to a lesser extent. Tumor-promoter-enhanced expression of these genes (mal 1 and 2) also showed similar kinetics compared with promoter-induced DNA synthesis. It is known that the labeling index of epidermal cells reaches a maximum at 18 and 30 h after tumor-promoter application[22]. Thus, we have correlative data suggesting that tumor-promoter-induced expression of mal 1 and 2 might be associated with proliferative responses which are not tumor promoter specific. In contrast, expression of mal 4 (ß-actin) and transin was induced earlier and for a shorter time. In addition, these genes, which were overexpressed only in malignant but not in benign tumors[7,9] were only inducible by tumor promoters, but not by EPP. These observations suggest that the stimulated expressions of mal 4 and transin might be associated with tumor-promoter-specific processes. Of interest was the fact that enhanced expression of the mal 4 sequence (found preferentially expressed in undifferentiated cells of the normal epidermis) was found in both basal and undifferentiated cells of the TPA-treated adult epidermis. Mal 4 enhanced expression in the differentiated cell population could represent a reprogramming of suprabasal cells to dedifferentiate, albeit transiently. It is also possible that post-transcriptional effects, such as a change in mRNA stability could explain the differences in steady-state levels of mal 4 transcripts. In contrast, the enhanced expression of the transin sequence was only found in the basal cells of the TPA-treated epidermis.

In conclusion we have identified and partially characterized a series of cellular genes (mal 1, 2, 3, 4 and transin) whose pattern of overexpression during mouse skin tumor formation help to define the boundaries between tumor promotion and progression. We are currently investigating the possible functional role of these genes in the formation of mouse skin tumors. In addition we have found that the transient TPA and RPA induced expression of the tumor associated genes, ß-actin and transin, in mouse epidermis is specific for tumor promoters. This differential gene expression may help to define the stage of tumor promotion and distinguish it from tumor progression.

ACKNOWLEDGEMENT

This work was supported in part by grants CA-42239 and CA-40584 awarded to G.T.B. from the National Institutes of Health. We also thank Ms. Sally Anderson for her expert secretarial assistance.

REFERENCES

1. T. J. Slaga, Overview of tumor promotion in animals, Environ. Health Perspect. 50:3-20 (1983).
2. J. F. O'Connell, A. J. P. Klein-Szanto, D. M. DiGiovanni, J. W. Fries, and T. J. Slaga, Malignant progression of mouse skin papillomas treated with ethylnitrosourea, N-methyl-N'-nitro-N-nitrosoguanidine, or 12-0-tetradecanoylphorbol-13-acetate, Cancer Lett. 30:269-274 (1986).
3. H. Hennings, R. Shores, M. L. Weick, E. F. Spangler, R. Tarone, and S. H. Yuspa, Malignant conversion of mouse skin tumors is increased by tumor initiators and unaffected by tumor promoters, Nature 304:67-69 (1983).
4. D. R. Jaffe, J. F. Williamson, and G. T. Bowden, Ionizing radiation enhances malignant progression of mouse skin tumors, Carcinogenesis 8:1753-1755 (1987).
5. G. J. Patskan, A. J. P. Klein-Szanto, J. L. Philips, and T. J. Slaga, Metastasis from squamous cell carcinomas of Sencar mouse skin produced by complete carcinogenesis, Cancer Lett. 34:121-127 (1987).
6. K. Melber, P. Krieg, G. Furstenberger, and F. Marks, Molecular cloning of sequences activated during multi-stage carcinogenesis in mouse skin, Carcinogenesis 7:317-322 (1986).
7. L. M. Matrisian, G. T. Bowden, P. Krieg, G. Furstenberger, J. P. Briand, P. Leroy, and R. Breathnach, The mRNA coding for the secreted protease transin is expressed more abundantly in malignant than in benign tumors, Proc. Natl. Acad. Sci. 83:9413-9417 (1986).
8. A. Balmain, R. Krumlauf, J. K. Vass, and G. D. Birnie, Cloning and characterization of the abundant cytoplasmic 7S RNA from mouse cells, Nucleic Acid Res. 10:4259-4262 (1982).
9. P. Krieg, K. Melber, G. Furstenberger, and G. T. Bowden, in: "Growth Factors, Tumor Promoters and Cancer Genes" (UCLA Symposium on Molecular and Cellular Biology), N. Colburn, H. Moses, E. Stanbridge, eds., Alan R. Liss, New York, 1987.
10. G. I. Goldberg, S. M. Wilhelm, A. Kornberger, E. A. Bauer, G. A. Grant, A. Z. Eisen, Human fibroblast collagenase: complete primary structure and homology to an oncogene transformation-induced rat protein, J. Biol. Chem. 261:6600-6605 (1986).
11. S. E. Whitman, G. Murphy, P. Angel, N. J. Rahmsdorf, B. J. Smith, A. Lyons, T. J. R. Harris, J. J. Reynolds, P. Herrlich, and J. P. Docherty, Comparison of human stromelysin and collagenase by cloning and sequence analysis, Biochem. J. 240:913-916 (1986).

12. P. Mignatti, E. Robbins, and D. B. Rifkin, Tumor invasion through the human amniotic membrane: requirement for proteinase cascade, Cell 47:487-498 (1986).
13. L. E. Ostrowski, J. Finch, P. Krieg, L. Matrisian, G. Patskan, J. F. O'Connell, J. Phillips, T. J. Slaga, R. Breathnach, and G. T. Bowden, Expression pattern of a gene for a secreted metalloproteinase during late stages of tumor progression, Molecular Carcinogenesis 1:13-19 (1988).
14. R. Breathnach, L. N. Matrisian, M.-C. Gesnal, A. Staub, and P. Leroy, Sequences coding for part of oncogene-induced transin are highly conserved in a related rat gene, Nucleic Acid Res. 15:1139-1151 (1987).
15. P. Krieg, J. Finch, G. Furstenberger, K. Melber, L. Matrisian, and G. T. Bowden, Tumor promoters induce a transient expression of tumor specific genes in both basal and differentiated cells of the mouse epidermis, Carcinogenesis 9:95-100 (1988).
16. G. Furstenberger, D. L. Berry, B. Song, and F. Marks, Skin tumor promotion by phorbol esters is a two-stage process, Proc. Natl. Acad. Sci. USA 78:7722-7726 (1981).
17. J. E. Paulsen and E. G. Astrup, Effects of single applications of 12-O-tetradecanoylphorbol-13-acetate, merzerein or ethylphenylpropiolate on DNA synthesis and polyamine levels in hairless mouse epidermis, Cancer Res. 43:4126-4131 (1983).
18. J. R. Chen, G. Murphy, and Z. Werb, Stromelysin, a connective tissue-degrading metalloendopeptidase secreted by stimulated rabbit synovial fibroblasts in parallel with collagenase, J. Biol. Chem. 260:12367-12376 (1985).
19. S. Garbisa, R. Pozzatti, R. J. Muschel, U. Saffiotti, M. Ballin, R. H. Goldfarb, G. Khoury, and L. A. Liotta, Secretion of type IV collagenolytic protease and metastatic phenotype: induction by transfection with c-Ha-ras but not c-Ha-ras plus Ad2-Ela, Cancer Res. 47:1523-1528 (1987).
20. R. Morrier, L. Daza-Grosjean, and A. Sarasin, The effect of 12-O-tetradecanoyl-phorbol-13-acetate (TPA) on cell transformation by simian virus 40 mutants, in "Carcinogenesis: Fundamental Mechanisms and Environmental Effects", B. Pullman, P. Tso, and H. Gelboin, eds., Reidel, Dordrecht (1980).
21. P. Angel, A. Poting, U. Mallick, H. J. Rahmsdorf, M. Schorpp, and P. Herrlich, Induction of metallothionein and other mRNA species by carcinogens and tumor promoters in primary human skin fibroblasts, Mol. Cell Biol. 6:1760-1766 (1986).
22. L. Krieg, I. Kuhlmann, and F. Marks, Effect of tumor-promoting phorbol ester and of acetic acid on mechanisms controlling DNA synthesis and mitosis (Chalones) and on the biosynthesis of histidine-rich protein on mouse epidermis, Cancer Res. 23:3135-3146 (1974).

DISCUSSION

Moderator (Lance A. Liotta): Questions?

Oscar Sudilovsky: Just a quick question. Is the expression of transin evident in other organs besides the skin?

G. Tim Bowden: We are now doing some work with a human cDNA clone which is a truncated version of transin, and we are presently looking at human prostate cancer using both northern and in situ hybridization. We do find the expression of this particular human transin gene in prostate cancer. Transin is not expressed in benign hyperplastic lesions of the prostate. We also are starting to look at skin lesions, including actinic keratosis and squamous cell carcinomas with these human probes, but I don't have answers yet for those studies.

Sandy Markowitz: I'm struck by what you don't find in this screening: the traditional nuclear oncogenes like myc and fos that tend to go up in transformed states. I was wondering whether you have looked with probes for those. Is it something about this kind of screening that you don't find them or are we mistaken in thinking that they should be elevated?

Bowden: I should point out the fact that we screened a relatively limited cDNA library of only about 5,000 to 10,000 clones. I can tell you that in collaborative work with Allan Balmain we have found evidence for increased expression of Ha-ras in both papillomas and squamous cell carcinomas. Work from Yuspa's laboratory, where they screened for the overexpression of a whole series of oncogenes, has not really shown consistently the overexpression of a number of different oncogenes, including myc, fos, and as I remember, erbB. So the data would suggest that there is not consistent overexpression of some of these oncogenes in the mouse skin model.

Peter Duesberg: Do you have any functional analysis?

Bowden: Yes, we are doing functional studies. We are presently doing this in terms of transin. We are using vectors to overexpress the rat as well as mouse transin cDNAs. We did it in Yuspa's 308 papilloma cell lines, but we do not have an answer to that yet. We also are in the process of making antisense vectors to try to turn down the expression of this transin gene in, for instance, a squamous cell carcinoma producing cell lines. I don't have an answer to that either.

ONCOGENES AND BREAST CANCER PROGRESSION

Robert Callahan

Chief, Oncogenetics Section
Laboratory of Tumor Immunology and Biology
National Cancer Institute
National Institutes of Health
Bethesda, MD

INTRODUCTION

The etiology of breast cancer is thought to involve a complex interplay of genetic, hormonal, and dietary factors that are superimposed on the physiological status of the host. Extensive studies have been undertaken to determine the relationship between these factors and tumor development in humans and experimental rodent models. Attempts to develop a cohesive picture of how these factors participate in mammary tumorigenesis have been hampered, in part, by a lack of information on the specific genetic lesions that contribute to the initiation and/or evolution of tumor development.

The study of experimentally induced mammary tumors has focused primarily on various strains of inbred mice with a high tumor incidence that are infected with the mouse mammary tumor virus (MMTV) (reviewed by Callahan, 1987). Chemical carcinogen also induces mammary tumors in mice and rats (reviewed by Weiss et al., 1982; Welsch, 1985). The demonstration that MMTV, like other retroviruses, can act as an insertion mutagen (Varmus, 1982) represents a significant advance, since the viral genome, unlike chemical carcinogens, can be used as a tag to identify the relevant mutagenic events. There is substantial evidence in neoplasias of several tissues that the expression of certain cellular genes has either been induced or the gene product altered by mutation (Klein, 1981). In many cases, these cellular genes are related to the acute tumorigenic retroviruses and are referred to as proto-oncogenes (Bishop, 1985). The particular proto-oncogene and type of genetic alteration (chromosomal translocation, gene amplification, point mutation, or deletion) are often specific to the type of tumor (Aaronson and Tronick, 1985). The high frequency with which certain genetic alterations occur in a particular type of cancer has been taken as evidence that they either provide the tumor cell with a selective growth advantage that contributes to the evolution of the tumor or deregulate normal tissue development (Klein, 1981). Recently, transgenic strains of mice have been developed that contain recombinant myc and ras oncogenes linked to different transcriptional regulatory sequences. Certain of these strains also have a high mammary tumor incidence. The purpose of this chapter is to describe the genes that are frequently activated or mutated in breast tumors and their potential role in the evolution of the tumor.

Boundaries between Promotion and Progression during Carcinogenesis
Edited by O. Sudilovsky *et al.*, Plenum Press, New York, 1991

The int Genes

Spontaneous mammary tumorigenesis in mice is frequently associated with a chronic infection of the host mammary tissue by the MMTV (Teich et al., 1982). MMTV is a biological mutagen (Varmus, 1982). As a part of its infectious cycle, the viral genome becomes integrated at numerous, perhaps random, sites in the host cellular genome. In this manner, the viral genome has the potential to act as an insertional mutagen. The association between MMTV and mammary tumor development was first recognized in inbred strains of mice that have a high incidence of tumors (Bitner, 1942). In these mice, MMTV is transmitted congenitally through the milk. The development of the disease appears to involve a multistep process. In the high incidence C3H strain, the earliest detectable stage of tumor development is the appearance of hyperplastic alveolar nodules (De Ome et al., 1959; Nandi, 1963). These lesions are considered premalignant precursors to the pregnancy-independent adenocarcinomas that develop later (8-12 months). The BR-6 inbred mouse strain contains an independent strain of milk-borne MMTV [designated MMTV(RIII)] that induces pregnancy-dependent mammary tumors (Foulds, 1949). These tumors regress at parturition and, after two or more pregnancies, progress to the hormone-independent type (Lee, 1968).

Three cellular genes (int-1, int-2, and int-3) have been identified that frequently contain a viral genome integrated in adjacent cellular sequences in MMTV-induced mammary tumor DNA (Nusse and Varmus, 1982; Peters et al., 1983; Gallahan and Callahan, 1986). In each case, integration of a viral genome at one of these loci activated the expression of the respective int gene. The int genes are unrelated to each other and are located on different mouse chromosomes (15, 7, and 17, respectively) (Nusse et al., 1984; Peters et al., 1984a; Gallahan et al., 1986).

The int genes are not expressed in normal adult mammary tissue but are expressed during mouse embryonic development. The int-1 gene is expressed between days 9 and 14.5 of development and is restricted to specific regions of the neural plate and its derivatives (Jakobovits et al., 1986; Shackleford and Varmus, 1987; Wilkinson et al., 1987). In the adult, int-1 expression is restricted to the testis in round spermatids, which are postmeiotic cells that undergo morphological conversion to become mature sperm (Shackleford and Varmus, 1987). The int-2 RNA is expressed in preimplantation embryos and is abundant in derivatives of the primitive endoderm lineage (Jakobovits et al. 1986). RNA for this gene has not been detected at later gestational stages or in any normal adult tissues. Therefore, it seems paradoxical that int-1 and int-2, which are normally expressed at different times and places in embryonic and adult tissue, are both activated in mammary tumors. This suggests that the phenotypic effect of their gene products is strongly dependent on cell type and the context of expression. Consistent with that possibility, the Drosophila homologue of the int-1 gene is mutated in the wingless developmental mutant (Rijsewijk et al., 1987b).

The nucleotide sequence of the int-1 and int-2 genes have been determined (Van Ooyen and Nusse, 1984; Fung et al., 1985; Moore et al., 1986). The int-1 gene is transcribed into a 2.6-kb RNA species that encodes a primary product of 370 amino acids. This gene has no appreciable homology to other known proto-oncogenes. In cells expressing int-1, the protein is glycosylated and ranges in size from 38,000-44,000 M_r (Brown et al., 1987). The biochemical properties of the int-1 protein and its association with the cell membrane are consistent with its being a secretory protein (Papkoff et al., 1987). Four species of int-2 RNA (3.2, 2.9, 1.8, and 1.4 kb) have been detected (Dickson et al., 1984; Jakobovits et al., 1986; Peters et al., 1984b). The open reading frame of these RNA species encodes a protein of 245 amino acids (Moore et al., 1986). Recently, Dickson and Peters (1987) have found a 38% amino acid sequence homology between the bovine basic fibroblast growth factor (bFGF) and the NH_2-terminal 180 amino acids of the int-2 protein. The COOH-terminal of the int-2 protein, encoded by the third exon, appears to be unique. This is an intriguing finding, since the family of FGFs not only stimulates the division of certain cell types but also acts as a potent angiogenic agent that promotes the growth of new blood vessels (Gospodarowicz et al., 1986). Presently, however, there has been no direct demonstration of the biological activity of the int-2 protein.

The thesis that activation of the int loci by MMTV contributes to mammary tumorigenesis has been supported primarily by the frequency with which it occurs in mammary tumors. Introduction of the int-1 gene into NIH3T3 fibroblast cells did not lead to morphological transformation of the cells (Rijsewijk et al., 1986) or other fibroblast cell lines (Brown et al., 1986). However, when the gene was introduced into murine mammary epithelial cell lines C57MG (Brown et al., 1986) or cuboidal RAC311C cells (Rijsewijk, 1987a), morphological transformation was observed. In the latter case, these cell lines also formed tumors in syngeneic BALB/c mice. Recently male and female transgenic mice containing recombinant int-1 linked to the MMTV long terminal repeat (LTR) enhancer element have been found to develop mammary hyperplasias that later give rise to focal tumors (H. Varmus, personal communication).

The myc and ras Oncogenes

The activation of the myc and ras proto-oncogene families has been implicated in the etiology of several neoplasias. The studies of Land et al. (1983) demonstrated that activated ras and myc oncogenes are both required for the malignant transformation of primary rat embryo fibroblasts in tissue culture. The notion that these genes can act in concert to malignantly transform tissues in vivo has been investigated through the development of transgenic mice containing the activated oncogenes. In these studies, the recombinant c-myc and ras genes have been linked to either the MMTV LTR (Leder et al., 1986; Sinn et al., 1987; Stewart et al., 1984) or the lactogenic, hormone-dependent regulatory sequence of the whey acidic protein (WAP) gene (Andres et al., 1987; Schonenberger et al., 1988). Since the MMTV LTR promotes transcription of associated genes earlier than the WAP promoter in mammary gland development, it was possible to evaluate the effect of the activated oncogene on tumorigenesis at different stages of mammary gland differentiation. The MMTV-myc transgene was associated with an increased incidence of tumors in various tissues, including a high frequency of focal mammary adenocarcinomas in female mice. Mice containing the WAP-myc transgene had an 80% incidence of mammary adenocarcinomas. In contrast, only the MMTV-ras transgene was associated with a high incidence of mammary adenocarcinomas. In this strain of mice, focal tumors developed in both males and females. Taken together, these studies show that the transforming potential of the activated oncogenes is a function of the differentiated state of the cell and the cell type in which it is active. Thus, activated c-myc is capable of contributing to the malignant transformation of mammary epithelial cells in early and late stages of differentiation, whereas the ras oncogene appears to be active only in early stages.

Although activated myc and ras are necessary for mammary tumorigenesis in the transgenic mice, they are not sufficient. This is indicated by the clonal nature of the tumors that arise. Recalling the experiment of Land et al. (1983), which demonstrated the concerted action of activated myc and ras oncogenes in the transformation of primary embryo fibroblast in culture, Sinn et al. (1987) mated MMTV-myc and MMTV-ras transgenic mice. The F_1 generation had a high incidence of mammary tumors with a decreased latency before tumor development. However, as with the parental strains, the tumors were monoclonal in origin. Thus, additional somatic mutations are required in the mammary epithelium of these mice for progression to malignancy.

The neu (c-erbB-2) Oncogene

The neu or c-erbB-2 oncogene was originally isolated from a chemically induced rat neuroglioblastoma (Shih et al., 1981). It encodes a transmembrane phosphoprotein that is related to the epidermal growth factor (EGF) receptor (Schachter et al., 1984). The transforming version of c-erbB-2 differs from the normal allele at amino acid 664 (Val to Glu) (Bargmann et al., 1986). Muller et al. (1988) have developed strains of transgenic mice in which the transforming c-erbB-2 gene is driven by the MMTV LTR. Virgin females of one strain (designated TG.NF) contain multiple hyperplastic and dysplastic nodules of mammary epithelium infiltrating the entire mammary fat pad. By 95 days, multiparous females develop polyclonal mammary adenocarcinomas that involve the entire epithelium of each gland. The age at tumor onset was independent of parity and sex of the animal. Thus, expression of the transforming c-erbB-2 gene appears to be sufficient to induce malignant transformation of mammary epithelium in one step. The importance of the combination of activated oncogenes and tissue context is again indicated by the fact that in these same mice overexpression of activated c-erbB-2 in the parotid gland or epididymis leads to bilateral hypertrophy and hyperplasia but not to malignant transformation.

Frequent Genetic Alterations in Primary Human Breast Carcinomas

Identification of specific genetic mutations in human breast carcinomas has been constrained by the heterogeneous nature of the disease. Multiple factors figure in the etiology of breast cancer, including genetic, physiological and environmental factors. Neither epidemiologic nor cytogenetic studies have identified any consistent genetic abnormalities in human breast cancer. Other approaches which have been pursued for the identification of genetic defects in breast cancer included surveys of proto-oncogenes in breast tumor cell lines and the transfection of primary breast tumor DNA into cultured cells. Although these methods provided clues about the irregularities of certain proto-oncogenes in breast carcinomas, they suffer from the serious drawback of the artificiality of in vitro culture systems.

To circumvent this problem, more recent studies have taken a direct approach in the comparative molecular analyses of primary human breast tumor DNAs and the constitutional genotypes of patients (reviewed by Ali et al., 1988a; Ali and Callahan, 1988). It is significant that many of the same genes that were found to contribute to mammary tumorigenesis in experimental murine models are also frequently affected in primary human breast tumors. Thus, several laboratories, which have collectively examined as many as 500 primary breast tumors, found that 16-56% of them contain an amplification of c-myc (Escot et al., 1986; Cline et al., 1987; Varley et al., 1987; Bonilla et al., 1988). High-level expression of c-myc RNA was detected by in situ hybridization in tumor cells of biopsy material where gene amplification had been demonstrated (Mariani-Costantini et al., 1988). The c-erbB-2 gene has also been found to be amplified in 10-40% of primary breast tumors (Slamon et al., 1987; van de Vijver et al., 1987; Vener et al., 1987; Zhou et al., 1987; Ali et al., 1988c; Berger et al., 1988c). In one study, the closely linked proto-oncogene c-erbA-1 was found to be frequently co-amplified with c-erbB-2; however, only c-erbB-2 RNA was expressed in these tumors (van de Vijver et al., 1987). The int-2 gene was amplified in 9-23% of the primary tumors examined (Lidereau et al., 1988; Varley et al., 1988; Zhou et al., 1988). More recently, int-2 has been shown to be co-amplified with two other closely linked loci (bcl-1 and hst) on chromosome 11q13 in 17 of 18 tumors (Ali et al., 1988d). The bcl-1 locus corresponds to a frequent breakpoint in t(11;14) translocations that occur in chronic lymphocytic leukemia (B-CLL) Tsujimoto et al., 1984). The oncogene hst was isolated from primary and metastatic human cancers (Sakamoto et al., 1986) and also from a Kaposi's sarcoma (Delli Bobi et al., 1987) by transfection into NIH3T3 cells. It also

encodes a protein with significant homology with basic and acidic FGF as well
as int-2 (Yoshida et al., 1987). Currently, there is little information on
the frequency with which amplification of the region within chromosome 11q3
leads to the expression of the int-2 or hst genes.

The idea that recessive mutations might underlie the tumorigenic phenotype, at least in some cancers, is based on two observations. First, the
neoplastic phenotype is often suppressed by the fusion of cancer cells with
normal cells, suggesting the presence of tumor suppressor genes in the normal
genome (Stanbridge et al., 1981; Kaelbling and Klinger, 1985). Second,
karyotypic analysis of a variety of human tumors has shown consistent deletions of specific chromosomal regions. This is consistent with more recent
demonstrations of loss of heterozygosity for regions of specific chromosomes
in several human malignancies. In these situations, an effective loss of the
normal allele of a tumor suppressor gene with a possible regulatory function
is believed to unmask an independent recessive mutation on the other homologous chromosome. This hypothesis was originally proposed by Knudson (1971)
to provide an explanation for the same genetic lesions occurring in both the
hereditary and sporadic forms of the childhood malignancy retinoblastoma.

Loss of heterozygosity for three different regions of the human cellular
genome has been observed in different subsets of primary breast tumor DNAs
(reviewed by Ali and Callahan, 1988). Allelic deletions specific for chromosome 11p were observed in approximately 20% (11 of 56) of the tumors from
patients who were heterozygous at multiple loci on chromosome 11 (Ali et al.,
1987). By characterizing overlapping deletions in 11 primary human breast
tumors, the shortest region of hemizygosity was mapped between the gamma-
globin and parathyroid hormone (PTH) loci on chromosome 11p13.

Deletions of sequences on the short arm of chromosome 3 have also been
detected by restriction fragment length polymorphism analysis (Ali et al.,
1988e). These occurred in a different subset of tumors from those that contain the 11p deletion. This region of chromosome 3p harbors a fragile site
located at 3p14-3p21 (Yunis and Soreng, 1984), the proto-oncogene c-raf-1 and
at least two members of the c-erbA gene family (Weinberger et al., 1986;
Rider et al., 1987; Ali et al., 1988f). The c-raf-1 proto-oncogene is a
cellular homolog of the mil/raf oncogene present in the MH_2 avian virus
(Rapp et al., 1983), which induces predominantly kidney and liver epitheli-
omas in chickens (Alexander et al., 1979). The c-erbA proto-oncogenes belong
to a superfamily of regulatory genes that code for DNA binding proteins
having hormone receptor function (Green and Chambon, 1986). Allelic deletions of genes on chromosome 3p, especially of c-erbA genes, occurring in a
significant number of breast tumors may be relevant to breast cancer.

In a study of 14 primary human breast tumors, 4 suffered loss of heterozygosity at multiple loci on chromosome 13 (Lundberg et al., 1987). More
recently, Lee et al. (1988) have shown that the retinoblastoma susceptibility
(RB) gene is inactivated in two of nine human breast cancer cell lines. One
of these cell lines, MDA-MB-468, had a homozygous deletion of the gene beyond
exon 2.

Attempts have been made to determine whether there is a link between the
mutations found in primary human breast tumors and the clinical course of the
disease, the patient's history, and characteristics of the tumor (reviewed by
Ali and Callahan, 1988). At the present time, the results are controversial.
For instance, Slamon et al. (1987) reported that amplification of c-erbB-2
was significantly associated with the lymph node status of the patient, but
other laboratories did not find this association (van de Vijver et al., 1987;
Zhou et al., 1987; Ali et al., 1988b,c). Similarly, a link between c-erbB-2
amplification and poor prognosis has been reported in some studies (Slamon et
al., 1987; Varley et al., 1987) and not others (Ali et al., 1988b,c). Ampli-

fication of c-myc was reported to be associated with the age of the patient (Escot et al., 1986) and in another study with poor short-term prognosis (Varley et al., 1987). The chromosome 11p deletion and int-2 amplification have been reported to be associated with relapse of the disease (Ali et al., 1987; Lidereau et al., 1988).

There are some plausible explanations for these discrepancies. Given the variable ratios of tumor cells to normal stromal tissue and infiltrating lymphocytes in breast tumor biopsies, the number of tumors amplified for a particular gene could be underestimated. On the other hand, polyploidy of chromosomes, a common feature of malignant cells, may tend to exaggerate the number of tumors with the amplified gene. Several studies suffer due to the fact that either relatively small numbers of patients were analyzed or follow-up information on the patients was unavailable. Even the analysis of a few hundred patients with long-term follow-up information can, only at best, identify potentially useful factors of prognostic significance.

DISCUSSION

The development of the mammary gland is a complex sequence of differentiation stages in which the interaction between mammary epithelium and basal cells is governed by steroid and peptide hormones. After massive cell proliferation, the endpoint in development is realized when functioning ducts and alveoli fill the fat pad at lactation. At weaning, involution or regression of the gland occurs with the loss of mammary epithelial cells. The number of different activated genes found in MMTV-induced mouse mammary tumors and primary human breast tumors suggests that there are probably multiple pathways of somatic mutation that contribute to malignancy. Some of these mutations may cryptically activate genes whose contribution to malignancy awaits additional mutagenic or epigenetic events. This is clearly the case in the MMTV-myc and MMTV-H-ras mice in which normal, nonmalignant mammary epithelium expressing these genes was found adjacent to tumor cells. The putative recessive mutations on chromosomes 3p, 11p13, and 13q14 represent similar situations in which the second mutagenic event is the deletion of the corresponding normal suppressor gene.

It seems probable that the cellular context and period of mammary gland development in which a gene is activated are important determinants in the gene's contribution to malignant transformation. The oncogenic potential of v-H-ras is dependent on its expression early in mammary gland development, whereas the oncogenic potential of c-myc is present at both early and late stages of development. Proto-oncogenes like c-myc, while playing a role in cell division, may also have other effects that are dependent on the cellular context. In this regard, the lactogenic hormone regulation of milk protein expression is interfered with in mammary tumors arising in WAP-myc transgenic mice. In these tumors, the milk proteins are expressed constitutively in the absence of hormonal stimulation. Similarly, Nusse (1988) has suggested that the int-1 and perhaps the int-2 gene products deregulate the interplay between differentiating epithelial cells and basal cells which, after secondary events, lead to tumorigenesis. Alternatively, the int gene products may interfere with the involution process, providing the epithelial cells with a prolonged life span. In subsequent rounds of gland development, these cells may be more prone to mutagenic events. In this regard, it may be pertinent that basic FGF delays cell senescence (Gospodarowicz et al., 1987), however, whether the int-2 gene product has a similar activity is unknown. The MMTV-neu transgenic mice represent an apparent exception to the notion that mammary tumorigenesis results from a series of stochastic events. The neu oncogene contains a point mutation that results in a single amino acid change relative to the normal gene product. Presently, the effect of this mutation on the biological activity of the protein in vivo is unknown. Similarly,

148

it is not known whether activation of expression of the normal neu (c-erbB-2) gene product is sufficient for tumorigenesis in vivo. It seems likely that this issue will be resolved in the near future through the development of new transgenic strains of mice in which the normal c-erbB-2 proto-oncogene is activated.

In primary human breast tumors, the c-myc, c-erbB-2, and int-2 proto-oncogenes are frequently activated by gene amplification. However, unlike the MMTV-c-myc and MMTV-neu (c-erbB-2) transgenic mouse mammary tumors, there appears to be tumor cell heterogeneity with respect to expression of the activated gene (Mariani-Costantini et al., 1988; van de Vijver et al., 1988; unpublished data). This suggests that activation of these mutations may occur late in tumor progression. In this regard, it is of interest that activation of c-myc, c-erbB-2, and int-2 has been reported to be associated with disease relapse (Varley et al., 1987; Slamon et al., 1987; Lidereau et al., 1988).

REFERENCES

S. A. Aaronson, and S. R. Tronick, 1985, The role of oncogenes in human neoplasia, in: "Important Advances in Oncology 1985," V. T. DeVita, S. Hellman, S. A. Rosenberg, eds., J. B. Lippincott Company, Philadelphia.

R. W. Alexander, C. Moscovici, and P. K. Vogt, 1979, Avian oncovirus Mill Hill No. 2: pathogenicity in chickens, J. Natl. Cancer Inst. 62:359-366.

I. U. Ali, R. Lidereau, C. Thiellet, and R. Callahan, 1987, Reduction to homozygosity of genes on chromosome 11 in human breast neoplasia, Science 238:185-188.

I. U. Ali, R. Lidereau, and R. Callahan, 1988a, Heterogeneity of genetic alterations in primary human breast tumors, in: "Breast Cancer: Cellular and Molecular Biology," M. E. Lippman and R. B. Dickson, eds., Martinus Nijhoff Publishers, Boston.

I. U. Ali, G. Campbell, R. Lidereau, and R. Callahan, 1988b, Amplification of c-erbB-2 and aggressive human breast tumors?, Science 240:1795-1796.

I. U. Ali, G. Campbell, R. Lidereau, and R. Callahan, 1988c, Lack of evidence for the prognostic significance of c-erbB-2 amplification in human breast carcinoma, Oncogene Res. 3:139-146.

I. U. Ali and R. Callahan, 1988, Prognostic significance of genetic alterations in human breast carcinoma, in: "Molecular Genetics and the Diagnosis of Cancer," J. Cossman, ed., Elsevier Science Pub., New York (In press).

I. U. Ali, G. Merlo, R. Lidereau, and R. Callahan, 1988d, The amplification unit on chromosome 11q13 in aggressive primary human breast tumors contains the bcl-1, int-1, and hst loci, Oncogene, (In press).

I. U. Ali, S. Meissner, R. Lidereau, and R. Callahan, 1988e, Allelic deletion of c-erbA-2 proto-oncogene in human breast carcinoma signifies a possible recessive mutation in member(s) of steroid/thyroid hormone receptor family, (Manuscript in preparation).

I. U. Ali, S. Miessner, N. Spurr, and R. Callahan, 1988f, Mapping of the c-erbA-2 proto-oncogene to chromosome 3p and its homology with the thyroid hormone receptor gene, (Manuscript in preparation).

A. C. Andres, C. A. Schonenberger, B. Groner, L. Hennighausen, M. LeMaur, and P. Gerlinger, 1987, Ha ras oncogene expression directed by a milk protein gene promoter: tissue specificity, hormonal regulation, and tumor induction in transgenic mice, Proc. Natl. Acad. Sci. USA 84:1299-1303.

C. I. Bargmann, M. C. Hung, and R. A. Weinberg, 1986, Multiple independent activations of the neu oncogene by a point mutation altering the transmembrane domain of p185, Cell 45:649-657.

M. S. Berger, G. W. Locher, S. Saurer, W. J. Gullick, M. D. Waterfield, B. Groner, and N. E. Hynes, 1988, Correlation of c-erbB-2 gene amplification and protein expression in human breast carcinoma with nodal status and nuclear grading, Cancer Res. 48:1238-1243.

J. Bishop, 1985, Viral oncogenes, Cell 42:23-38.

J. J. Bitner, 1942, The milk influence of breast tumors in mice, Science 94:462-463.

M. Bonilla, M. Ramirez, J. Lopez-Cueto, and P. Gariglio, 1988, In vivo amplification and rearrangement of c-myc oncogene in human breast tumors, J. Natl. Cancer Inst. 80:665-671.

A. M. C. Brown, J. Papkoff, Y. K. T. Fung, G. M. Shackleford, and H. E. Varmus, 1987, Identification of protein products encoded by the proto-oncogene int-1, Mol. Cell. Biol. 7:3971-3981.

R. Callahan, 1987, Retrovirus and proto-oncogene involvement in the etiology of breast neoplasia, in: "The Mammary Gland," M. C. Neville and C. W. Daniel, eds., Plenum Publishing Company, New York.

M. J. Cline, H. Battifora, and J. Yokota, 1987, Proto-oncogene abnormalities in human breast cancer: correlation with anatomic features and clinical course of disease, J. Clin. Oncol. 5:999-1010.

P. Delli Bovi, A. M. Curatola, F. G. Kern, A. Greco, M. Ittmann, and C. Basilico, 1987, An oncogene isolated by transfection of Kaposi's sarcoma DNA encodes a growth factor that is a member of the FGF family, Cell 50:729-740.

K. B. DeOme, L. J. Faultein, H. H. Bern, and P. B. Blair, 1959, Development of mammary tumors from hyperplastic alveolar nodules transplanted into gland free mammary fat pads of female C3H mice, Cancer Res. 19:515-520.

C. Dickson and G. Peters, 1987, Potential oncogene product related to growth factors, Nature 326:833-836.

C. Dickson, R. Smith, S. Brookes, and G. Peters, 1984, Tumorigenesis by mouse mammary tumor virus: proviral activation of a cellular gene in the common integration region int-2, Cell 37:539-550.

C. Escot, C. Theillet, R. Lidereau, F. Spyratos, M. H. Champeme, J. Gest, and R. Callahan, 1986, Genetic alteration of the c-myc proto-oncogene in human primary breast carcinoma, Proc. Natl. Acad. Sci. USA 83:4834-4838.

L. Foulds, 1949, Mammary tumors in hybrid mice: The presence and transmission of the mammary tumor agent, Br. J. Cancer 3:230-239.

Y. K. T. Fung, G. M. Shackleford, A. M. C. Brown, G. S. Sanders, and H. E. Varmus, 1985, Nucleotide sequence and expression in vitro of cDNA derived from mRNA of int-1, a provirally activated mouse mammary oncogene, Mol. Cell. Biol. 5:3337-3348.

D. Gallahan and R. Callahan, 1986, A new common integration region (int-3) for the mouse mammary tumor virus on mouse chromosome 17, J. Virol. 61:218-220.

D. Gospodarowicz, G. Neufeld, and L. Schweigerer, 1986, Fibroblast growth factor, Mol. Cell. Endocrinol. 46:187-204.

D. Gospodarowicz, N. Ferrara, L. Schweiger, and G. Newfeld, 1987, Structural characterization and biological functions of fibroblast growth factor, Endocrine Rev. 8:95-114.

S. Green and P. Chambon, 1986, A superfamily of potentially oncogenic hormone receptors, Nature 324:615-620.

A. Jakobovits, G. M. Shackleford, H. E. Varmus, and G. R. Martin, 1986, Two proto-oncogenes implicated in mammary carcinogenesis, int-1 and int-2, are independently regulated during mouse development, Proc. Natl. Acad. Sci. USA 83:7806-7810.

M. Kaebling and H. P. Klinger, 1985, Suppression of tumorigenicity in somatic cell hybrids. III. Cosegregation of human chromosome 11 of a normal cell and suppression of tumorigenicity in intraspecies hybrids of normal diploid x malignant cells, Cytogenet. Cell Genet. 41:65-70.

G. Klein, 1981, The role of gene dosage and genetic transposition in carcino-genesis, <u>Nature</u> 294:290-293.

A. G. Knudson, 1971, Mutation and cancer: statistical study of retinoblasto-ma, <u>Proc. Natl. Acad. Sci USA</u> 68:820-823.

H. Land, L. Parada, and R. A. Weinberg, 1983, Cellular oncogenes and multi-step carcinogenesis, <u>Nature</u> 304:596-602.

A. Leder, P. K. Pattengale, A. Kuo, T. A. Stewart, and P. Leder, 1986, Conse-quences of widespread deregulation of the c-<u>myc</u> gene in transgenic mice: multiple neoplasms and normal development, <u>Cell</u> 45:485-495.

E. Lee, Y.-H. P. H. To, J.-Y. Shew, R. Bookstein, P. Scully, and W.-H. Lee, 1988, Inactivation of the retinoblastoma susceptibility gene in human breast cancers, <u>Science</u> 241:218-222.

A. B. Lee, 1968, Genetic and viral influences on mammary tumors in BR6 mice, <u>Br. J. Cancer</u> 22:77-82.

R. Lidereau, R. Callahan, C. Dickson, G. Peters, C. Escot, and I. U. Ali, 1988, Amplification of the <u>int</u>-2 gene in primary human breast tumors, <u>Oncogene Res.</u> 2:285-291.

C. Lundberg, L. Skoog, W. K. Cavenee, and M. Nordenskjöld, 1987, Loss of heterozygosity in human ductal breast tumors indicates a recessive mutation on chromosome 13, <u>Proc. Natl. Acad. Sci. USA</u> 84:2373-2376.

R. Mariani-Costantini, C. Escot, C. Theillet, A. Gentile, G. Merlo, R. Lidereau, and R. Callahan, 1988, In situ <u>myc</u> expression and genomic status of the c-<u>myc</u> locus in infiltrating ductal carcinomas of the breast, <u>Cancer Res.</u> 48:199-205.

R. Moore, G. Casey, S. Brookes, M. Dixon, G. Peters, and C. Dickson, 1986, Sequence, topography and protein coding potential of mouse <u>int</u>-2: a putative oncogene activated by mouse mammary tumor virus. <u>EMBO J</u> 5:919-924.

W. J. Muller, E. Sinn, P. K. Pattengale, R. Wallace, and P. Leder, 1988, Single step induction of mammary adenocarcinoma in transgenic mice bearing activated c-<u>neu</u> oncogene, <u>Cell</u> 54:105-115.

S. Nandi, 1963, New method for detection of mouse mammary tumor virus. I. Influence of foster nursing on the incidence of hyperplastic mammary nodules on BALB/c Crg 1 mice, <u>J. Natl. Cancer Inst.</u> 31:57-73.

R. Nusse, 1988, The <u>int</u> genes in mammary tumorigenesis and in normal develop-ment, <u>Trends in Genetics</u> (In press).

R. Nusse, and H. Varmus, 1982, Mammary tumor induced by the mouse mammary tumor virus: evidence for a common region for provirus integration in the same region of the host genome, <u>Cell</u> 31:99-109.

J. Papkoff, A. M. C. Brown, and H. E. Varmus, 1987, The <u>int</u>-1 proto-oncogene products are glycoproteins that appear to enter the secretory pathway, <u>Mol. Cell Biol.</u> 7:3978.

G. Peters, S. Brookes, R. Smith, and C. Dickson, 1983, Tumorigenesis by mouse mammary tumor virus: evidence for a common region for provirus in-tegration in mammary tumors, <u>Cell</u> 33:369.

G. Peters, C. Kozak, and C. Dickson, 1984a, Mouse mammary tumor virus in-tegration region <u>int</u>-1 and <u>int</u>-2 map on different mouse chromosomes, <u>Mol. Cell Biol.</u> 4:375.

G. Peters, A. E. Lee, and C. Dickson, 1984b, Activation of cellular gene by mouse mammary tumor virus may occur early in mammary tumor develop-ment, <u>Nature</u> 309:273.

T. H. Rabbitts, 1985, The c-<u>myc</u> proto-oncogene: involvement in chromosomal abnormalities, <u>Trends In Genetics</u> 1:327-331.

U. R. Rapp, F. H. Reynolds, and J. R. Stephenson, 1983, New mammalian trans-forming retrovirus: demonstration of a polyprotein gene product, <u>J. Virol.</u> 45:914-922.

S. H. Rider, P. A. Gorman, J. M. Shipley, G. Moore, B. Vennström, E. Solomon, and D. Sheer, 1987, Localization of the oncogene c-<u>erbA</u>-2 to human chromosome 3, <u>Ann. Hum. Genet.</u> 51:53-156.

F. Rijsewijk, M. van Lohuizen, A. van Ooyen, and R. Nusse, 1986, Construction of a retroviral cDNA version of the int-1 mammary oncogene and its expression in vitro, Nucleic Acids Res. 14:693-670.

F. Rijsewijk, L. van Deemter, E. Wagenaar, A. Sonnenberg, and R. Nusse, 1987a, Transfection of the int-1 mammary oncogene in cuboidal RAC mammary cell line results in morphological transformation and tumorigenicity, EMBO J. 6:127-131.

F. Rijsewijk, M. Schuerman, L. Wagenaar, P. Parren, D. Weigel, and R. Nusse, 1987b, The Drosophila homolog of the mouse mammary oncogene int-1 is identical to the segment polarity mutant wingless, Cell 50:649-659.

H. Sakamoto, M. Mori, M. Taira, T. Yoshida, S. Matsukawa, K. Shimizu, M. Sekiguchi, M. Terada, and T. Sugimura, 1986, Transforming gene from human stomach cancers and a non-cancerous portion of stomach mucosa, Proc. Natl. Acad. Sci. USA 83:3997-4001.

A. L. Schechter, D. F. Stern, L. Valdyanathan, S. J. Decker, J. A. Drebin, M. E. Greene, and R. A. Weinberg, 1984, The neu oncogene: an erb B related gene encoding a 185,000-M tumor antigen, Nature 312:513-516.

C. A. Schoenenberger, A. C. Andres, B. Groner, M. van der Valk, M. LeMeur, and P. Gerlinger, 1988, Targeted c-myc gene expression in mammary glands of transgenic mice induces mammary tumors with constitutive milk protein gene transcription, EMBO J. 7:169-179.

G. M. Shackleford and H. Varmus, 1987, Expression of the proto-oncogene int-1 is restricted to postmeiotic male germ cells and the neural tube of mid-gestational embryos, Cell 50:89-100.

C. Shih, L. Padney, M. Murray, and R. A. Weinberg, 1981, Transforming genes of carcinomas and neuroblastomas introduced into mouse fibroblasts, Nature 290:261-264.

E. Sinn, W. Muller, P. Pattengale, I. Tepler, R. Wallace, and P. Leder, 1987, Coexpression of MMTV/v-H-ras and MMTV/c-myc genes in transgenic mice: synergistic action of oncogenes in vivo, Cell 49:465-475.

D. J. Slamon, G. M. Clark, S. G. Wong, W. S. Levin, A. Ullrich, and W. L. McGuire, 1987, Human breast cancer: correlation of relapse and survival with amplification of the HER-2/neu oncogene, Science 235:177-182.

E. J. Stanbridge, R. R. Flandermeyer, D. W. Daniels, and W. A. Nelson-Rees, 1981, Specific chromosome loss associated with the expression of tumorigenicity in human cell hybrids, Somatic Cell Genet. 7:699-712.

T. Stewart, P. Pattengale, and P. Leder, 1984, Spontaneous mammary adenocarcinomas in transgenic mice that carry and express MMTV/myc fusion genes. Cell 38:627-637.

N. Teich, J. Wyke, T. Mak, A. Bernstein, and W. Hardy, 1982, Pathogenesis of retrovirus induced disease, in: "Molecular Biology of Tumor Viruses, RNA Tumor Viruses," R. Weiss, N. Teich, H. E. Varmus, and J. Coffin, eds., Cold Spring Harbor Laboratory, New York.

Y. Tsujimoto, J. Yunis, L. Onorato-Shouie, J. Erikson, P. C. Nowell, and C. M. Croce, 1984, Molecular cloning of the chromosomal breakpoint of B-cell lymphomas and leukemias with the t(11;14) chromosome translocation, Science 224:1403-1406.

M. van de Vijver, R. van de Bersselaar, P. Devilee, C. Cornelisse, J. Peterse, and R. Nusse, 1987, Amplification of neu (c-erbB-2) oncogene in human mammary tumors is relatively frequent and is often accompanied by amplification of the linked c-erbA oncogene, Mol. Cell Biol. 7:2019-2023.

M. J. van de Vijver, J. L. Peterse, W. J. Mooi, P. Wisman, J. Lomans, and R. Nusse, 1988, Overexpression of the neu (or c-erbB-2 or HER-2) protein is very frequent in comedo type ductal carcinoma in situ but not of prognostic value in stage II breast cancer, N. Engl. J. Med. 319:1239-1245.

A. van Ooyen and R. Nusse, 1984, Structure and nucleotide sequence of the putative mammary oncogene int-1: proviral insertions leave the protein-encoding domain intact, Cell 39:233-240.

J. M. Varley, J. E. Swallow, W. J. Brammar, J. L. Whittaker, and R. A. Walker, 1987, Alterations to either c-erbB-2 (neu) or c-myc proto-oncogenes in breast carcinomas correlate with poor short-term prognosis, Oncogene 1:423-430.

J. M. Varley, R. A. Walker, G. Casey, and W. J. Brammar, 1988, A common alteration to the int-2 proto-oncogene in DNA from primary breast carcinomas, Oncogene 3:87-90.

H. E. Varmus, 1982, Recent evidence for oncogenesis by insertion mutagenesis and gene activation, Cancer Surv. 1:309-319.

D. J. Venter, N. L. Tuzi, S. Kumar, and W. J. Gullick, 1987, Overexpression of the c-erbB-2 oncoprotein in human breast carcinomas: immunohistological assessment correlates with gene amplification, Lancet ii:69-72.

C. Weinberger, C. C. Thompson, E. S. Ong, R. Lebo, D. J. Gruol, and R. M. Evans, 1986, The c-erbA gene encodes a thyroid hormone receptor, Nature 324:641-646.

R. N. Weiss, N. Teich, H. Varmus, and J. Coffin, 1982, Origins of contemporary RNA tumor virus research, in: "Molecular Biology of Tumor Viruses, RNA Tumor Viruses," R. Weiss, N. Teich, H. Varmus, and J. Coffin, eds., Cold Spring Harbor Laboratory, New York.

C. W. Welsch, 1985, Host factors affecting the growth of carcinogen-induced rat mammary carcinomas: a review and tribute to Charles Brenton Huggins, Cancer Res. 45:3415-3443.

D. G. Wilkinson, J. A. Bailes, and A. P. McMahon, 1987, Expression of the proto-oncogene int-1 is restricted to specific neural cells in the developing mouse embryo, Cell 50:79-89.

T. Yoshida, K. Miyagawa, H. Odagiri, H. Sakamoto, P. F. R. Little, M. Terada, and T. Sugimura, 1987, Genomic sequence of hst, a transforming gene encoding a protein homologous to fibroblast growth factors and the int-2 encoded protein, Proc. Natl. Acad. Sci. USA 84:7305-7309.

J. J. Yunis and A. L. Soreng, 1984, Constitutional fragile sites and cancer, Science 226:1199-1204.

D. Zhou, H. Battifora, J. Yokota, T. Yamamoto, and M. J. Cline, 1987, Association of multiple copies of the c-erbB-2 oncogene with spread of breast cancer, Cancer Res. 47:6123-6125.

D. J. Zhou, G. Casey, and M. J. Cline, 1988, Amplification of human int-2 in breast cancers and squamous carcinomas, Oncogene 2:279-282.

DISCUSSION

Unidentified speaker: Did you see any histologic differences in the type of tumors, once the tumors grew out in the transgenic mice?

Robert Callahan: This is work by Phil Leder at Bernd Groner. It is my impression that all of these tumors appeared to be alveolar carcinomas, similar to the kind that MMTV induces.

Helene Smith: We, too, find that there aren't any ras mutations in human tumors in breast cancers, and there are two possible interpretations. One, that the mouse model isn't the same as human perhaps because human breast cancer is not caused by carcinogens, whereas in carcinogen-induced tumors you find the mutations. So the question is, if you look at spontaneous mouse tumors do you also find that there are ras mutations?

Callahan: We haven't looked, and I don't know that anybody has really looked seriously for point mutations in spontaneous mouse mammary tumors. It's known, however, that tumors derived from hyperplastic outgrowth lines in mice which had been treated with DMBA frequently have, maybe even always, a point mutation in Harvey ras.

Philip Frost: The only thing that concerns me is the 60 or more odd patients that you screened that don't have any of these changes. How do you explain the fact that they do not show them? You have an incidence of a particular change in an onc gene in 30% at most, I think in one series. What about the other 70% of the patients that don't show the change.

Callahan: The point I tried to make is that there were probably multiple combinations or pathways of mutations which will provide you with the common phenotype, a tumor.

Frost: But if there are multiple, then there are inumerable ones.

Callahan: Potentially, yes.

Frost: How would you ever expect to really get to the bottom of it?

Callahan: We would hope that there are a few common pathways.

Frost: You're saying that in the future common pathways will appear, and that we haven't seen them yet.

Callahan: Yes. This is what we are searching for right now. It is an attempt to identify the affected genes. What I didn't tell you was the number of genes we've looked at for which there are no apparent alterations, which at this point is probably 30-50 genes. With respect to human breast tumors, the only option that I see is an exhaustive walk through the genome looking for mutations. Our rationale for using the mouse model system was the hope that it may provide us with clues for genes which are important to look at in human tumors. This, I believe, is the importance of the int loci. It provides us a starting point to begin looking in human breast tumors.

Unidentified speaker: You show gene activation in higher grade tumors; so could you postulate that the earlier tumors, the ones that haven't progressed so far, are negative because they haven't...

Callahan: That's a good question, actually. There's a logistical problem here. The problem is to have enough tissue to extract enough DNA and RNA to be able to do the experiment. Operationally, many times larger tumors have been taken. When one looks in the literature, there has not been really an adequate study of Stage I tumors. It's something we're very interested in doing, but the logistics of it have not made it an easy...

Unidentified speaker: That's a sensible answer to the question why don't all of the carcinomas show activation, because it occurs late in progression.

José Russo: The idea that int I may play a role in tumor evolution is a very attractive one. Did you localize the int I by in situ hybridization in hyperplastic nodules?

Callahan: That experiment has not been done, that I'm aware of. But what we're attempting to do now is to use retroviral vectors to introduce these different genes into primary mammary epithelium, reintroduce the infected cells back into cleared fat pads, to ask the kinds of questions that you've just posed.

Peter Duesberg: What is the rationale of using these LTR ras or LTR myc constructs? It seems to me like whoever did this, (is it Leder?) is about to rediscover retroviruses. Isn't it just one more LTR and you have the retrovirus. You make the retroviral oncogene and then you get a tumor. It's nice that things are being rediscovered slowly, but that would transform...

Callahan: It hasn't been rediscovered insofar as the mammary gland, because there are no acute transforming viruses for the mammary gland.

Duesberg: Well, so we make one now. But what does that tell us about cancer?

Callahan: You're correct. Why not make an acute transforming MMTV genome; is that the point?

Duesberg: Yes. That's what we did.

Callahan: The problem is first, that MMTV is very poorly infectious in tissue culture. Secondly there's something about the viral genome that does not lend itself to growth in bacteria, which is necessary in the preparation of the various constructs that you would have to make. For that reason, investigators have taken another tack, two other tacks actually. One is to introduce the various genes of interest into type C retroviral vectors -these viruses do grow in mammary epithelial cells- and do the types of experiments that you are suggesting. The other tack, which I think is equally valid although it has some problems, is to introduce the gene into the germ line under the control of an LTR or (in the case of Bernd Groner) the WAP promoter. The problem with the transgenic mouse studies, apart from the fact that the normal c-Harvey ras gene has not been introduced, is that it's difficult to control the effect of expression of these genes in other tissues of the same animal because of the contribution it may have to the hormonal or growth peptide milieu that the mammary gland would see.

Duesberg: Have you looked in any of those tumors for virus production?

Callahan: This work that I've summarized is from Leder's group. They say that the mice they use as their host do not have an incidence of mammary tumors, which leads me to believe that they probably do not have a virus either.

Peter Nowell: I think one other possible explanation for the 70% of tumors without known oncogene involvement is simply that we have yet to identify most of the growth regulatory genes that are important in human neoplasia. A number of these may turn out to be lineage specific. I think the recent work in the lymphomas certainly suggests both of those aspects.

Callahan: In that respect, most of these studies have been carried out primarily on invasive ductal carcinomas while other histopathological types of breast tumors have not been examined to any great extent.

Nowell: To get back to the early question on Harvey ras and your comments about the 11p deletions, do any of those involve Harvey ras?

Callahan: Yes, in some cases Harvey ras is deleted, but there are other cases where it isn't.

Lance A. Liotta: Could you comment on increased expression of ras, because in experimental systems increased normal p21 will cause transformation compared to activated.

Callahan: There are two comments. One is that Harvey ras is expressed probably in 60% of primary human breast tumors, yet we have found no point mutations to indicate that there would be an expression of it as an oncogenic product. In the mouse, it's interesting that neither ras nor myc have been found to be activated as an int locus for MMTV. In that particular case it may be that activation of those genes occurs in cells which are restricted from infection by MMTV.

155

Russo: The fact that not all the tumors have increased expression of ras could be a methodological problem, because in reading all these papers, in general they say that they plate the tumors and they correlate that with the histology of the tumor. But everybody that is studying mammary tumors realizes that there is a tremendous heterogeneity in the tumors.

Callahan: Yes. The study of ras expression in primary human breast tumors was done by the immunoperoxidase technique.

Russo: Yes, but I am talking about expression of the other oncogene that you were talking about. It is quite possible that the portion of the tumor that was analyzed didn't contain any tumor at all.

Callahan: That's absolutely correct.

Russo: I believe that before reaching any conclusions as to the percentage of expression, we should have a better methodological approach to be sure that the tumor is analyzed. I think that we may have different results.

Callahan: I didn't have time to go into it, but as a matter of fact, those genes that had been activated by amplification have also been looked at for expression, using techniques which allow one to look at the tumor cells, such as by immunoperoxidase or in situ RNA hybridization. In those cases where it has been looked at extensively, amplification seems to correlate with expression of the gene. But you're correct. In looking for amplification one has to do it by grinding up a tumor. So, at best one can only say that this is a minimum number of tumors in which the gene is amplified, and it makes it almost impossible to make any sort of statement about the degree of amplification in association with other parameters.

GENE AMPLIFICATION DURING STAGES OF CARCINOGENESIS

Joseph Locker

Department of Pathology
University of Pittsburgh
Pittsburgh, PA

INTRODUCTION

Models of carcinogenesis have developed from extensive observation of clinical and experimental systems. (This summary is based on Pitot, 1981.) The related concepts of promotion and progression describe an essential feature of carcinogenesis, the gradual change from a normal cell to a tumor composed of proliferating, invading, metastasizing cells. Promotion is the gradual conversion of undetectable initiated cells to tumors without the continuing action of a carcinogen. Promotion requires distinct agents and involves both cell proliferation and gradually changing cell properties. In other words, promotion is the process by which an initiated cell evolves from a state of homeostatically regulated proliferation to a state of inadequately regulated proliferation. However, the appearance of a tumor is not really an endpoint in this process, and the more general term, progression, refers to the continuous evolution in characters both before and after a tumor has formed.

Reports of abnormal genes in cancer cells are now too numerous to list. These have presented a conceptual problem because mutations are abrupt and carcinogenesis is gradual and evolving. I do not mean to imply that these observations of abnormal genes are wrong, or unimportant. Oncogenes are central to contemporary studies of tumor biology, and proto-oncogenes are crucial genes that regulate proliferative, behavioral, and developmental responses. I imply only that we have yet to learn how the generation of genetic abnormalities and their effect on cell behavior are synthesized into the progressive evolution of lesions in carcinogenesis.

Like many laboratories, mine has been engaged in the study of oncogenes in cancer. Our studies have been surveys, correlating abnormal gene structure and abnormal gene expression, in both experimental and clinical systems. We frequently observe gene amplification, which has been reported by many groups in a variety of tumors. Our studies have been based on the premise that gene amplification starts out small, as a single extra gene copy. Therefore our experimental rationale has been that sensitive analysis of small changes will be required to survey the full importance of gene amplification in both experimental and clinical systems.

EXPERIMENTAL ANALYSIS OF myc GENE AMPLIFICATION

Using quantitative hybridization systems and specific gene probes, we detected excess myc gene hybridization in several experimental systems. I will not go into much detail, but Figure 1 illustrates representative Southern blot analysis. The figure shows matched blots from 6 breast cancers, internally controlled by hybridization to a probe for the mos gene. Mos and myc are both on chromosome 8. Both myc bands vary in hybridization intensity independent of mos hybridization. The differences are small but clearly visible. We use these Southern blots for confirmation, although most of our data analysis is based on less variable hybridization systems with multiple controls. In principle, a single extra gene copy should give a myc/mos ratio of 1.5; however, cancers, particularly breast cancers, contain

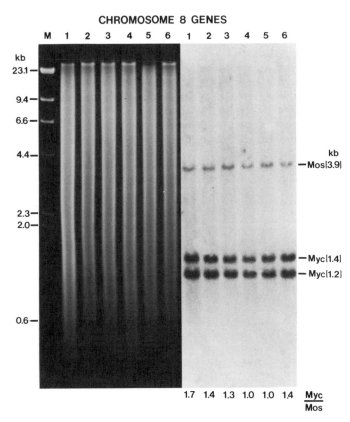

Fig. 1. Myc gene amplification in breast cancer DNA. 5 µg of DNA from six different breast carcinomas was digested with PvuII, blotted, and hybridized simultaneously with myc and mos gene probes. The stained gel is included to show the equivalence of DNA concentration in the gel lanes. The DNA of lane 1 is moderately degraded, a typical finding in cancer specimens, and thus appears brighter towards the bottom of the gel lane; however, this lane contains the same amount of DNA by optical density, and the mos gene intensity matches the other lanes.

Table 1. Myc Gene Amplification Studies

System	Incidence %		Amplification Size	Other Genes
Human Ductal Breast CA	10/21	48	1.4-1.9	Neu, N-myc
Rat HCC Cell Lines	2/ 7	29	1.6-1.7	-
Human Hepatic Tumors				
Adult HCC	9/15	60	1.4-2.3	-
Pediatric and Other	0/14	0	-	N-myc
Rat Choline Deficiency				
HCC	14/14	100	1.7-70	Endog. Retrovirus

variable amounts of stromal cells. A tumor which is 50% stromal cells and has one extra myc gene will give a signal of 1.25, a value usually below the statistically significant range of our quantitative hybridizations. We probably underestimate the incidence of myc gene amplification.

Table 1 summarizes studies, some preliminary, of myc gene amplification recently carried out in my laboratory (Chandar et al., 1989; Locker and Crawford, 1989; Contento and Locker, in preparation). Some of these will be discussed in detail below. These include two sets of data from clinical specimens and 2 sets from research systems. In human breast cancers, we observed small myc gene amplifications in almost half of the specimens (Contento and Locker, in preparation). Other groups have reported incidences ranging from 15% to 56% (Bonilla et al., 1988; Varley et al., 1987). Myc gene amplification occurs independently of neu gene amplification, which we studied in parallel. The other three studies deal with hepatocellular carcinoma. We characterized small myc gene amplifications in two of seven rat hepatocellular carcinoma lines (Locker and Crawford, 1989). These are our most rigorous characterizations. We used these cell lines to study the relationship between myc gene amplification and myc gene expression. On the basis of these observations of rat cell lines, we then studied human liver carcinomas. We found similar gene amplifications in more than half of the adult hepatocellular carcinomas (HCC), but not in the pediatric HCC or other types of liver tumors (Contento and Locker, in preparation). These observations, which will not be further discussed, provide a link between our experimental systems and human hepatic carcinogenesis.

In collaboration with Nalini Chandar and Benito Lombardi, we studied the choline-devoid (CD) diet experimental system (Yokoyama et al., 1985; Chandar et al., 1987). Rats fed this diet without chemical carcinogens, develop liver damage and hepatocellular carcinomas. We found myc gene amplification in all of these tumors. This experimental system has enabled us to formulate working hypotheses about the role of gene amplification in carcinogenesis.

We first studied livers and tumors of animals fed only the CD diet. Tumors appeared in about 1/3 of the animals after an average of 14 months. All of the tumors (eventually 8 were studied) showed small myc gene amplifications. Chandar and Lombardi (1988) later observed that feeding a CD diet and then replacing it with a choline supplemented (CS) diet caused a much higher incidence of carcinomas. In these sequential diets, CD was fed for periods of 3, 6, 9, or 12 months, followed by CS for a total of 16 months. The incidence of carcinomas ranged from 13% in the 3 month CD group to 73% in the 12 month group. We studied these tumors, expecting to find the same changes as in the tumors produced by the continuous CD-diet. However, to our surprise, these tumors had larger myc gene amplifications. Further, gene amplification could also be detected in non-tumorous portions of these livers. These livers show very little damage, but they have hepatocytic nuclear pleomorphism suggesting active cell proliferation and possibly abnormal ploidy.

THE PHASES OF GENE AMPLIFICATION

Our observations on the CD-diet system imply two separate phases of gene amplification. A model, presented in Figure 2, is derived from standard models of gene amplification presented by Schimke and others (Schimke, 1984; Schimke, et al., 1986; Van Hoff et al., 1988). Amplification obviously occurs as a result of the CD diet, presumably as a consequence of hepatocyte damage and regeneration. This chromosomal lesion is detected even in animals that have had the diet for as short a time as three months. However, the fact that significant enlargement occurs only in animals that have been taken off the CD diet indicates that the induction and the enlargement of gene amplification are different processes that occur under significantly different conditions.

The initial phase of gene amplification probably involves an accident during DNA replication. Schimke has stressed overreplication of DNA, which results from the disruption of coordinated replication along the length of the chromosome, although my personal view is that double strand breaks at replication forks are also significant. These would become prominent if DNA replication and DNA repair were not adequately coordinated (Edwards and Taylor, 1980). The CD diet kills hepatocytes and thus induces compensatory cell proliferation in the survivors. Further, many of these dividing cells are probably injured. The CD diet causes abnormal DNA methylation, and choline deficiency interferes with many pathways associated with DNA replication and repair, including polyamine and purine biosynthesis. The combination of injury, biochemical alteration, and cell division predisposes to this particular type of chromosomal lesion and probably others. Small amplifications differ from chromosomal lesions induced by the classical DNA-modifying carcinogens and might not be detected by standard mutagenesis assays. However, many such carcinogens also cause necrosis and could induce gene amplification along with point mutations.

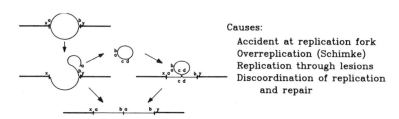

Causes:

Accident at replication fork
Overreplication (Schimke)
Replication through lesions
Discoordination of replication
and repair

I. Initial Gene Duplication

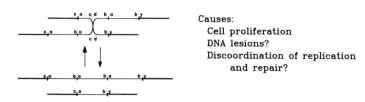

Causes:

Cell proliferation
DNA lesions?
Discoordination of replication
and repair?

II. Enlargement

Fig. 2. Two-stage model of gene amplification.

The second phase of gene amplification does not require the same kind of DNA damage near the replication fork. It can occur by unequal crossing over, between two chromosomes, between chromosome and episome, or between two episomes. Note that enlargement by unequal crossing over can be a geometrical progression. Unequal crossing over may be related to cell division and enhanced by DNA damage and repair. Incidentally, this process is reversible, and duplicated gene segments can be excised as well as expanded. If a chromosome normally contains a duplicated element, then the first phase of gene amplification could also be caused by unequal crossing over. Cell proliferation and consequent DNA replication appear to be the common factors of both stages of gene amplification; the first stage occurs in a setting of damaged cells and may require both replication and broken DNA molecules.

Some chromosomal regions may be more susceptible because of specific sequences they contain. Distance from the telomere would determine the probability of overreplication. A repetitive sequence might predispose to unequal crossing over. Thus, a gene may be very susceptible to amplification in one species and resistant, because of a different chromosomal location in another. Myc has similar locations in human chromosome 8 and rat chromosome 7.

EFFECTS OF GENE AMPLIFICATION ON myc EXPRESSION

The discussion so far has established the presence of myc gene amplifications and considered the processes that caused them. Do these amplifications affect myc gene expression?

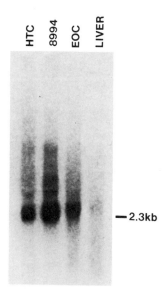

Fig. 3. Myc transcript levels in hepatocellular
carcinoma cell lines. Cell lines EOC
and HTC have myc gene amplification,
while line 8994 does not. Line 8994 has
the highest and line HTC the lowest
level of myc transcripts that was observed in hepatocellular cell lines.
The blots represent total cellular RNA,
except for the liver control, which is
cytoplasmic RNA.

We used hepatocellular carcinoma cell lines, as well as the CD livers and tumors, to study the relationship between myc gene expression and amplification. This analysis proved difficult, because of the low intrinsic levels of myc expression in normal hepatocytes compared to most other tissues (Zimmerman et al., 1986). Fausto and his co-workers (Thomson et al., 1986) have shown that myc transcripts increase approximately 5-fold during liver regeneration. Our own studies confirm this range: hepatocytes have a basal level of myc transcripts that increases 3 to 5 fold when they are stimulated to divide. We surveyed a series of rat hepatocellular carcinoma cell lines and characterized two with myc gene amplification (Locker and Crawford, 1989). These amplifications have remained small despite years in culture. We quantified myc transcripts (Figure 3) and found that all of the cell lines had higher levels than normal liver. This figure does not show all of the cell lines, but it includes the two with gene amplification (HTC and EOC), and the line with the highest level of myc expression (8994). During liver regeneration, fos transcript levels rise before myc, a sequence present in other tissues following stimulation by growth factors. Most, but not all, of the cell lines showed both fos and myc elevation. Elevated myc expression appears to be required for normal cell proliferation and is also found in most hepatocellular carcinomas. When increased expression is caused by normal cellular regulatory pathways, it should be accompanied by elevated fos expression. Alternatively, a primary abnormality in the myc gene could bypass the normal cellular pathway. Line HTC has no significant fos elevation. Gene amplification appears to be the major cause of increased myc expression in HTC cells.

In the CD-diet experimental system (Figure 4), we observed modest myc transcript elevations in livers and tumors. The CD diet causes liver damage and hepatocyte proliferation. When gene amplification is present, we can calculate the transcript level per gene, and find that there is only a five-

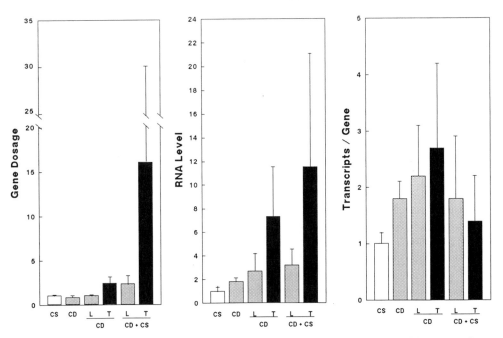

Fig. 4. Relative myc gene and transcript levels in CD-diet livers and tumors.

Table 2. Gene Alterations and Selections

---

1) Lethal change --> cell death
2) Neutral change --> accumulated damage, karyotypic abnormalities
3) Selective advantage --> cell will outsurvive its companions
4) Competitive advantage --> cell will outcompete its companions
5) Proliferative advantage --> cell will outgrow its companions

---

fold range of activity over all of these samples, the range observed in liver regeneration. This contrasts with some of the cell lines, which have as much as a 20-fold increase over basal hepatocyte levels. The expression per gene is higher in specimens from the continuous CD diet, which have active liver damage, than in livers and tumors from the sequential diet, where the liver damage has mostly stopped. Since the amplified genes function within the normal range of expression and can be stimulated normally, the extra copies must account for most of the increased expression.

ROLE OF myc IN PROLIFERATION CONTROL

Extra myc genes thus caused increased myc gene expression, but the effect is relatively weak. How do these small increases affect the cell? Several lines of evidence show that the myc protein participates in the control of cell proliferation (reviewed in Cole, 1986). 1) Transfected, constitutively controlled, myc genes cause autonomous cell proliferation. 2) Myc appears to act by triggering the G_o-G_1 transition in the cell cycle (Schweinfest et al., 1988). 3) Stimulation of regulated cell division by growth factors, including TSH in thyroid (Colletta et al., 1986) or PDGF in fibroblasts (Kelly et al., 1983) causes a sequential elevation of several proteins, including fos, myc, and p53. This suggests a replication control pathway. Myc may act as a mitotic trigger in this pathway. 4) Myc gene abnormalities causing overexpression are found in a large variety of cancers.

In normal cells, myc modulates from a basal to a stimulated level. The level of myc is tightly regulated by at least four levels of control, at initiation of transcription, transcription attenuation, mRNA turnover, and protein turnover (reviewed in Fahrlander and Marcu, 1986; Peichachzyk, 1987). This extensive regulation implies that cells are very sensitive to small changes in myc expression.

From these lines of evidence, I hypothesize that even a single extra myc gene significantly alters the control of cell division. Specifically, a cell with an extra myc gene would be triggered to divide by a smaller stimulus than a normal cell. This cell would behave like a normal cell, but would divide at a lower level of growth stimulating factor. Since there appears to be a basal level of expression from unstimulated myc genes in some cell types, as myc genes increased this basal level would increase until mitosis would be constitutively activated. Thus, as gene amplification enlarges, cell control progresses through altered homeostasis to autonomy. Several other forms of mutation deregulate the myc gene in cancer cells, including translocation and mutation of the first exon. These occur by different mechanisms, but could also deregulate cell division.

SELECTION

Can these observations about gene amplification be integrated into our understanding of carcinogenesis? When a gene becomes amplified, or any other stable DNA lesion is introduced into a cell, this alteration may affect the behavior and survival of the cell. Since a lesion could presumably occur at many genomic locations, we must account for the selection process that leads to recurrent amplification of specific genes like myc. I have considered 5 kinds of selection (Table 2). 1) Obviously, many kinds of genetic damage are lethal. Amplification could disrupt the function of a crucial gene. 2) Other changes should be neutral and have little effect on cell function. However, as these accumulate, chromosomal abnormalities may become apparent. 3) Some changes may impart a selective advantage, such as amplification of the dhfr gene during selection with methotrexate (Schimke, 1984). 4) Some changes will impart a competitive advantage. The altered behavior of the cell may enable it to obtain nutrients more efficiently than surrounding cells, or to interfere with surrounding cells, etc. This may be the most important consideration in acquisition of the invasive phenotype. 5) Some alterations will cause cells to outgrow their companions. The myc gene alterations could fall into this fifth category.

Myc GENE ABNORMALITIES IN PROMOTION AND PROGRESSION

To conclude this discussion, I would like to use our observations about CD-diet carcinogenesis, and about breast cancer, to synthesize a view of promotion and progression.

Our findings in CD-diet carcinogenesis suggest that myc gene amplification is a precancerous event in the genesis of these liver tumors. Myc gene amplification was clearly present in non-tumorous portions of some livers of rats fed the sequential diet, and by correlating histopathology, we excluded the possibility that these specimens contained enough tumor cells to account for the observed amplification.

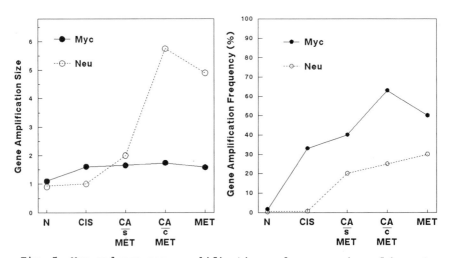

Fig. 5. Myc and neu gene amplification and progression of breast cancers.

In our analysis of 21 breast cancers, we were able to compare to the incidence and size of myc and neu gene amplifications in the same set of tumors (Figure 5). Myc amplification could be detected in carcinoma in situ as well as all later stages. Incidence increased with stage and tumor size. However, there was little progressive change in the size of myc gene amplifications. In contrast, neu gene amplification occurred in fewer tumors and they enlarged with tumor progression. Several tumors contained both myc and neu amplification. Neither gene showed a significant difference between primary tumors and their metastases. The slightly lower values for the metastases probably reflects increased stroma, hence fewer tumor cells with gene amplifications.

These observations suggest general features of myc gene amplification that are common to other gene amplifications, and special features of the myc gene that cause its abnormalities to be selected at an early stage in carcinogenesis. These myc gene abnormalities can lead to further progression. Deregulation of proliferation appears to be the first requirement in the sequence of promotion. Not only does proliferation expand the altered cell population, but its DNA replication is the process that gives rise to chromosomal abnormalities. Thus, an autonomously dividing cell would be more or less self-promoting. Why, then, do myc gene amplifications often remain small? In one sense, it is because they have not been selected to be any bigger. That is, having one or two extra myc genes may be enough to deregulate cell proliferation within the environment in which selection takes place. For example, in the continuous CD-diet system, there is ongoing liver damage, cell death and cell replacement. Factors that stimulate hepatic regeneration are present, and we observed that both liver and tumor myc genes function at near maximal levels. In the sequential CD-CS diet, the damage has nearly ceased. The myc gene amplifications are larger but the activity per gene is less.

Myc is probably not the only gene that could have this role in carcinogenesis. N-myc and L-myc appear to function similarly in certain cell types; p53 abnormalities could also have a similar effect; and, a number of oncogenic viruses have genes that stimulate cell division. Finally, all forms of myc and related gene deregulation should act synergistically with other kinds of gene alteration, particularly gene activation by direct mutation.

ACKNOWLEDGEMENTS

I would like to thank Dr. Benito Lombardi for encouragement and helpful discussions; Drs. Hector Tobon and Joseph Amenta, Magee Women Hospital, for contributing breast cancer specimens; and Drs. Suzanne Taylor and Eduardo Yunis, Pittsburgh Children's Hospital, for contributing pediatric liver tumors. This work was supported by National Cancer Institute Grant CA43909 and the Pathology Education and Research Foundation.

REFERENCES

Bonilla, M., Ramirez, M., Lopez-Cueto, J., and Gariglio, P., 1988, In vivo amplification and rearrangement of c-myc oncogene in human breast tumors, J. Natl. Cancer Inst., 80:665-671.
Chandar, N., Amenta, J., Kandala, J. C., and Lombardi, B., 1987, Liver cell turnover in rats fed a choline-devoid diet, Carcinogenesis, 8:669-673.
Chandar, N., and Lombardi, B., 1988, Liver cell proliferation and incidence of hepatocellular carcinomas in rats fed consecutively a choline-devoid and a choline-supplemented diet, Carcinogenesis, 9:259-263.
Chandar, N., Lombardi, B., and Locker, J., 1989, C-Myc gene amplification in liver tumors of rats fed a choline devoid diet, Proc. Natl. Acad. Sci., U.S.A., in press.

Cole, M.D., 1986, The myc oncogene: its role in transformation and differentiation, Ann. Rev. Genet., 361-384.

Colletta, G., Cirafici, A. M., and Vecchio, G., 1986, Induction of the c-fos oncogene by thyrotropic hormone in rat thyroid cells in culture, Science 233:456-460.

Edwards, M. J., and Taylor, A. M. R., 1980, Unusual levels of (AFP-ribose)n and DNA synthesis in ataxia telangiectasia cells following gamma-ray irradiation, Nature, 287:745-747.

Fahrlander, P. D., and Marcu, K. B., 1986, Regulation of c-myc gene expression in normal and transformed mammalian cells, in: "Oncogenes and Growth Control," P. Kahn, and T. Graf, eds., Springer, Berlin.

Kelly, K., Cochran, B. H., Stiles, C. D., and Leder, P., 1983, Cell-specific regulation of the c-myc gene by lymphocyte mitogens and platelet-derived growth factor, Cell, 35:603-610.

Locker, J., and Crawford, N., 1989, Nuclear oncogenes in hepatoma cell lines, Submitted for publication.

Piechaczyk, M., Blanchard, J.-M., and Jeanteur, P., 1987, c-myc gene regulation still holds its secret, Trends Genet., 3:47-51.

Pitot, H.C., 1981, "Fundamentals of Oncology," Dekker, New York.

Schimke, R. T., 1984, Gene amplification, drug resistance, and cancer, Cancer Res., 44:1735-1742.

Schimke, R. T., Sherwood, S. W., Hill, A. B., and Johnston, R. N., 1986, Overreplication and recombination of DNA in higher eukaryotes: Potential consequences and biological implications, Proc. Natl. Acad. Sci., U.S.A., 83:2157-2161.

Schweinfest, C. W., Fijiwara, S., Lau, L. F., and Papas, T. S., 1988, c-myc can induce expression of G0/G1 transition genes, Mol. Cell. Biol., 8:3080-3087.

Thompson, N. L., Mead, J. E., Braun, L., Goyette, M., Shank, P. R., and Fausto, N., 1986, Sequential protooncogene expression during rat liver regeneration, Cancer Res., 46:3111-3117.

Van Hoff, D. D., Needhan-VanDevanter, D. R., Yucel, J., Windle, B. E., and Wahl, G. M., 1988, Amplified human MYC oncogenes localized to replicating submicroscopic circular DNA molecules, Proc. Natl. Acad. Sci., U.S.A., 85:4804-4808.

Varley, J. M., Swallow, J. E., Brammer, W. J., Whittaker, J. L., and Walker, R. A., 1987, Alterations to either c-erbB-2 or c-myc proto-oncogenes in breast carcinomas correlate with poor short-term prognosis, Oncogene 1:423-430.

Yokoyama, S., Sells, M. A., Reddy, T. V., and Lombardi, B., 1985, Hepatocarcinogenic and promoting action of a choline-devoid diet in the rat, Cancer Res., 45:2834-2842.

Zimmerman, K. A., Yancopoulos, G. D., Collum, R. G., Smith, R. K., Kohl, N. E., Denis, K. A., Nau, M. M., Witte, O. N., Toran-Allerand, D., Gee, C. E., Minna, J. D. and Alt, F. W., 1986, Differential expression of myc family genes during murine development, Nature, 319:780-783.

DISCUSSION

Oscar Sudilovsky: I thought it was interesting that the regulation or amplification of myc increased after you stopped the choline deficient diet. How many weeks after you discontinued the choline deficiency did you detect that increase?

Joseph Locker: The interval was generally several months. The experimental endpoint in these animals was usually when a tumor was palpated. Typically the combination of the two diets was maintained for more than a year and the animals were studied at least three and as long as nine months after stopping the choline deficiency.

Sudilovsky: That eliminates the possibility that the increase in myc could be attributed to cell death. How do you explain, then, the increase of myc if cell death is not the cause?

Locker: When liver cells are dying, proliferation is stimulated by hepatic regeneration pathways. In the livers we studied, the liver cells are no longer dying. Presumably they are still proliferating and myc is elevated because it has become autonomous. I would propose that these myc gene amplifications have raised the basal level of myc expression, and that cell division is triggered although the stimuli have either reduced or disappeared. It evolved from a situation where the cells had to be stimulated to one where they no longer require stimulation.

Unidentified Speaker: Do you see any example where your cells lose amplification and reverted back to a different biological phenotype?

Locker: We haven't studied it in that kind of detail. We have studied a few examples where amplifications have been lost, but just at the level of observing that it has occurred. We don't have any direct correlation.

George Michalopoulos: Are you sure that you observed gene amplification in non-tumorous liver areas?

Locker: We were very concerned about this because there was always the possibility that the liver was just poorly sampled, for example, if it was contaminated with tumor cells. The livers are abnormal, but the myc gene amplification is clearly not in a population of tumor cells. It seems to be in abnormal hepatocytes. We would interpret that to mean that myc gene amplification is a precancerous change in this system, and as far as I know it is permanent. I would be happy to show you the morphology.

Unidentified Speaker: If you give a limited number of cycles of this treatment so that you really don't get tumors, could you have a situation where the liver remains non-neoplastic, non-tumor bearing, and yet have widespread myc amplification?

Locker: I would presume that to be the case, but we didn't study it. We took the livers when they had tumors. Our rationale was that the longer you let these processes continue, the bigger the changes and the easier they will be to find. If cells had a ten-fold gene amplification and were one tenth of the population of the liver, such gene amplification would be undetectable. So, we think it's very likely that these changes are there, but undetectable even a relatively short time interval before we sampled. I think the studies are most powerful in the very late stages.

Peter Duesberg: If I understood correctly, you proposed at the end that amplification of myc would explain or possibly help to explain proliferation of liver cells, that hyperplastic proliferation in turn would lead to cancer.

Locker: To me, hyperplasia is a regulated process. Hyperplasia would be the proliferation of liver cells in regeneration, because they've been stimulated to do so.

Duesberg: So in liver you have the perfect control system to study liver regeneration. If you take half of it out you can watch the remaining half grow back in a couple of days, I understand.

Locker: Yes. That's right.

Duesberg: Now, do these have amplified myc genes.

Locker: They do not, but they have elevated _myc_ expression.

Duesberg: So, what's the basis for your proposal then, if the cells can proliferate without amplification and probably grow 100 times faster than a tumor will ever grow?

Locker: Without amplification they do it by a controlled pathway. Presumably there are hormones, hepatocyte proliferation factors, like the ones that Dr. Michalopoulos has been characterizing, which stimulate the cells to proliferate. This proliferation pathway is probably just like the thyroid or the fibroblast pathway. In thyroid, TSH binds to a receptor. There is an elevation of _fos_ followed by an elevation of other gene products. There is then elevation of _myc_ and the cells are triggered to divide. In this regulated system, _myc_ would be a trigger. As you get a few extra copies of _myc_, proliferation would still have features of regulation. You would get more _myc_ from the same signal, and you would see a cell which still proliferates under stimulation, but which is easier to trigger than a normal cell. And as you increase the _myc_ levels beyond that point, you would see an autonomous cell which no longer needs stimulation. This would be an evolving process as gene amplifications are progressively selected to larger size.

Duesberg: Is there any reason to believe that the hyperplastic liver doesn't have the mechanism that makes it grow fast during regeneration? These would be normal genes, regular genes that are in it.

Locker: For me this is the definition of promotion, that a cell starts out having only its normal pathways for regulating proliferation, and this is why a promoting agent is needed. It has to be stimulated to proliferate because the control systems do not tell it divide. In the case of the liver, it's very clear that manipulations like partial hepatectomy are promoting systems. The end stage of promotion is a tumor that grows without that stimulation; it has to evolve through stages where it proliferates with stimulation and eventually loses its need for that stimulation. What I should point out is that gene amplification is one process that has these properties. There should be a number of other processes that work the same way. _Myc_ gene amplification is very prevalent, surprisingly prevalent, but it's by no means the only way to deregulate cell division.

Lance A. Liotta: It would seem to me that when you went through your five different categories, selective advantage and competitive advantage were one and the same. I wasn't clear why you divided those up into separate mechanisms.

Locker: I think I probably didn't describe them adequately. Selective advantage refers to conditions like antibiotic selection. Methotrexate kills cells, but cells that have gene amplification of the _dhfr_ gene can deal with methotrexate and survive.

Liotta: You mean in response to an outside selective pressure?

Locker: Yes, in response to exogenous selective pressure. Competitive advantage might be the same except we don't know what the selection procedure is. It's a cell that is aggressive and can obtain nutrients in a more efficient fashion than its surrounding cells, by the way it aligns itself along a blood vessel or because it kills other cells around it. I think they're quite different, although all these categories could overlap.

Liotta: Passive versus active, that is what you're saying.

Harry Rubin: When cells are stimulated to multiply by whatever mechanism, either hepatectomy, various growth factors or nonspecific factors, almost

every process in the cell changes its rate. It occurs in every synthetic
process, every aspect of intermediary metabolism. As one investigates in
detail, one finds more and more changes. Nowadays you look at genes and find
different changes. There is this question of causality. If you are trying
to go in and focus on a particular gene or set of genes, or glycolysis or the
TCA cycle, sooner or later you're going to find something, but in fact the
cell responds as a whole to some kind of signal. It seems to me that there
ought to be more emphasis placed on what kind of signal the cell is responding to. I think it's the old lamppost analogy. You look at the kind of
things that you can measure. But that doesn't tell you whether it plays any
causal role. You could amplify myc by itself, without increasing the expression of many other genes or even without increasing the expression of genes.
The mere fact that you can get tumorigenesis with choline deficiency, which
presumably affects something about phospholipid synthesis in the cell, sort
of raises the critical question: at what level do we think about causality
in systems like this? Why should we fasten on any particular genetic or
metabolic change in the cell when the cell is responding as a unit? Any one
of those specific changes by themselves, if we had a probe to do that,
perhaps would do nothing to the cell.

Locker: I think it's a mistake to consider them all individually. First of
all, all the changes mean something. A liver cell is a good example. In its
normal state, it's in a resting G_o compartment, a cell whose function is to
make secretory proteins and to regulate metabolic pathways. When you trigger
that cell to divide, there is a major reordering of priorities because the
synthetic machinery has to be shifted into making cellular components, replicating DNA and making cytoskeleton. Triggering a cell to divide which is
normally resting will alter virtually every synthetic pathway in the cell.
Anything you study will be changed: albumin will go down, actin will go up,
DNA synthesis will go up. These are not random changes, although they may
appear to be random if you're sampling half a dozen enzymes.

Rubin: I didn't imply they were. I suggest the opposite.

Locker: You may not be able to interpret them. Now myc, from many studies
in many laboratories, seems to be a very special gene. I won't get involved
in whether to call it an oncogene or not, but myc clearly has a role in the
control of cell proliferation. There is substantial evidence that myc, in
fact, may be a trigger for cell proliferation.

Rubin: I have never been able to see that.

Peter Nowell: I don't know how much you want to get into this now, but I
would like to follow up the myc story for a moment from the cytogenetic stand-
point. I think it is impressive that in every single Burkitt lymphoma you
have a chromosome translocation that puts a myc gene into juxtaposition with
a rearranged immunoglobulin gene. As a result there is a constitutive abnormal expression, not necessarily elevated, but abnormal regulation of expression of the myc gene. That chromosome translocation appears in every cell of
the tumor, which suggests that it was there from early on and is at least an
important part of the clonal expansion of the tumor. Since you see it in
every tumor, it does suggest very strongly that altered expression of the myc
gene is playing a major role in the pathogenesis of this tumor. It doesn't
say it's the only factor, but it does say, to me at least, that it's an
important factor.

Duesberg: I just want to comment briefly on this point. It is often said in
literature based on the Burkitt lymphoma cell lines which have been studied;
but if you look at the literature for primary Burkitt lymphomas, the evidence
is by far not as clear. There are papers, one from the NIH, about six years
old from Jackie Whang Peng, and one from France, that also came out a few

years ago, that showed that up to 50% of primary Burkitt lymphomas do not
show a rearrangement of chromosome 8. It's only when you put these cells in
culture that you select for variants where the chromosome shows a rearrangement. Often the breakpoint is downstream of myc. The myc gene isn't even
affected by it, and sometimes it's a 3'-5' opposite orientation, where the
myc gene cannot be transcribed. The simplification that the myc gene is
always "activated", rearranged, is simply not true for primary tumors. Even
in cell lines it is frequently in a configuration where the rearrangement cannot affect transcription. In fact, this configuration decapitates the gene
and inactivates it entirely.

RECESSIVE MUTATIONS IN CANCER PREDISPOSITION

AND PROGRESSION

Webster K. Cavenee

Ludwig Institute for Cancer Research
Royal Victoria Hospital
687 Pine Avenue West
Montreal, Quebec
H3Z 1A1
Canada

INTRODUCTION

Cancer is widely considered to represent the phenotypic manifestation of
the accumulation of genetic damage[1], and this notion is the subject of much
of the present volume. There is a great deal of circumstantial evidence in
its support which arises from examination of human populations[2]. There has
been, for example, extensive documentation of familial aggregation of specific histological types of tumors, sometimes developing with the formal behaviour of an autosomal dominant Mendelian trait. At the level of cytogenetics, various chromosomal aberrations of the germline appear to result in
increased propensities for the development of tumors. Tumors often have
specific chromosomal rearrangements and, sometimes, such aberrations resemble
those which, when inherited, predispose to similar disease.

The major question which arises from these observations is whether such
chromosomal derangements are causal of or caused by the neoplastic process.
The challenge is to sort through the bewildering array of such changes and to
determine which are common amongst neoplastic diseases at the microscopic and
molecular levels and then to categorize them in the context of the large
amount of knowledge we have about the process of experimental carcinogenesis.

In this way, we can have some reasonable hope to gain insight into the
process of human tumorigenesis by having taken advantage of experiments that
Nature has already performed. In this chapter I shall cite two examples of
studies from this laboratory which demonstrate the power of molecular biology
to this end.

PREDISPOSITION

In the parlance of experimental chemical carcinogenesis, the earliest
event in tumorigenesis is termed the initiating event. In the human population such initiations may be transmitted as inherited predisposition. At
least fifty different forms of human cancer have been observed to aggregate
in families as well as to have corresponding sporadic forms[3]. In many of

Boundaries between Promotion and Progression during Carcinogenesis
Edited by O. Sudilovsky *et al.*, Plenum Press, New York, 1991

171

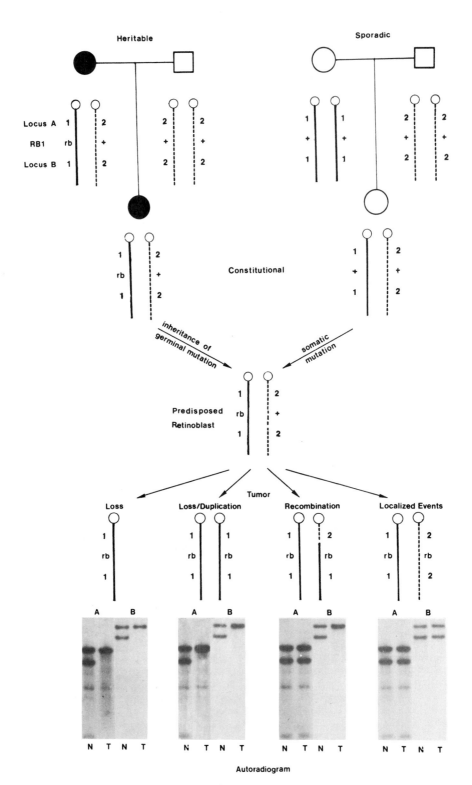

Figure 1

these cases the aggregation occurs with a pattern consistent with the transmission of an autosomal dominant Mendelian trait.

One important line of evidence indicating the genetic origin of certain neoplastic diseases is the frequent finding of constitutional chromosomal abnormalities in patients with specific types of tumors. Perhaps the best characterized of these are deletions involving the chromosome region 13q14 which are found in normal tissues of 3-5% of children with bilateral retinoblastoma, a tumor of embryonic neural retina[4]. Such deletions may occur de novo, be inherited, or be unbalanced segregants of insertion/deletion translocations. In the latter case, the balanced carriers and family members with duplications do not develop tumors, whereas those with unbalanced deletions of chromosome 13q do[5].

In contrast, tumors that appear as inherited forms often show abnormal segregation of chromosomes, resulting in monosomy and/or duplication of a whole chromosome complement, although structural rearrangements are also frequent. Deletions are mechanistically related to monosomy since both events result in loss of genetic information and due to these events, the tumor cells are hemizygous for part of the genome. Additionally, although retinoblastoma tumor cells often carry 13q-deletions involving chromosome region 13q14, their most frequent[6] rearrangements are structural alterations of chromosomes 1 and 6.

In the context of the requirement for multiple events in tumorigenesis[7], such deletions could act as the first "hit" and, when they are germinal, they could confer a risk of tumor formation in an autosomal dominant manner. Evidence that the same locus is involved in retinoblastoma cases that lack an apparent chromosomal deletion was provided through the demonstration of tight genetic linkage between retinoblastoma and the esterase D locus, the latter being a moderately polymorphic isozymic enzyme whose encoding locus also map to 13q14[8]. This is particularly important since cytogenetically detectable deletions of chromosome band 13q14 are found in only 3-5% of all bilateral retinoblastoma patients and, in most familial forms of retinoblastoma cases, the primary mutation has not been characterized. However, a model has been proposed[9], which makes specific mechanistic predictions concerning the nature of the chromosomal rearrangements which could serve to unmask the initial predisposing recessive mutation through the somatic attainment of homozygosity for the mutation. This model is illustrated in Figure 1 and includes the following chromosomal mechanisms: mitotic recombination between the chromosomal homologues with a breakpoint between the tumor locus and the centromere, which would result in heterozygosity at loci in the proximal region of the chromosome and homozygosity throughout the rest of the chromosome including the tumor locus; mitotic nondisjunction with loss of the wild-type chromosome, either without or with duplication of the mutant chromosome, resulting in hemizygosity or homozygosity at all loci on

Fig. 1. Chromosomal mechanisms that could unmask predisposing recessive germ-line or somatic mutations. Top left, inheritance of a chromosome 13 that carries a recessive defect at the TB1 locus designated "rb", results in a child who is genotypically rb/+ in all her cells. A retinoblastoma might develop by eliminating the dominant wild-type allele at the RB1 locus by one of the indicated mechanisms. Top right, a recessive mutation occurring in a single retinal cell could also be unmasked by one of the same mechanisms. Bottom, the chromosomal mechanism involved in tumor formation can be discerned by comparing genotypes at loci A and B on chromosome 13 in normal (N) and tumor (T) tissues. The use of marker loci in this way does not allow the detection or discrimination of the other regional mechanisms shown at bottom right.

Table 1. Loss of Heterozygosity for Loci on Chromosome 13q in Retinoblastomas

Patient	Tissue	Alleles Present at Locus					Mechanism
		D13S1	D13S7	D13S4	D13S5	D13S53	
A. Sporadic:							
Retin LA69	N	1,2	1,1	1,2	1,2	1,1	Chromosome Loss
	T	2	1	1	2	1	
Retin 409	N	2,2	1,2	2,2	1,1	1,2	Isotrisomy
	T	2,2,2	1,1,1	2,2,2	1,1,1	2,2,2	
Retin 412	N	1,2	1,1	1,2	1,1	2,2	Mitotic Recombination
	T	1,2	1,1	2,2	1,1	2,2	
B. Heritable:							
Retin KS2H	N	2,2	1,2	2,2	1,1	1,1	Isodisomy
	T	2,2	2,2	2,2	1,1	1,1	
Retin 462F	N	1,2*	1,1	1,2	1,2	1,1	Mitotic Recombination
	T	1,2	1,1	2,2	1,1	1,1	

* D13S6 examined, not D13S1

the chromosome; and, mitotic or regional second events such as gene conversion or point mutation. Sporadic disease could also arise through the appearance of nullizygosity at the tumor locus, the difference being, in this instance, two somatic events as compared to one germinal and one somatic event in heritable cases.

Chromosome specific, single copy segments of the human genome, isolated in recombinant DNA form can be used to recognize polymorphisms at the corresponding chromosomal locus. Sequence variation in restriction endonuclease recognition sites, giving rise to restriction fragment length polymorphism (RFLP) in the locus defined by the probe, are revealed as distinct bands on an autoradiogram and represent alleles of the locus; one from the paternally-derived and one from the maternally-derived chromosomal homologue and behave as Mendelian codominant alleles in family studies. These RFLP markers can be used as linkage markers in inherited disorders, including retinoblastoma. If a disease locus is located close to a polymorphic RFLP marker locus, they are likely to segregate together in a family. Therefore, the genotype of DNA markers can be used to infer the genotype at the retinoblastoma locus, and thus to predict if the offspring has inherited the predisposition. Chromosome segregation during tumorigenesis can also be determined in each patient by comparing the child's constitutional and tumor genotypes at each of these marker loci. Recombinant DNA segments have been isolated from human chromosome 13 and used to determine somatic changes in the germline genotypes in several such cases in the manner illustrated in Figure 1.

Detailed analyses of many retinoblastomas have shown that such events are common and are detected in about 3/4 of all retinoblastoma tumors. These rearrangements fall into four different classes as illustrated in Figure 1. In 20 of 33 tumors, one constitutional allele was missing at all informative loci along the whole chromosome and 19 of these tumors contained two intact chromosomes 13 as determined either by cytogenetic analysis of the tumor cells or densitometric quantitations of the autoradiographic signal of the remaining alleles. Therefore, the loss of alleles along the chromosome must involve two separate events; a nondisjunction resulting in loss of one chromosomal complement and either a duplication of the remaining homologue or an abnormal mitotic segregation of the chromosome resulting in isodisomy as shown for Retin 409 in Table 1A. In one case, data consistent with the sole loss of chromosome 13 were obtained. Evidence for mitotic recombination between the chromosome homologues was provided in 4 of the 33 tumors (one example is Retin 412, Table 1A). The constitutional genotype was maintained at all informative loci in 9 of the 33 tumors, and therefore, in these cases, the mechanism of attainment of homozygosity could not be determined. These studies strongly suggest that the second event in tumor initiation is comprised of a specific chromosomal rearrangement involving physical loss of the balancing wild type allele at the RB1 locus. This inference was corroborated by examining cases of heritable retinoblastoma and showing that the chromosome 13 homologue retained in these tumors was derived from the affected parent as would be predicted. Two examples are shown in Table 1B.

It is noteworthy that, although the unmasking of predisposing mutations at the RB1 locus occurs in mechanistically similar ways in sporadic and heritable retinoblastoma cases, only the latter carry the initial mutation in each of their cells. Heritable cases also seem to be at greatly increased risk for the development of second primary tumors, particularly osteogenic sarcomas[11]. This high propensity may not be merely fortuitous but may be genetically determined by the predisposing RB1 mutation. This notion of a pathogenetic causality in the clinical association between these two rare tumor types was tested[12] by determining the constitutional and osteosarcoma genotypes at RFLP loci on chromosome 13 and representative data are shown in Table 2. Osteosarcomas arising in retinoblastoma patients had become specifically homozygous around the chromosomal region carrying the RB1 locus

Table 2. Loss of Heterozygosity for Loci on Chromosome 13q in Osteosarcomas

Patient	Tissue	Alleles Present at Locus					Mechanism
		D13S1	D13S7	D13S4	D13S5	D13S3	
A. Sporadic:							
Osteo 03	N	1,2	1,1	2,2	-	1,2	Isodisomy
	T	2,2	1,1	2,2	-	1,1	
Osteo 06	N	1,2	1,1	1,1	1,2	1,1	Isodisomy
	T	1,1	1,1	1,1	1,1	1,1	
Osteo 09	N	2,2	1,1	2,2	2,2	1,2	Isodisomy
	T	2,2	1,1	2,2	2,2	1,1	
B. Second primary to retinoblastoma:							
Rb1-1	N	1,2	1,1	1,2	1,2	2,2	Translocation
	T	1,1	1,1	1,1,2	1,2,2	2,2,2	Isodisomy
Rb108	N	1,2	1,1	2,2	1,1	1,2	Translocation
	T	1,1,2	1,1	2,2	1,1	1,1	

- Not determined

(Table 2B). Furthermore, these same chromosomal mechanisms were observed in sporadic osteosarcomas (Table 2A), suggesting a genetic similarity in pathogenetic causality. These findings are of obvious relevance to the interpretation of human mixed cancer families as they suggest differential expression of a single pleiotropic mutation in the etiology of clinically associated cancers of different histological types.

A likely explanation for the association between retinoblastoma and osteosarcoma is that both tumors arise subsequent to chromosomal mechanisms which unmask recessive mutations in either one common locus that is involved in normal regulation of differentiation of both tissues, or in separate loci that are located closely within chromosome region 13q14. In either case, germline deletions of the retinoblastoma locus may also affect the osteosarcoma locus. Deletions are likely to be an important form of predisposing mutation at the RB1 locus since a considerable fraction of bilateral retinoblastoma cases carry visible constitutional chromosome deletions and submicroscopic deletions have been detected by reduction of esterase D activity and by molecular analyses using a cDNA for a gene which is, in all likelihood, the transcription product of the retinoblastoma locus[13]. The data are also consistent with the chromosomal mechanisms attaining nullizygosity for mutation in a tumor suppressor locus.

The information derived from these studies raises two points relevant to familial predisposition to cancer. Chromosomal mechanisms capable of unmasking predisposing recessive mutations occur in more than one tumor and, at least for chromosome 13, clinically associated tumors share this mechanism of pathogenesis. This latter point suggests that these loci have pleiotropic tissue specificity; however, this pleiotropy appears to be restricted to a small number of tissue types.

PROGRESSION

The foregoing section suggests a general approach to identifying the chromosomal positions of loci whose recessive alleles predispose to human cancer. The approach takes advantage of specific and frequent somatic chromosomal alterations in tumors and draws its power from the conjoint use of such information and familial genetic analysis. Clearly, one would anticipate that a segregating tumor trait will be genetically linked to the predisposing mutation which elicits it[14]. This single characteristic should provide a means of distinguishing predisposing from progressionally acquired genetic damage since the latter would be unlikely to be genetically linked to the former in families.

Our first efforts in attempting to utilize genotypic distinctions for tumor staging have been with glial tumors[15]. Tumors of astrocytic origin are the most frequently occurring neoplasms of the central nervous system and are identified using conventional histopathology in combination with histochemistry or immunohistochemistry[16]. Biological aggressiveness characteristics in these tumors have been empirically determined and the tumors are now grouped under four malignancy grades. Prognoses vary accordingly with five-year postoperative survival rates reaching a nadir of < 5% for patients whose tumors are diagnosed as grade IV astrocytomas (also referred to as glioblastomas). Clinical recurrence of astrocytoma is frequent and histologic examination commonly reveals that the recurrent tumor is less well differentiated than its predecessor; data which suggest that astrocytoma is a progressive disease.

Cytogenetic analyses of direct preparations and short term cultures of malignant astrocytomas have provided information regarding the gross chromosomal changes taking place in these tumors. For example, monosomy for

Table 3. Loss of Heterozygosity for Loci on Chromosome 10 in Gliomas

Patient	Clinical Grade	Tissue	Alleles Present at Locus		
			D10S1	D10S4	PLAU
Glio 03	AIV	N	1,2	1,2	1,2
		T	1	2	1
Glio 12	AIV	N	1,2	1,2	1,2
		T	2	2	2
Glio 16	AIII	N	1,2	1,2	1,2
		T	1,2	1,2	1,2
Glio 19	AIII	N	1,2	1,2	1,2
		T	1,2	1,2	1,2
Glio 24	AII	N	1,2	1,2	1,2
		T	1,2	1,2	1,2
Glio 25	AII	N	1,2	1,2	1,2
		T	1,2	1,2	1,2

Alleles shown are composited from various enzyme digestions for maximal illustration.

chromosome 10 has been detected in about half of grade IV astrocytomas[17]. Since the genotypic changes in other tumor types can be determined even in the absence of gross chromosomal aberration by comparing restriction fragment length alleles in DNA from constitutional and neoplastic tissues from affected individuals as described above, we applied this type of analysis to astrocytomas to assess whether such data distinguish tumors histologically classified by malignancy grade.

We compared constitutional and tumor genotypes at loci on chromosome 10 for DNA samples from 26 adult cases of astrocytoma, histologically representing a continuum of malignancy grades. Allelic combinations were determined with probes homologous to three different chromosome 10 loci: D10S1, D10S4 and PLAU. Representative data obtained with samples of various histological grades are shown in Table 3. Each of 15 grade IV tumors examined showed loss of constitutional heterozygosity at one or more of the chromosome 10 loci as exemplified by Glio 03 and Glio 12 and these losses appeared to be elicited by nondisjunction resulting in monosomy. In sharp distinction, none of the eleven tumors of lower malignancy grades showed a loss of alleles at any of the chromosome 10 loci examined as exemplified by Glio 16, 19, 24 and 25 (Table 3). Further several of the grade IV astrocytomas displayed areas of varying cellularity and diverse morphology with or without necroses and, in three cases, such areas are dissected and analyzed separately. In each instance, DNA from these different areas showed loss of constitutional heterozygosity for chromosome 10 loci consistent with the high malignancy grade classification of the tumor. Central nervous system tumors of four other types were also analyzed for allelic combinations at these loci: 6 oligodendrogliomas, 5 oligoastrocytomas, 1 primitive neuroectodermal tumor, and 3 medulloblastomas. All were informative for at least one marker locus and none showed loss of constitutional heterozygosity on chromosome 10.

These findings have two major implications. The first of these is the demonstration of a clonal origin of the cells comprising these tumors. The cellular pleomorphism of malignant astrocytomas and karyotypic heterogeneity of in vitro derived cell subpopulations arising from primary tumors have complicated attempts to determine the nature of any relationship between the cells which constitute this type of neoplasm. These data show that grade four astrocytomas arise from the expansion of cells deficient in all or part of chromosome 10. The second issue involves the histopathological evidence that astrocytomas progress and become more malignant with time. Since the losses of heterozygosity for chromosome 10 loci were restricted to tumors of the highest malignancy grade, it may be that this aberration is an event of tumor progression, rather than of initiation. Conversely, it is possible that the etiologies and ontogenies of astrocytomas exhibiting low and high degrees of cellular differentiation have no interrelating molecular pathways. However, the postoperative, post-therapeutic recurrence and histological progression of astrocytoma is well documented, providing clinical support for an ontogenic relationship[18]. While these data are only correlative at present, the identification of these nonrandom events may serve as the beginning of a genotypic, rather than phenotypic, approach to the definition of the molecular underpinnings of tumor progression.

CONCLUSIONS

The studies described in this chapter demonstrate the utility of identifying specific genotypic alterations in human tumors. Information so derived bears on each phase of carcinogenesis. When it is used in concert with familial aggregation, inferences concerning predisposition can be drawn and strategies for gene isolation devised. Of course, the applications the information can have viz à viz workforce screening and assignment, actuarial table revision, early disease detection and vigilance in periodic medical examination are easily envisaged. When the analyses are applied to a dissection of the pathway of evolution of tumors[1], the opportunities are great as well. For example, one may be able to examine low grade tumors with the object of determining which among them will likely recur or progress. Such information, although only speculative at present would have great medical impact, for example, on therapy selection. All of these possibilities are so tantalizing that the future of genotypic research into the bases of tumor initiation and progression seems bright and exciting indeed.

REFERENCES

1. P. C. Nowell, The clonal evolution of tumor cell populations, Science 194:23-28 (1976).
2. M. Nordenskjold, and W. K. Cavenee, Genetics and the etiology of solid tumors, in: "Important Advances in Oncology 1988," V. T. DeVita, S. Hellman, and S. A. Rosenberg, eds., J. B. Lippincott, Philadelphia (1988).
3. J. J. Mulvihill, Genetic repertory of human neoplasia, in: "Genetics of Human Cancer," J. J. Mulvihill, R. W. Miller, and J. F. Fraumeni, eds., Raven Press, New York (1977).
4. U. Francke, Retinoblastoma and chromosome 13, Cytogenet. Cell Genet. 16:131-134 (1976).
5. L. C. Strong, V. M. Riccardi, R. E. Ferrell, and R. S. Sparkes, Familial retinoblastoma and chromosome 13 deletion transmitted via an insertional translocation, Science 213:1501-1503 (1981).
6. J. Squire, B. L. Gallie, and R. A. Phillips, A detailed analysis of chromosomal changes in heritable and nonheritable retinoblastoma, Hum. Genet. 70:291-301 (1985).
7. A. G. Knudson, Jr., Mutation and cancer: Statistical study of retinoblastoma, Proc. Natl. Acad. Sci., U.S.A. 68:820-823 (1971).

8. R. S. Sparkes, A. L. Murphree, R. W. Lingua, M. C. Sparkes, L. L. Field, S. J. Funderburk, and W. F. Benedict, Gene for hereditary retinoblastoma assigned to chromosome 13 by linkage to esterase D, Science 219:971-973 (1983).

9. W. K. Cavenee, T. P. Dryja, R. A. Phillips, W. F. Benedict, R. Godbout, B. L. Gallie, A. L. Murphree, L. C. Strong, and R. L. White, Expression of recessive alleles by chromosomal mechanisms in retinoblastoma, Nature 305:770-784 (1983).

10. W. K. Cavenee, M. F. Hansen, E. Kock, M. Nordenskjold, I. Maumenee, J. A. Squire, R. A. Phillips, and B. L. Gallie, Genetic origins of mutations predisposing to retinoblastoma, Science 228:501-503 (1985).

11. F. D. Kitchin, and R. M. Ellsworth, Pleiotropic effects of the gene for retinoblastoma, J. Med. Genet. 11:244-246 (1974).

12. M. F. Hansen, A. Koufos, B. L. Gallie, R. A. Phillips, O. Fodstad, A. Brogger, T. Gedde-Dahl, and W. K. Cavenee, Osteosarcoma and retinoblastoma: a shared chromosomal mechanism revealing recessive predisposition, Proc. Natl. Acad. Sci., U.S.A. 82:6216-6220 (1985).

13. S. H. Friend, R. Bernards, S. Rogelj, R. A. Weinberg, J. M. Rapoport, D. M. Albert, and T. P. Dryja, A human DNA segment with properties of the gene that predisposes to retinoblastoma and osteosarcoma, Nature 323:643-646 (1986).

14. M. F. Hansen, and W. K. Cavenee, Retinoblastoma and the progression of tumor genetics, Trends in Genet. 4:125-128 (1987).

15. C. D. James, E. Carlbom, J. P. Dumanski, M. Hansen, M. Nordenskjold, V. P. Collins, and W. K. Cavenee, Clonal genomic alterations in glioma malignancy stages, Cancer Res. 48:5546-5551 (1988).

16. W. R. Shapiro, Treatment of neuroectodermal tumors, Ann. Neurol. 12:231-237 (1982).

17. S. H. Bigner, J. Mark, M. S. Mahaley, and D. D. Bigner, Patterns of the early gross chromosomal changes in malignant human gliomas, Hereditas 101:103-113 (1984).

18. P. C. Burger, F. S. Vogel, S. B. Green, and T. A. Strike, Glioblastoma multiforme and anaplastic astrocytoma. Pathologic criteria and prognostic implications, Cancer 56:1106-1111, 1985.

DISCUSSION

Philip Frost: Dr. Cavenee, I certainly think that what you've shown is very interesting. I have some general questions. The deletions you've described are different for different tumors, which means that for a colon tumor there may be a deletion in chromosomes 5, 17 and 18, in glioblastoma in 10, and yet they each produce a tumor. The deletions are occurring in totally different chromosomes, which means that there must be many, many of these so-called suppressor genes. I'd like to hear what your views are in particular with relation to deletions in particular organs and in the development of tumors in those organs.

Webster Cavenee: I think there are two possibilities. The first is that the retinoblastoma paradigm holds in these tumors. That is, we're achieving nullizygosity, so if a deletion eliminates a wild side copy, a mutant copy is left; there's really nothing there. That's one interpretation, but that's not demanded by this data. The data, of course, is consistent also with a dosage effect of something which is encoded by a locus on chromosome 10. So whether or not we're looking at one or the other of those two sorts of events, I don't know. In terms of the colon cancer situation the interpretation is murky because of several changes which occur either sequentially or concurrently. But sequentially is not a very good argument, because you can have any sequence you like as long as you end up with the aggregate number of changes that are required. Whether those are tumor suppressors or not is subject to question. The reason that it is though to be tumor suppressors is because of this behavior, but that's in the lack of functional evidence. Obviously what you'd like to do is to be able to put in combinations of these chromosomes into the stages and see if that happens. That's what we're doing, in fact, with Buddy Weisman.

Peter Duesberg: Also regarding the retinoblastoma paradigm, recently we had a pleasure in Berkeley that our friend and distinguished cancer researcher Dr. Weinberg, from MIT, came to talk about the retinoblastoma. He reported that in over 50% of the retinoblastomas that he studied, the suppressor RNA was expressed. How does that fit with the paradigm? Are you familiar with that data? Could you comment on it?

Cavenee: I am familiar with that data, Peter. I think it fits just marvelously with the paradigm.

Moderator (Lance A. Liotta): We're going to hear more about suppressor genes from the next speaker.

Duesberg: But this is not the paradigm. Isn't it appropriate to discuss this at least briefly?

Cavenee: No.

Duesberg: You don't think so.

Cavenee: I'll give you a brief answer. I think it's entirely possible that the expression of that protein which you're talking about...

Duesberg: RNA, it was. I think he measured RNA.

Cavenee: Presumably RNA would be translated into protein, as I remember the textbook.

Duesberg: I'm not so sure about all of them. Some of them do.

Cavenee: Some retinoblastomas in fact have retinoblastoma gene product, let's put it that way, regardless of whether it's RNA or protein. But since there is no functional assay for that gene at the moment, as you know, there are two questions. The question is the authenticity of the gene; whether that's a question or not I don't know, but it's certainly one explanation for your data. The second is whether or not the expression of protein implies the activity of that protein, and that I don't know either. There's certainly plenty of precedence in biology for cross-reactive materials which are not active.

George Michalopoulos: Because of the prolific angiogenic response induced by astrocytomas and glioblastomas there was a speculation in the past that a model for their progression may involve fusion of normal endothelial cells with astrocytoma neoplastic cells. Is that model still tenable with your data?

Cavenee: That model is not tenable (at least not with the high frequency in those data) because we would pick up the normal nuclei in this way, but only to the level of resolution of that sort of analysis, which is probably 5-10%. So although it certainly could happen, it doesn't have to be. The other thing that I should say is that Peter Collins spent a lot of time painstakingly pulling out pieces which were as highly proportionate with tumor cells as possible, so there may be a bias in there as well.

ROLE OF TUMOR SUPPRESSOR GENES IN

A MULTISTEP MODEL OF CARCINOGENESIS

Jeff A. Boyd and J. Carl Barrett

Laboratory of Molecular Carcinogenesis
National Institute of Environmental Health Sciences
P. O. Box 12233
Research Triangle Park, NC 27709

Most cancers are clonal in origin, arising from a single aberrant cell[1]. The multistep nature of this process is now a widely accepted paradigm[2], and is supported by several lines of evidence from both clinical oncology and experimental models. Among these are: 1) histopathological observations of human tumors reveal multiple stages of tumor progression such as dysplasia and carcinoma in situ[3]; 2) mathematical models based on age-specific tumor incidence curves are consistent with four to seven independent hits required for tumor formation[4]; 3) studies of genetic predisposition to cancer in some families suggest that one step in the carcinogenic process is a germline mutation and that additional somatic events are required for full neoplastic development[5]; 4) well documented experimental systems, including the two-stage model of chemical carcinogenesis in rodent skin[6,7] and liver[8], show that different chemicals affect qualitatively different stages in the carcinogenic process; 5) cell culture studies reveal that chemical carcinogen-induced neoplastic transformation of normal cells involves a progressive process[9]; 6) cell culture studies with viral and tumor-derived oncogenes show that the neoplastic conversion of normal cells generally requires multiple cooperating oncogenes[10]; and 7) studies on the development of tumors in transgenic mice carrying activated proto-oncogenes in their germline indicate that the tumors which arise are clonal in origin, suggesting that additional somatic events besides activation of two cooperating oncogenes are required for full malignant development[11].

The stages of carcinogenesis have historically been broadly defined as initiation, promotion, and progression. Initiation is generally considered to result from an irreversible genetic alteration, which is insufficient in itself to induce tumorigenesis. In regard to the stages of carcinogenesis after initiation, which are the subject of this symposium, there exists considerably greater disparity in their definition. Promotion may be defined as the clonal expansion of the initiated cells into a benign tumor or preneoplastic focus, usually resulting from reversible, epigenetic phenomena. Progression may be defined as the events necessary for the conversion of the benign or preneoplastic population to the malignant state. These definitions were developed in large part to accommodate the observed actions of certain chemical agents found to affect or induce the various stages. While this model has proven the basis for great advances in the field of experimental carcinogenesis, the limitations of the model are becoming increasingly apparent. Promotion is a phenomenon and may occur by any one of a number of

Boundaries between Promotion and Progression during Carcinogenesis
Edited by O. Sudilovsky *et al.*, Plenum Press, New York, 1991

Table 1. Two Classes of Genes Involved in Carcinogenesis

Proto-oncogenes	Tumor Suppressor Genes
1. Involved in cellular growth and differentiation	1. Function unknown, but possibly involved in cellular growth and differentiation (negative regulators of cell growth?)
2. Family of genes exists	2. Family of genes exists
3. Must be activated (quantitatively or qualitatively) in cancers	3. Must be inactivated or lost in cancers
4. Mutational activation by point mutation, chromosome translocation, or gene amplification	4. Mutational inactivation by chromosome loss, chromosome deletion, point mutation, somatic recombination, or gene conversion
5. Little evidence for involvement in hereditary cancers	5. Clear evidence for involvement in hereditary and non-hereditary cancers

mechanisms, including further genetic alterations of the initiated cell. Furthermore, progression may occur in the absence of any tumor promoting treatment or without the detection of a preneoplastic or benign lesion. Therefore, the potential exists for an overlap between the stages of promotion and progression. We prefer to define progression as a series of qualitative, heritable changes in a subpopulation of initiated cells, resulting in malignancy or an increased potential to progress to malignancy. Tumor promotion, the clonal expansion of initiated cells, would accelerate progression to malignancy in this model. The ensuing discussion will hopefully provide greater insight into the rationale for this type of definition.

Implicit in most multistep models of carcinogenesis is the assumption that multiple genes are involved in the process, and this is supported by experimental findings (vide infra). Only through the identification and characterization of these genes will it be possible to fully define the temporal order, interrelationships, and "boundaries" between the various stages of neoplastic development, if they in fact exist. It will be the purpose of this monograph to describe one such class of genes, i.e. tumor suppressor genes, and to present data from our laboratory which supports the involvement of these genes in the multistage development of cancer.

There are two distinct classes of genes believed to play a role in cancer development (Table 1); the first, oncogenes, were originally discovered through the study of genetic transduction by the acutely transforming retroviruses[12]. Tremendous progress has been made during the last decade in our understanding of oncogenes and their normal cellular counterparts, proto-oncogenes[12-13]. Evidence for links between the products of oncogenes and growth factors, growth factor receptors, and various components of signal transduction pathways is increasingly apparent[12-14], and confirms the hypothesis that oncogenes and proto-oncogenes are involved in the growth control of normal cells and in the abnormal growth control of neoplastic cells.

There is currently no conclusive evidence, however, that the activation of a single cellular oncogene is sufficient to effect the neoplastic transformation of an otherwise normal cell. Studies in which this phenomenon was reported have employed either aneuploid or immortalized rodent cell lines as activated oncogene recipients[15], or have utilized highly altered oncogene constructs[16]. Two cooperating oncogenes have been shown to neoplastically transform normal primary rodent cells[17]; however, in these experiments the possibility of additional genetic changes during clonal expansion and tumor formation still exists. Studies from this laboratory have demonstrated that in early passage Syrian hamster embryo fibroblasts transfected with two cooperating oncogenes, i.e., v-myc and v-Ha-ras, an additional change, the nonrandom loss of one copy of chromosome 15, is required for neoplastic transformation[18,19]. Similar chromosomal changes have also been observed in Chinese hamster embryo fibroblasts, regardless of the transfected gene or method of transformation[20,21]. These and numerous other studies underscore the importance of chromosome losses in neoplastic transformation. The nonrandom loss of genetic material in cancers suggests that normal cellular genes must be lost or inactivated for a tumor to arise.

The essential role of a second family of cellular genes in carcinogenesis, termed tumor suppressor genes, anti-oncogenes, or recessive oncogenes, is becoming increasingly evident. (The term tumor suppressor gene reflects the function of these genes without implying a mechanism, and will therefore be used in this monograph.) Tumor suppressor genes are normal cellular genes that act as negative regulators of tumor cell proliferation in vivo and must be lost or inactivated for neoplastic transformation to occur. In contrast, proto-oncogenes are a distinct class of normal cellular genes that are activated by point mutation, gene amplification, or rearrangement to become oncogenes which act as positive proliferative signals in neoplastic cells.

The existence of tumor suppressor genes is supported by several lines of evidence, which include: 1) suppression of tumorigenicity in cell hybrids; 2) genetic predisposition to cancer in animals and humans; 3) nonrandom chromosome loss or deletion in specific tumor types; 4) loss of heterozygosity of specific chromosomal regions in tumors; and 5) reversion of tumor cells to a non-malignant state by interactions with normal cells, or by treatment of tumor cells with certain differentiation-inducing chemicals or growth factors.

The earliest evidence that normal cells possess genes which may suppress tumorigenicity was provided by studies showing that hybrids between highly tumorigenic and nontumorigenic mouse cells were often nontumorigenic (reviewed in 22). The original studies of this type suggested that the tumorigenic phenotype was dominant[23,24]; however, this was due to the instability of the hybrid cells, and Harris and colleagues later established that hybrids are usually nontumorigenic when they maintain the full chromosome complement of both parental cells[25,26]. Definitive studies by Stanbridge[27] using normal human fibroblasts and human cervical carcinoma (HeLa) cells showed that normal cells can suppress the tumorigenicity of highly malignant cells. The non-tumorigenic hybrids produced in these experiments were extremely stable[28], but eventually variant hybrids reexpressed tumorigenicity which was shown to correlate with the loss of one copy each of chromosomes 11 and 14[29]. Likewise, the reversion to tumorigenicity of hybrids between normal human fibroblasts and tumorigenic Chinese hamster ovary cells was shown by Klinger and Shows to correlate with the loss of specific human chromosomes[30]. The presence of a presumed tumor suppressor gene(s) on human chromosome 11 was elegantly demonstrated recently by Stanbridge, Weissman, and colleagues by the introduction of this chromosome via microcell transfer into HeLa and Wilms' tumor cells[31,32]. These were the first experiments to demonstrate that a single human chromosome is capable of completely suppressing the tumorigenic phenotype. These studies have now been extended to other tumor cells (Koi and Oshimura, unpublished data).

The tendency for certain cancers to cluster in families has been recognized for over a century, and at least 50 dominantly inherited cancers have now been identified[33]. The prototypical tumors of this class generally manifest in early childhood and include retinoblastoma and Wilms' tumor, the former being most thoroughly studied (reviewed in 5). The pioneering work of Knudson[34-36] led to more recent molecular studies which established the hypothesis that a germline mutation in one allele of a tumor suppressor gene constitutes the genetic predisposition to these cancers. The locus of the inherited mutation frequently becomes homozygous or hemizygous through a secondary somatic event, either nondisjunction leading to elimination of the normal chromosome, recombination, chromosome deletion, or point mutation[37]. The implication of these findings is that cancer is suppressed unless a mutation and/or deletion arises in both alleles of the suppressor gene. It is important to note, however, that retinoblastoma tumors may also contain other critical genetic changes elsewhere in the genome, for example, oncogene amplifications[38,39], consistent with the concept that loss of a tumor suppressor gene is only one of multiple genetic events essential for tumorigenesis[40].

The retinoblastoma gene (Rb) has now been cloned by at least three groups[41-43], and is located on chromosome 13q14. The Rb gene is over 200 kb in size and contains at least 20 exons; a portion or all of the gene is deleted in several retinoblastomas and osteosarcomas. The corresponding 4.7 kb mRNA was originally reported to be altered in size or absent from many retinoblastoma and osteosarcoma cell lines[41-43], but a more recent study by Gallie and coworkers[44] concluded that a normal sized transcript was present in a majority of retinoblastoma tumors and cell lines tested. These workers utilized an RNase protection assay to demonstrate that several retinoblastoma tumors contain point mutations in the Rb gene[45]. The predicted protein of 816 amino acids[42,46] is unrelated to other known proteins, but contains a potential metal-binding domain similar to that found in several nucleic acid-binding proteins. This putative Rb gene product is a nuclear phosphoprotein, possesses DNA-binding activity, and is expressed in many normal cells but not those from retinoblastoma tumors, suggesting that the Rb protein may be involved in the regulation of gene expression in normal cells[46]. Consistent with the role of the Rb gene in suppression of other tumor types are recent reports describing the structural rearrangement of the Rb gene in human breast carcinoma[47] and small cell lung carcinoma[48].

Additional evidence for the existence of tumor suppressor genes comes from the study of nonrandom chromosome deletions or losses in specific types of tumors[18,49,50]. The development of molecular probes for specific chromosomal regions which reveal restriction fragment length polymorphisms (RFLP) on Southern blot analyses have been used to demonstrate that heterozygous alleles on specific chromosomes frequently become homozygous or hemizygous in certain human tumors[5]. RFLP analysis has been used to study the loss of tumor suppressor genes in hereditary cancers such as retinoblastoma and Wilms' tumor[5,37,40] as well as a number of adult onset cancers including lung cancer[49], breast cancer[50], and colorectal cancer[51]. The locations of these genes encoding possible tumor suppressor activity for each tumor type are on different chromosomes, implying that a family of tumor suppressor genes exists. The number of tumor suppressor genes in this family is currently unknown, but studies of the above type have localized potential human tumor suppressor genes to at least eight different chromosomes.

The mechanism of action of tumor suppressor genes is also currently unknown, but several possibilities exist[52]; among these are: 1) regulation of oncogene expression or function; 2) negative regulation of cell growth through transcriptional and post-transcriptional control of normal gene expression; 3) induction of terminal differentiation[53,54] or cellular senescence[55-57]; or 4) production of growth regulatory substances[58,59]. As is the case for oncogene function, it is very likely that members of the tumor suppressor gene family will exhibit different mechanisms of action.

Research in our laboratory has taken several approaches toward the identification, cloning, and elucidation of the mechanism of action of tumor suppressor genes. The experimental system employed consists of two Syrian hamster embryo (SHE) cell lines, designated 10W and DES4, immortalized by treatment with chemical carcinogens, asbestos or diethylstilbestrol, respectively. These cell lines at early passage, like normal diploid SHE cells, have the ability to suppress anchorage independence and tumorigenicity of a highly tumorigenic, benzo(a)pyrene-transformed cell line (BP6T) when the two cell types are fused together[55]. At later passage, however, before the immortal cell lines acquire anchorage independence or tumorigenicity, the cells lose the ability to suppress the transformed phenotype in cell hybridization experiments. Loss of this tumor suppressive function also results in an increased susceptibility of the immortal cell lines to transformation following transfection with oncogene or tumor cell DNA. These results suggest that chemically induced neoplastic progression of SHE cells involves at least three steps: 1) induction of immortality, 2) loss of tumor suppressive function, and 3) activation of a transforming oncogene[55].

A comparative analysis of immortal, nontumorigenic SHE cell clones which have either retained (sup$^+$) or lost (sup$^-$) the tumor suppressor phenotype (see Fig. 1) should yield insight into the properties and function of the relevant tumor suppressor gene(s); our current strategy is directed toward both molecular cloning of the gene and analysis of the cellular phenotypes controlled by the gene. The screening of cDNAs for differential expression has proven a powerful tool for the identification of genes for a specific phenotype, and the application of this methodology to the sup$^+$/sup$^-$ cells has revealed at least three genes that are differentially expressed (Cizdziel

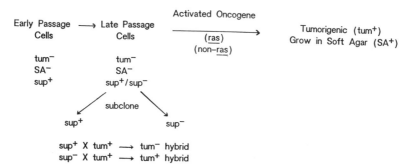

Neoplastic Progression of Immortal, Syrian Hamster Embryo Cells

Fig. 1. SHE fibroblasts were immortalized by treatment with the chemical carcinogens diethylstilbestrol or asbestos. At early passage, these immortalized cells were nontumorigenic, failed to produce colonies in soft agar, and retained the tumor suppressor phenotype as determined in cell hybridization studies with BP6T tumor cells. At later passage, some cells lost the abililty to suppress tumorigenicity in cell hybrids with BP6T tumor cells; the variants were cloned and designated sup$^+$ or sup$^-$. Both sup$^+$ and sup$^-$ clones were still nontumorigenic and anchorage-dependent. Further activation of oncogenes in sup$^-$ cells resulted in the completely tumorigenic phenotype, and ability to form colonies in soft agar.

and Hosoi, unpublished data). Two of these cDNAs have been sequenced, and were found to correspond to the collagen type II and type IX genes. These genes, which are chondrocyte differentiation markers, are expressed in the sup[+] cells but not in the sup[-] cells. The significance of this finding is unclear at present, and the remaining cDNA is being sequenced.

A similar comparative approach was used at the protein level in order to identify differentially expressed cellular gene products[60]. Total protein was isolated from sup[+] and sup[-] cell lines and subjected to quantitative two-dimensional gel analysis, utilizing a SHE cell protein database to compare the levels of over 1000 proteins. A six-fold reduction of tropomyosin-1 was observed in the sup[-] variants relative to the sup[+] clones or normal SHE cells. Tropomyosins are abundant cellular proteins, and at least six isoforms have been identified in mammalian cells[61]. The role of tropomyosins in skeletal muscle cells is to mediate the effects of calcium on the actin-myosin interaction in muscle contraction[62], and although tropomyosin is a major structural component of cytoskeletal microfilaments in nonmuscle cells, its function in these cells is still largely unknown[63]. The suppression of high molecular weight isoforms of tropomyosin has been observed in fibroblasts transformed by a variety of oncogenes[64]. Since these isoforms demonstrate the highest binding affinity to actin, it has been proposed that their decreased synthesis leads to destabilization of microfilaments, resulting in the cytoskeletal alterations commonly associated with the transformed phenotype[65].

We have pursued this hypothesis and studied the organization of actin microfilaments in sup[+] and sup[-] cell lines, using fluorescence microscopy[66]. The sup[-] clones, like tumorigenic BP6T cells, invariably demonstrate a low number of actin stress fibers, while the sup[+] clones, like normal SHE cells, consistently exhibit many ordered, well-defined actin stress fibers (Fig. 2). Molecular data indicate that actin mRNA as well as actin protein levels are similar in both sup[+] and sup[-] variants, consistent with differences in actin protein organization rather than decreased transcription or translation. Taken together, we believe that the above data are consistent with the hypothesis that decreased synthesis of high molecular weight isoforms of tropomyosin, resulting in destabilization of the actin cytoskeleton, is an early event in the chemical transformation of SHE cells, correlating with loss of the tumor suppressor phenotype. Interestingly, several previous studies have implied an association between the tumor suppressor phenotype and actin organization. In normal fibroblasts from patients with inherited colonic

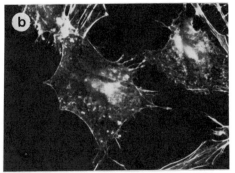

Fig. 2. Comparison of actin microfilament organization in chemically immortalized SHE cell variants which have retained (A, DES4sup[+]) or lost (B, DES4sup[-]) the tumor suppressor phenotype. Fixed cells were stained with rhodamine-phalloidin and photographed under a fluorescent light source at 630x.

Table 2. Effect of Growth Surface on Growth Factor Stimulation
of C-fos and C-myc Transcription in Quiescent SHE Cell
Derivatives

Cell Line	Plastic		Agar	
	c-fos	c-myc	c-fos	c-myc
Normal SHE	+	+	+	-
10Wsup$^+$	+	+	+	-
10Wsup$^-$	+	+	+	+
DES4sup$^+$	+	+	+	-
DES4sup$^-$	+	+	+	+
BP6T	-	+	-	+

Growth factors: insulin, 1 μg/ml; EGF, 50 ng/ml; PDGF, 5 ng/ml
+ = transcriptional activation
- = no discernable transcriptional activation

cancer, actin organization is disrupted[67], and the actin protein turnover
rate is 10-fold higher in normal fibroblasts from patients with dominantly
inherited retinoblastoma, polyposis coli, or nevoid basal cell carcinoma[68].
Furthermore, actin organization was found to correlate with the suppression
of the anchorage-independent phenotype of cell hybrids between chemically-
transformed baby hamster kidney cells and normal human fibroblasts; angiogenesis and the presence of human chromosome 1 also correlated[69].

We have also shown that loss of tumor suppressor gene function in the
SHE cell system is correlated with enhanced growth factor responsiveness of
sup$^-$ cells in agar[70]. In initial studies, suppressed cell hybrids were
formed between a tumor cell line (BP6T) which produces autocrine growth factors and normal SHE cells which do not. One such hybrid still produced these
growth factors, but was unable to respond to them with growth in soft agar.
This led to the observation that the ability of immortal hamster cell clones
to suppress tumorigenicity correlated with their response in agar to BP6T
cell-conditioned media. The cosegregation of these two phenotypes in further
subcloning experiments strongly suggested that they are interrelated. The
ability of defined growth factors alone or in combination to replace BP6T-
conditioned medium was then examined. Although no single growth factor alone
was effective, various combinations of insulin, EGF, and PDGF stimulated
colony formation of sup$^-$ but not sup$^+$ cell clones in soft agar. The number
of growth factor receptors as well as their affinity for these growth factors
was similar in both cell types. These findings suggest that this tumor suppressor phenotype is involved in negatively regulating the responsiveness of
cells under selective conditions to growth factors.

In order to investigate further the possible relationships between loss
of tumor suppressor function, response to growth factors, and cytoskeletal
and cell shape alterations, we examined various aspects of signal transduction and cell growth of sup$^+$ and sup$^-$ cells on different growth surfaces[71].
The purpose of these experiments was to examine the effects of cell shape and
attachment on the ability of the cells to respond to growth factors. The
response of cells either attached and spread on a plastic substratum or sus-
pended on top of a solid agar substratum to growth factor induced DNA synthesis, cell growth, and the transcriptional activation of the "early-immediate"
genes c-fos and c-myc was measured. There was little difference between
sup$^+$ and sup$^-$ clones in either DNA synthesis or cell growth in response to
growth factors, when the cells were cultured on a plastic substratum. In
contrast, when the cells were cultured in suspension on an agar surface, significant differences were observed between sup$^+$ and sup$^-$ lines in

Table 3. Similarities Between sup$^+$ and sup$^-$ Variants

---

1. Nontumorigenicity and anchorage-dependent growth
2. Growth rates and colony forming efficiencies on plastic
3. Growth factor response of cells on plastic
4. Growth factor receptors (number and affinity)
5. Tropomyosin IV levels
6. Amounts of actin mRNA and protein
7. Induction by growth factors of c-myc and c-fos expression on plastic
8. Induction by growth factors of c-fos expression on agar

---

regard to cell growth and DNA synthesis in response to growth factors. Likewise, northern blot analyses demonstrated that when the cells were cultured on a plastic surface, growth factors stimulated the transcriptional activation of the c-myc and c-fos genes identically in quiescent normal SHE cells, sup$^+$ clones, and sup$^-$ clones. In contrast, growth factor stimulation of quiescent cells on an agar surface resulted in c-fos transcription in all cell lines, but c-myc transcription was significantly induced only in sup$^-$ variants, and not in sup$^+$ clones and normal SHE cells (Table 2).

In summary, the findings presented above (summarized in Tables 3 and 4) suggest that the loss of tumor suppressor gene function in two chemically transformed, preneoplastic SHE cell lines results in the decreased transcription of several genes, including tropomyosin-1, collagen types II and IX, and an unknown gene. One hypothesis is that the tumor suppressor gene is involved in the transcriptional regulation of these other genes. This is currently being tested. We propose furthermore that the reduced levels of tropomyosin-1 result in decreased formation of actin microfilaments in sup$^-$ cells; changes in cell shape by cytoskeletal alterations may allow these cells to respond to growth factor stimulation independent of cell shape. The altered signal transduction in sup$^-$ cells appears to involve transcriptional activation of the c-myc but not the c-fos gene. This hypothesis is consistent with the paradigm that certain tumor suppressor genes may be negative regulators of tumor growth by controlling gene expression.

Based on the preceeding discussion, it is obvious that the loss of tumor suppressor gene function may occur at various points in the process of neoplastic development, and may therefore be relevant to any of several stages of carcinogenesis. As an early/predisposing event in tumor development, tumor suppressor gene loss may be considered an initiating event, for example, in genetic predispositions to childhood cancers. On the other hand, altered expression of tumor suppressor genes and other cellular genes by chemicals may play a role in tumor promotion. Finally, loss of suppressor gene function subsequent to the activation of oncogenes could occur during

Table 4. Characteristics of sup$^-$ Phenotype

---

1. Loss of ability to suppress tumorigenicity of a chemically-transformed cell line (BP6T) in cell hybrids
2. Enhanced growth response to growth factors in soft agar
3. Decreased expression of tropomyosin I
4. Disorganized actin microfilaments
5. Decreased expression of collagen types II and IX
6. Decreased expression of an anonymous cDNA
7. Increased growth factor induction of c-myc expression on agar

---

tumor progression. In any case, the identification and cloning of specific tumor suppressor genes is required to define the temporal relationships of tumor suppressor gene loss to multistage carcinogenesis.

REFERENCES

1. P. C. Nowell, The clonal evolution of tumor cell populations, Science 194:23-28 (1976).
2. J. C. Barrett, A multistep model for neoplastic development: Role of genetic and epigenetic changes, in: J. C. Barrett, ed., "Mechanisms of Environmental Carcinogenesis, Vol. II, Multistep Models of Carcinogenesis," CRC Press, Boca Raton (1987).
3. L. Foulds, "Neoplastic Development," Academic Press, New York (1975).
4. R. Peto, Epidemiology, multistage models, and short-term mutagenicity tests. in: "Origins of Human Cancer," H. H. Hiatt, J. D. Watson, and J. A. Winsten, eds., Cold Spring Harbor Laboratory, Cold Spring Harbor, NY (1977).
5. M. F. Hansen and W. K. Cavenee, Genetics of cancer predisposition, Cancer Res. 47:5518-5527 (1987).
6. R. K. Boutwell, The function and mechanism of promoters of carcinogenesis, CRC Crit. Rev. Toxicol. 2:419-443 (1974).
7. T. J. Slaga, S. M. Fischer, K. Nelson, and G. L. Gleason, Studies on the mechanism of skin tumor promotion: evidence for several stages in promotion, Proc. Natl. Acad. Sci. USA 77:3659-3663 (1980).
8. E. Farber, The multistep nature of cancer development, Cancer Res. 44:4217-4223 (1984).
9. J. C. Barrett and W. F. Fletcher, Cellular and molecular mechanisms of multistep carcinogenesis in cell culture models, in: "Mechanisms of Environmental Carcinogenesis, Vol. II, Multistep Models of Carcinogenesis," J. C. Barrett, ed., CRC Press, Boca Raton (1987).
10. D. G. Thomassen, T. G. Gilmer, L. A. Annab, and J. C. Barrett, Evidence for multiple steps in neoplastic transformation of normal and preneoplastic Syrian hamster embryo cells following transfection with Harvey murine sarcoma virus oncogene (v-Ha-ras), Cancer Res. 45:726-732 (1985).
11. E. Sinn, W. Muller, P. Pattengale, I. Tepler, R. Wallace, and P. Leder, Coexpression of MMTV/v-Ha-ras and MMTV/c-myc genes in transgenic mice: synergistic action of oncogenes in vivo, Cell 49:465-475 (1987).
12. J. M. Bishop, The molecular genetics of cancer, Science 235:305-311 (1987).
13. R. A. Weinberg, The action of oncogenes in the cytoplasm and nucleus, Science 230:770-776 (1985).
14. M. B. Sporn and A. B. Roberts, Autocrine growth factors and cancer, Nature 313:745-750 (1985).
15. L. F. Parada, C. J. Tabin, C. Shih, and R. A. Weinberg, Human EJ bladder carcinoma oncogene is homologue of Harvey sarcoma virus ras gene, Nature 297:474-478 (1982).
16. D. A. Spandidos and N. M. Wilkie, Malignant transformation of early passage rodent cells by a single mutated human oncogene, Nature 310:469-475 (1984).
17. H. Land, L. F. Parada, and R. A. Weinberg, Tumorigenic conversion of primary embryo fibroblasts requires at least two cooperating oncogenes, Nature 304:596-602 (1983).
18. M. Oshimura, T. Gilmer, and J. C. Barrett, Nonrandom loss of chromosome 15 in Syrian hamster tumors induced by v-Ha-ras plus v-myc oncogenes, Nature 316:636-639 (1985).
19. M. Oshimura, M. Koi, N. Ozawa, O. Sugawara, P. W. Lamb, and J. C. Barrett, Role of chromosome loss in ras/myc-induced Syrian hamster tumors, Cancer Res. 48:1623-1632 (1988).

20. R. M. Kitchen and R. Sager, Genetic analysis of tumorigenesis: VI. Chromosome rearrangements in tumors derived from diploid premalignant Chinese hamster cells in nude mice, Somat. Cell Genet. 6:615-629 (1980).

21. C. C. Lau, I. K. Gadi, S. Kalvonjian, A. Anisowicz, and R. Sager, Plasmid-induced "hit and run" tumorigenesis in Chinese hamster embryo fibroblast (CHEF) cells, Proc. Natl. Acad. Sci. USA 82:2839-2843 (1985).

22. B. E. Weissman, Suppression of tumorigenicity in mammalian cell hybrids, in: "Mechanisms of Environmental Carcinogenesis, Vol. I, Epigenetic Changes," J. C. Barrett, ed. CRC Press, Boca Raton (1987).

23. B. Ephrussi, "Hybridization of Somatic Cells," Princeton University Press, Princeton (1972).

24. G. Barski, S. Sorieul, Fr. Cornefurt, Production dans des cultures in vitro de deux souches cellulaires en association de cellules de caractere "hybrid", C. R. Acad. Sci. 251:1825-1827 (1960).

25. H. Harris, O. J. Miller, G. Klein, P. Worst, and T. Tachibana, Suppression of malignancy by cell fusion, Nature 223:363-368 (1969).

26. H. Harris, Cell fusion and the analysis of malignancy, Proc. Roy. Soc. Lond. B. Biol. Sci. 179:1-20 (1971).

27. E. J. Stanbridge, Suppression of malignancy in human cells, Nature 260:17-21 (1976).

28. E. J. Stanbridge, C. J. Der, C. J. Doersen, R. Y. Nishimi, D. M. Peehl, B. E. Weissman, and J. Wilkinson, Human cell hybrids: analysis of transformation and tumorigenicity, Science 215:252-259 (1982).

29. E. J. Stanbridge, R. R. Flandermyer, F. W. Daniels, and W. A. Nelson-Rees, Specific chromosome loss associated with the expression of tumorigenicity in human cell hybrids, Somat. Cell Genet. 7:699-712 (1981).

30. H. P. Klinger and T. B. Shows, Suppression of tumorigenicity in somatic cell hybrids, II. Human chromosomes implicated as suppressors of tumorigenicity in hybrids with Chinese hamster ovary cells, J. Nat. Cancer Inst. 71:559-569 (1983).

31. P. J. Saxon, E. S. Srivarsan, and E. J. Stanbridge, Introduction of human chromosome 11 via microcell transfer controls tumorigenic expression of HeLa cells, EMBO J. 5:3461-3466 (1986).

32. B. E. Weissman, P. J. Saxon, S. R. Pasquale, G. R. Jones, A. G. Geiser, and E. J. Stanbridge, Introduction of a normal human chromosome 11 into a Wilms' tumor cell line controls its tumorigenic expression, Science 236:175-180 (1987).

33. A. G. Knudson, Hereditary cancer, oncogenes, and antioncogenes, Cancer Res. 45:1437-1443 (1985).

34. A. G. Knudson, Mutation and cancer: statistical study of retinoblastoma, Proc. Natl. Acad. Sci. USA 68:820-823 (1971).

35. A. G. Knudson, Retinoblastoma: a prototypic hereditary neoplasm, Semin. Oncol. 5:57-60 (1978).

36. A. G. Knudson, Model hereditary cancers of man, Prog. Nucl. Acids Res. Mol. Biol. 29:17-25 (1983).

37. W. K. Cavenee, T. P. Dryja, R. A. Phillips, W. F. Benedict, R. Godbout, B. L. Gallie, A. L. Murphree, L. C. Strong, and R. L. White, Expression of recessive alleles by chromosomal mechanisms in retinoblastoma, Nature 305:779-784 (1983).

38. W. G. Benedict, A. Banerjee, C. Mark, and A. L. Murphree, Nonrandom chromosomal changes in untreated retinoblastomas, Cancer Genet. Cytogenet. 10:311-333 (1983).

39. E. Chaum, R. M. Ellsworth, D. H. Abramson, B. G. Haik, F. D. Kitchin, and R. S. K. Chaganti, Cytogenetic analysis of retinoblastoma: evidence for multifocal origin and in vivo gene amplification, Cytogenet. Cell Genet. 38:82-91 (1984).

40. A. L. Murphree and W. F. Benedict, Retinoblastoma: clues to human oncogenesis, Science 223:1028-1033 (1984).

41. S. H. Friend, R. Bernards, S. Rogelj, R. A. Weinberg, J. M. Rapaport, D. M. Albert, and T. P. Dryja, A human DNA segment with properties of the gene that predisposes to retinoblastoma and osteosarcoma, Nature 323:643-646 (1986).

42. W.-H. Lee, R. Bookstein, F. Hong, L.-J. Young, J.-Y. Shew, and E. Y.-H. P. Lee, Human retinoblastoma susceptibility gene: cloning, identification, and sequence, Science 235:1394-1399 (1987).

43. A. D. Goddard, H. Balakier, M. Canton, J. Dunn, J. Squire, E. Reyes, A. Becker, R. A. Phillips, and B. L. Gallie, Infrequent genomic rearrangement and normal expression of the putative RB1 gene in retinoblastoma tumors, Molec. Cell. Biol. 8:2082-2088 (1988).

45. J. L. Dunn, R. A. Phillips, A. J. Becker, and B. L. Gallie, Identification of germline and somatic mutations affecting the retinoblastoma gene, Science 241:1797-1800 (1988).

46. W.-H. Lee, J.-Y. Shew, F. D. Hong, T. W. Sery, L. A. Donoso, L.-J. Young, R. Bookstein, and E.Y.-H.P. Lee, The retinoblastoma susceptibility gene encodes a nuclear phosphoprotein associated with DNA binding activity, Nature 329:642-645 (1987).

47. A. T'Ang, J. M. Varley, S. Chakraborty, A. L. Murphree, and Y.-K.T. Fung, Structural rearrangement of the retinoblastoma gene in human breast carcinoma, Science 242:263-266 (1988).

48. J. W. Harbour, S.-L. Lai, J. Whang-Peng, A. D. Gazdar, J. D. Minna, and F. J. Kaye, Abnormalities in structure and expression of the human retinoblastoma gene in SCLC, Science 241:353-357 (1988).

49. J. Yokota, M. Wada, Y. Shimasato, M. Terada, and T. Sugimura, Loss of heterozygosity on chromosomes 3, 13, and 17 in small-cell carcinoma and on chromosome 3 of adenocarcinoma of the lung, Proc. Natl. Acad. Sci. USA 84:9252-9256 (1987).

50. I. U. Ali, R. Lidereau, C. Theillet, and R. Callahan, Reduction to homozygosity of genes on chromosome 11 in human breast neoplasia, Science 238:185-188 (1987).

51. D. J. Law, S. Olschwang, J.-P. Monpezat, D. Lefrancois, D. Jagelman, N. J. Petrelli, G. Thomas, and A. P. Feinberg, Concerted nonsystemic allelic loss in human colorectal carcinoma, Science 241:961-965 (1987).

52. R. Sager, Genetic suppression of tumor formation: A new frontier in cancer research, Cancer Res. 46:1573-1580 (1986).

53. D. M. Peehl and E. J. Stanbridge, Characterization of human keratinocyte x HeLa somatic cell hybrids, Int. J. Cancer 27:625-635 (1981).

54. H. Harris and M. E. Bramwell, The suppression of malignancy by terminal differentiation: evidence from hybrids between tumor cells and keratinocytes, J. Cell Sci. 87:383-388 (1987).

55. M. Koi and J. C. Barrett, Loss of tumor suppression function during chemically induced neoplastic progression of Syrian hamster embryo cells, Proc. Natl. Acad. Sci. USA 83:5992-5996 (1986).

56. O. M. Pereira-Smith and J. R. Smith, Evidence for the recessive nature of cellular immortality, Science 221:963-966 (1983).

57. C. K. Lumpkin, J. K. McClung, O. M. Pereira-Smith, and J. R. Smith, Existence of high abundance antiproliferative mRNA's in senescent human diploid fibroblasts, Science 232:393-395 (1986).

58. J. Keski-Oja and H. L. Moses, Growth inhibitory peptides in the regulation of cell proliferation, Med. Biol. 65:13-20 (1987).

59. C. Knabbe, M. E. Lippman, L. M. Wakefield, K. C. Flanders, A. Kasid, R. Derynck, and R. B. Dickson, Evidence that transforming growth factor-beta is a hormonally regulated negative growth factor in human breast cancer cells, Cell 48:417-428 (1987).

60. R. W. Wiseman, M. E. Lambert, P. W. Lamb, J. I. Garrels, and J. C. Barrett, Alterations in gene expression at various stages of neoplastic transformation of Syrian hamster embryo (SHE) cells, J. Cell. Biochem. Supplement 12A:219 (1988).

61. F. Matsumura and S. Yamashiro-Matsumura, Purification and characterization of multiple isoforms of tropomyosin from rat cultured cells, J. Biol. Chem. 260:13851-13859 (1985).

62. L. B. Smillie, Structure and functions of tropomyosins from non-muscle sources, Trends Biochem. Sci. 4:151-155 (1979).

63. G. P. Cote, Structural and functional properties of non-muscle tropomyosins, Molec. Cell. Biochem. 57:127-146 (1983).

64. H. L. Cooper, N. Feuerstein, M. Noda, and R. H. Bassin, Suppression of tropomyosin synthesis, a common biochemical feature of oncogenesis by structurally diverse retroviral oncogenes, Molec. Cell. Biol. 5:972-983 (1985).

65. F. Matsumura and S. Yamashiro-Matsumura, Tropomyosin in cell transformation, Cancer Rev. 6:21-39 (1986).

66. J. C. Barrett, R. W. Wiseman, and J. A. Boyd, Ultrastructural changes in actin microfilaments which correlate with loss of the tumor suppressor phenotype in Syrian hamster embryo (SHE) cells, J. Cell Biol. 107:684a (1988).

67. L. Kopelovich, S. Conlon, and R. Pollack, Defective organization of actin in cultured skin fibroblasts from patients with inherited adenocarcinoma, Proc. Natl. Acad. Sci. USA 74:3019-3022 (1977).

68. M. H. Antecol, A. Darveau, N. Sonenberg, and B. B. Mukherjee, Altered biochemical properties of actin in normal skin fibroblasts from individuals predisposed to dominantly inherited cancers, Cancer Res. 46:1867-1873 (1986).

69. N. Bouck, A. Stoler, and P. J. Polverini, Coordinate control of anchorage independence, actin cytoskeleton, and angiogenesis by human chromosome 1 in hamster-human hybrids, Cancer Res. 46:5101-5105 (1986).

70. M. Koi, C. A. Jones, L. A. Annab, and J. C. Barrett, Loss of a tumor suppressor gene function results in enhanced growth responsiveness of hamster cells in agar, Submitted for publication (1989).

71. J. A. Boyd, C. A. Jones, P. A. Futreal, and J. C. Barrett, Alterations in signal transduction which correlates with loss of the tumor suppressor phenotype in Syrian hamster embryo (SHE) cells, J. Cell Biol. 107:264a (1988).

DISCUSSION

Moderator (Lance A. Liotta): I would like to begin with just a few questions. What do you know about the karyotypic differences between the suppressor⁻ and ⁺ cells? Do you get metastases at the end of your multistage progression? Are these cells metastatic in animals? And finally, I was puzzled by the role of type 2 collagen, the cartilage related collagen. How would that have anything to do with the tumor formation in this system?

Jeff A. Boyd: I'll start from the end. We have no idea what collagen 2 means in this system. My personal bias is that it's a somewhat artifactual result obtained due to the nature of these proteins. They are among the most abundant proteins in cells, and it may be that we're looking at a result based simply on the most abundant cDNAs present. Some of the more relevant genes we just haven't detected by this cDNA hybridization technology. As far as metastasis, we don't know. We have of course shown that they are tumorigenic in syngeneic hosts, but we haven't looked for metastases yet. In regards to karyotyping the cells, there are no differences at the gross karyotypic level, as far as number of chromosomes or gross deletions or rearrangements that we can detect among them. Both types of cells are aneuploid, but as far as we can tell, at the gross karyotypic level there are no differences.

Michael Lieberman: Not a question for Dr. Boyd but a more general comment, picking up his question about screening of cDNA libraries. In fact, as I've watched what has gone on this morning and a whole number of other sets of experiments, one of the problems with different libraries and differential screening is that in fact you pick up the most abundant molecules.

Boyd: I agree completely.

Lieberman: And, if you think about what Tim Bowden found with the actin and myelin and myosine and what you found, one of the problems is that you can look at something like 1,000 proteins, while the cell makes 10,000. What we really want to know is what are the differences in regulatory molecules that control cell behavior, and those are molecules of low abundance. So this technique is a very difficult one to try in order to get the answers to the questions we'd all like to know.

Boyd: Yes, that's why I didn't spend much time on it. We're not putting very much stock in this technology.

G. Tim Bowden: Through the technology of subtractive hybridization one can get to levels of expression of .001 percent, using either phenol emulsion hybridization or a new technique of avidin-biotin separation. So it is possible, using subtractive hybridization, to get to very low levels of expression, which might include regulatory molecules.

George Michalopoulous: Probably one of the best characterized protein suppressors of epithelial cell growth is TGF beta and the different types of TGF beta 1, 2, 3 and I think it's up to 5 at this point, as far as I know. Do any of these forms of TGF beta have any effect in your system?

Boyd: We've currently got several graduate students totally devoted to just that question. We're screening virtually every growth factor we can get our hands on and looking for their effects on those clones, but we don't have any data as yet to bear on that question.

Bowden: I was quite interested in the decreased expression of tropomyosin-1. Have you looked at the turnover rates of actin itself in the sup$^-$ and sup$^+$?

Boyd: That's the experiment I intend to do as soon as I can. I think that's going to be the critical factor. I don't think there's going to be a difference in transcription of actin. There's not going to be a difference at the cDNA level. I think the lack of tropomyosin-1 is going to result in an increased turnover in actin which one particular study has in fact shown, in some human fibroblasts from these predisposed individuals.

Meenhard Herlyn: Besides the response to growth factor, one also might look at how independent are they from those factors. So, what are the potential autocrine growth factors?

Boyd: We are also addressing that question now. This data is all very new and very recent and we just have not had time to do a lot of the obvious things that I'm sure you are all wondering about.

Liotta: Do they show a differential response in terms of growing in serum-free media or media without growth factors?

Boyd: None of these clones will grow in FBS deficient medium, and we're currently trying to develop a serum-free system for addressing that type of question.

Unidentified speaker: I'm sure that you are aware that the hamster is not a very sensitive species in detecting chemical carcinogens. In view of your talk, could you comment on what might be the reasons? Is there any reason to explain why the hamster is not very sensitive?

Moderator: Why did you use the hamster and why do you think it is less sensitive to carcinogenesis than mice?

Boyd: Carl Barrett started using this SHE system back in his post-doc days with Paul T'so at Johns-Hopkins and I'm not sure why they chose it. I can't answer why this particular cell type might be less sensitive.

Unidentified speaker: Aren't Syrian hamsters tetraploid?

Unidentified speaker: I believe they are, so that may explain why they're more resistant to chemical carcinogens.

Boyd: I wasn't aware of that.

Moderator (Lance A. Liotta): I don't think so either. Does anyone know the answer to that?

Oscar Sudilovsky: Cells in the golden hamster are diploid.

CANCER GENES BY NON-HOMOLOGOUS RECOMBINATION*

Peter H. Duesberg, David Goodrich
and Ren-Ping Zhou

Department of Molecular Biology
University of California
Berkeley, CA 94720

INTRODUCTION

The only proven cancer genes to date are the onc genes of directly transforming retroviruses[1-4]. These are autonomous transforming genes because they transform diploid cells in culture with single hit kinetics, and because all susceptible cells become transformed as soon as they are infected. Accordingly, tumors induced by such viruses in animals are all polyclonal. Such viruses have never been found in healthy animals, a statement that cannot be made for retroviruses without onc genes or DNA tumor viruses, which are commonly found in animals outside the laboratory and only transform cells indirectly and inefficiently[5-7].

However, retroviruses with onc genes play a very minor role as natural carcinogens[2]. This is because these viruses have an extremely low birth rate, less than one hundred have been isolated in over 80 years of tumor virus research[6], and because the transforming genes of these viruses are very unstable, having half-lives of only a few replicative cycles[8-10]. In fact, spontaneous deletion of the src gene of Rous sarcoma virus (RSV) was the original basis for the discovery of onc genes[8,11]. The short half-life of retroviral onc genes is also the reason that such viruses have never caused epidemics of cancer, and hence have never played a major role as natural carcinogens[2]. The unstable element of retroviral onc genes is a sequence that is not essential for the retrovirus, and that was originally termed transformation-specific[12]. Although this sequence is not really less stable than the essential retrovirus genes[9], deletion mutants of nonessential sequences remain viable and hence readily detectable[8,13]. By contrast, deletions of essential retrovirus genes are not detectable because they do not survive unless complemented by a helper virus[6,13].

Since the transformation-specific sequences of retroviruses have very short half-lives, they must be recent genetic acquisitions. Indeed, the most plausible source of nonessential genetic information for viruses is the host cell, from which viruses can transduce genetic information, as was originally demonstrated for the bacterial virus, lambda[14]. The cellular origin of transformation-specific sequences was first confirmed in the 70's for the transforming ras genes of Harvey and Kirsten sarcoma viruses, using liquid

*P.H.D. is supported by an Outstanding Investigator Grant #5-R35-CA39915-03 from the National Cancer Institute and grant #1547AR1 from the Council for Tobacco Research.

Boundaries between Promotion and Progression during Carcinogenesis
Edited by O. Sudilovsky et al., Plenum Press, New York, 1991

hybridization between viral cDNA sequences and cellular DNA[15,16]. This technique demonstrated that all transformation-specific sequences of retroviral onc genes were derived from one or more cellular genes, which have since been termed proto-onc genes[1,2].

Liquid hybridization is adequate to compare sequences, but is insufficient to compare genetic structures or functional homologies of genes that share homologous sequences. The genetic structure of the proto-onc genes could only be compared to viral onc genes much later, since molecular cloning and nucleic acid sequencing had not yet been discovered or licensed for work on oncogenic viruses. Nevertheless, the hybrid nature of some viral onc genes, which include retrovirus-derived and proto-onc-derived coding regions, like the delta-gag-myc gene of MC29 virus, provided the first clues that viral onc genes and proto-onc genes were not isogenic[17]. This was confirmed later when the corresponding proto-myc gene had been cloned and sequenced (see below)[18].

THE ONCOGENE CONCEPT POSTULATES CANCER VIA ACTIVATION OF LATENT CELLULAR ONCOGENES

The Oncogene Concept

On the basis of the homology between the transformation-specific sequences of retroviruses and cellular proto-onc genes, it was proposed that retroviral onc genes are cellular oncogenes (c-onc) transduced from the cell[6,19-21], analogous to the complete, bacterial genes transduced by bacterial viruses[14]. Thus, the retroviruses were viewed as the lambda phages of eukaryotes. This proposal, later termed the oncogene concept[22], predicted latent cancer genes in normal cells, hence termed "enemies within"[19]. The proponents of the oncogene concept postulate that these cellular oncogenes are not only converted to cancer genes from without the cell by transducing retroviruses, but also from within, either by direct "activation," or indirectly by inactivation of a suppressor gene or an anti-oncogene[23,24] like other latent or inactive cellular genes. According to the oncogene concept, five different mechanisms "activate" latent cellular oncogenes or proto-oncogenes to active cancer genes (Table 1)[6,22-28].

Three Classes of Cellular Oncogenes

In addition to proto-onc genes, related to retroviral onc genes, two other classes of cellular genes are now also termed cellular proto-onc genes or oncogenes, and thought to be subject to activation by the mechanisms listed in Table 1. One class consists of genes from tumors which -unlike their counterparts from normal cells- transform upon transfection the morphology and enhance the tumorigenicity of the highly aneuploid mouse NIH 3T3 cell line[6,21,25-28]. The other class consists of cellular genes that serve as preferential integration sites of retroviruses in mammary tumors of certain inbred strains of mice, but are not related to viral onc genes and do not transform 3T3 cells[6,21,25,26,29].

Promises and Problems of the Oncogene Concept

At first sight, the oncogene concept was appealing because known derivatives of some cellular oncogenes, namely the viral onc genes, are authentic cancer genes. Above all, it promised access to the long-sought cancer genes of virus-negative cancers -in the form of hybridization- or antibody-probes derived from either retroviral onc genes[30-32], or from cloned, 3T3-cell transforming tumor DNAs[6,25,26], or from specific retrovirus integration sites in tumors of retrovirus-infected animals[7,29]. Activation from within was the key to the enormous popularity of the oncogene-concept because most

Table 1. Mechanisms that Activate Cellular Oncogenes
According to the Oncogene Concept

---

1) Amplification
2) Translocation with and without gene rearrangement
3) Point mutation
4) Inactivation of suppressor or antioncogenes
5) Promoter or enhancer insertion from an integrating retrovirus

---

tumors are virus-negative; directly oncogenic viruses could only be isolated from a handful of animal cancers and have yet to be found in any human cancer[2,6,22].

However, a serious problem with the concept was that it postulated latent cancer genes in normal cells that could be activated by a multiplicity of mechanisms (Table 1), although activated cancer genes are the least desirable genes conceivable for multicellular eukaryotes -even a death gene would be preferable. If a single cell dies in a multicellular organism, its place can be taken by another. But if a single cell is converted to an autonomous cancer cell, it will inevitably kill the organism by authoring a clonal tumor[34,35].

Predictions of the Oncogene Concept Remain Unconfirmed

The oncogene concept makes five experimentally testable predictions, listed in Table 2. Despite numerous efforts in the last 6 to 8 years, these predictions have not been confirmed, with some possible exceptions of prediction 3.

Prediction 1. Structural comparisons between viral onc genes and corresponding proto-onc genes have revealed that viral onc genes and proto-onc genes are not isogenic. Instead, all viral onc genes are tripartite hybrids consisting of retroviral promoters and coding regions derived mostly from virus and proto-onc genes, or sometimes only from proto-onc genes, and of terminal retroviral control elements (Fig. 1)[1,2]. On the basis of such studies, the hypothesis emerged that substitution of the cellular promoter and of as yet poorly defined non-transcribed regulatory sequences by the strong, constitutive promoter of a retrovirus, termed long terminal repeat (LTR), is a structural alteration that is essential to convert a proto-onc gene to a viral onc gene[2,36-38].

In the cases of the conversion of proto-myc to the myc genes of the avian carcinoma/leukemia viruses MC29, MH2, OK10 and CMII[38,39], and of proto-ras to the ras genes of Harvey, Balb and other murine sarcoma viruses[36,37], we have demonstrated that promoter substitution is indeed sufficient to convert a proto-onc gene to a viral onc gene. This was proven with virus constructs in which the native proto-myc or proto-ras genes were linked to a retroviral LTR promoter in a retroviral vector. These synthetic retroviruses transformed diploid cells like wild type avian and murine tumor viruses with myc or ras genes[36-38]. However, it appears that multiple copies of a LTR-ras or LTR-myc onc gene are necessary to transform a diploid cell. This was deduced from comparisons in which the transforming function of synthetic LTR-ras and LTR-myc genes was compared to that of synthetic proviruses like LTR-ras-LTR or LTR-myc-LTR that can be replicated by helper viruses. Such studies indicate that replicating viruses transform cells that can only be very inefficiently transformed, if at all, by equivalent nonreplicating transforming genes[36-38]; unpublished data.

199

Table 2. Predictions Made by the Oncogene Concept

1) Cellular oncogenes and viral onc-genes are isogenic
2) Transcriptional activation of cellular oncogenes leads to cancer
3) Activated cellular oncogenes transform diploid cells
4) Diploid cancers exist with activated oncogenes as the only genetic distinction
5) The probability of oncogene activation ≤ the probability of cancer

Fig. 1. The generic, recombinant structures of retroviral onc genes and their relationship to viral onc genes (stippled) and cellular proto-onc genes (unshaded). The genes are compared as transcriptional units or mRNAs. All known viral onc genes are tripartite hybrids of a central sequence derived from a cellular proto-onc gene, which is flanked by 5' and 3' elements derived from retroviral genes. Actual size differences, ranging from about 2 to 7 kilobases (kb), are not faithfully recorded. The map order of the three essential retrovirus genes, gag, pol,

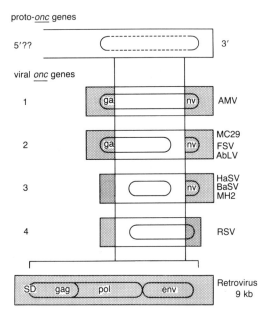

and env, and the splice donor (SD) are indicated. Four groups of viral onc genes are distinguished based on the origins of their coding sequence. Group 1: the coding unit has a tripartite structure of a central proto-onc-derived sequence that is initiated and terminated by viral coding sequences. Avian myeloblastosis virus (AMV) is an example[6]. Group 2: The coding unit is initiated by a viral sequence and terminated by a proto-onc sequence. The delta-gag-myc gene of avian carcinoma virus MC29 is an example[17,18]. The hybrid onc genes of Fujinami avian sarcoma virus (FSV) and Abelson murine leukemia virus (AbLV) are other examples[6]. Group 3: The coding unit of the viral onc gene is colinear with a reading frame of a cellular proto-onc gene. The ras gene of the Harvey and Balb murine sarcoma viruses (HaSV and BaSV)[36,37], and the myc gene of the avian carcinoma virus MH2 are examples (see Fig. 2)[38]. Group 4: The coding unit is initiated by a proto-onc derived domain and terminated by a viral reading frame. The src gene of Rous sarcoma virus (RSV) is an example[2,6]. The transcriptional starts and 5' nontranscribed regulatory sequences of most proto-onc genes are as yet not, or not exactly, known[2,36-38]. It is clear, however, that proto-onc-specific regulatory elements are always replaced by viral promoters and enhancers and that proto-onc coding sequences are frequently recombined with viral coding sequences. This figure is adapted from reference 2.

Prediction 2. Several, but not all, proto-onc genes are highly expressed in normal cells, for example, proto-myc, indicating that proto-onc
expression is not sufficient to transform cells[1,2,6].

Prediction 3. Not one of the proto-onc genes from tumor cells "activated" according to those mechanisms of the oncogene concept (Table 1), which
do not alter their germline configuration, has been shown to transform
diploid cells in culture[1,2,6,25,26,40]. On the contrary, point mutated
("activated") proto-ras genes like those found in some human cancers[30,31],
have been observed in clonal hyperplasias of mice. These hyperplasias subsequently differentiated into normal tissues[41,42]. Likewise, transgenic mice
have been generated that are tumor-free, yet carry in every cell recombinant
ras genes with a point mutation linked to a mammary tumor virus promoter[43],
or proto-myc genes "activated" by rearrangements with globin genes that were

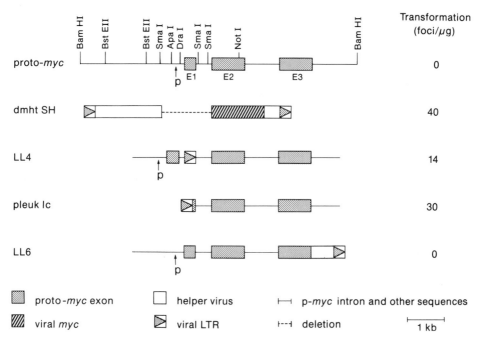

Fig. 2. Structure of cloned proto-myc genes from chicken lymphomas.
Three molecularly cloned proto-myc genes from retrovirus-
positive avian leukemias, termed LL4, p-leuk1C and LL6, are
analyzed. Their structures and transforming functions are
compared to native chicken proto-myc DNA and proviral DNA of MH2
virus with a deletion (dotted line) in the non-transforming mht-
sequence[38]. The MH2 virus carries a transformation-specific
myc sequence (hatched area). Open boxes indicate viral sequences; stippled boxes indicate p-myc exons 1 to 3; boxes with
arrowheads stand for retroviral LTRs, in which the arrowheads
point in the transcriptional direction; solid lines represent
other sequences; major restriction enzyme sites of proto-myc are
indicated. The distance between the two BamH1-sites of chicken
proto-myc is about 9.5 kb. The p-myc promoter is signaled by
p. Transfection of cloned cellular proto-myc DNAs onto quail
embryo cells has been described[38]. The transforming
efficiencies, measured in focus-forming units per microgram of
DNA (foci per µg) are the average of at least two independent
assays. The figure is adapted from reference 38.

derived from murine or human B-cell lymphomas[44,45]. Since some of these mice developed clonal tumors, the "activated" oncogenes were proposed to be necessary but not sufficient for tumorigenesis (see below).

However, proto-myc genes isolated from retrovirus-induced chicken leukemias were recently shown to transform diploid cells in vitro like proviral DNAs from retroviruses with onc genes[38]. The native promoters of these proto-myc genes had been substituted in the animal by that of a retrovirus. Fig. 2 shows two such hybrid genes from chicken lymphomas with a 5' retroviral LTR promoter linked to a proto-myc coding region which had retained its native 3' terminus. Upon transfection, these genes transformed primary quail cells like the corresponding proviral DNAs of retroviruses with myc genes. Nevertheless, preliminary evidence indicates that multiple copies of these genes are necessary to transform a cell (R.-P. Zhou and P. H. Duesberg, unpublished data). These are as yet the only known genes derived from tumor cells that, upon transfection, transform diploid cells in culture[38].

The oncogene concept further proposes that the LTR of an integrated retrovirus could convert a proto-onc gene to an active oncogene, not only by promoter substitution as originally postulated[32], but also as a position-independent enhancer[25,26,33]. However, neither a proto-myc gene with a retrovirus integrated downstream that was isolated from a chicken lymphoma (Fig. 2), nor similar constructs made in vitro were found to transform quail embryo cells upon transfection[38]. Therefore, it was concluded that retrovirus integration can convert a proto-myc gene only to a transforming gene if the viral LTR functions as a classical promoter as in viral myc genes, but not as a position-independent enhancer. Accordingly, it was suggested that the lymphoma from which the nontransforming proto-myc hybrid gene was isolated may not have been caused by this altered proto-myc gene, although it cannot be excluded that a proto-myc gene, which cannot transform embryo cells in vitro, may transform lymphocytes in vivo[38].

Prediction 4. As yet, no diploid tumors have been described that differ from normal cells only in a cellular oncogene "activated" by a mechanism that does not alter its genetic structure, like point mutation, amplification or transcriptional activation (Table 1)[2]. The only exceptions are tumors caused by retroviruses with onc genes, which are initially diploid[2]. Other possible exceptions may be tumors like the chicken lymphomas induced by retroviruses that substitute the promoter of a proto-onc gene, although the karyotype of such tumors has yet to be determined.

Indeed, most, if not all, animal and human cancers have chromosome abnormalities[2,35,46-48]. For example, the human bladder carcinoma cell lines from which proto-ras with a point mutation was first isolated[30,31] contain over 86 chromosomes, including several abnormal marker chromosomes[2]. In view of such fundamental genetic alterations, the point mutation of proto-ras said to have caused the tumor[30,31] seems like a very minor event.

Prediction 5. Clearly, if activation of a proto-onc gene according to the oncogene concept were directly relevant to cancer, the probability of cancer should be at least as high as the sum of the probabilities of activation of all cellular oncogenes. Since about 50 putative cellular oncogenes have been described[6,21-28], cancer should occur at least 50 times more often than the activation of a given oncogene by any of the five mechanisms listed above (Table 1).

However, there is an astronomical discrepancy between the real probabilities of cancer in man and animals and the probabilities predicted by the oncogene concept -even if one takes into consideration that not every cell lineage is susceptible to transformation by every oncogene. For example, the probability of activating proto-ras by a point mutation according to the

Table 3. Probability of proto-ras gene activation by point mutation

---

The net probability of a point mutation in prokaryotic
or eukaryotic cells per nucleotide per mitosis after
repairs is[2]: 10^{-9}

Since at least 50 different mutations are said to
"activate" proto-ras[2,25], and eukaryotes are diploid,
the probability of proto-ras activation per mitosis is: 10^{-7}

---

oncogene concept is about 10^{-7} per mitosis (Table 3). Since proto-ras with
a point mutation is a "dominant oncogene" according to the 3T3 mouse cell
assay[6,25,28,30,31], one mutation should be sufficient for transformation.

By contrast, the probability of spontaneous conversion of a human (or
animal) cell to the malignant parent cell of a clonal cancer is only 2 x
10^{-17} per mitosis (Table 4). Thus, the probability of proto-ras activation
according to the oncogene concept is about 10^{10} times higher than the probability of malignant transformation of a human cell to a cancer cell in
vivo, although such a mutation may not transform every cell in which it
arises. Moreover, only a small minority of human cancers contain proto-ras
mutations[2,25].

Other events that do not affect the genetic structure of genes, such as
amplification, translocation, elevated expression, or inactivation and deletion of transacting suppressor genes, also each occur at probabilities be
tween 10^{-4} and 10^{-9} per mitosis[2,4]. Hence, there is a discrepancy between
the probabilities of cancer predicted by the oncogene concept and reality of
at least 10^8 or more, depending on how many of the 50 postulated cellular
oncogenes -and of the five different mechanisms said to activate them- are
included in the estimate, and depending on how many cell lineages are assumed
to be susceptible to transformation by a given oncogene.

It follows that the oncogene concept has basically failed in predicting
the reality of cancer because the probability of generating cancer genes by
most of the postulated mechanisms is much higher than the reality of cancer.
Moreover, a gene or combination of genes that transform diploid cells in
vitro have yet to be isolated from a human tumor. Only one class of such
genes, namely the retrovirus-proto-myc recombinant genes from chicken lymphomas (Fig. 2), has been isolated from animals.

ACTIVATED PROTO-ONC GENES NECESSARY FOR CANCER?

In view of these difficulties with the original oncogene concept that
postulated functional equivalence between viral and cellular oncogenes, it
has been argued more recently that activated cellular oncogenes are necessary, but not sufficient for carcinogenesis[6,26,40,49]. For example, it was
proposed that activated proto-myc would complement activated proto-ras to
transform cells, because these two activated oncogenes were found in the same
tumor cell line[40]. Hence, one "activated" oncogene by itself might occur
much more frequently than cancer, and would not be expected to transform
diploid cells upon transfection.

This proposal is a significant revision of the original oncogene concept, which held that, upon activation, cellular oncogenes were like viral
onc genes, and that their effects can be observed when they are introduced

Probability of a Human Cell to Initiate a Clonal
Precursor of a Malignant Cancer

The probability of spontaneous malignant transformation of a cell per
mitosis, based on the reality of human cancer is the product of the
following factors:

a)	Cancers are monoclonal[2,34,35]:	1
b)	1 in 5 humans develop cancer[2]:	2×10^{-1}
c)	Adult humans consist of about 10^{14} cells that go through an average of 10^2 mitoses per lifetime[2]:	$10^{-14} \times 10^{-2}$

Thus, the probability for a human cell
to originate a clonal tumor is: $\qquad$ 2×10^{-17}

---

into non-malignant cells[20,24-28,30,31]. This proposal redefines "activated"
oncogenes and sets them apart, from both viral onc genes, which are sufficient, and normal proto-onc genes, which are not known to be (directly)
necessary for carcinogenesis.

To support this proposal, it would be necessary to demonstrate a consistent correlation between an "activated" oncogene or proto-onc gene and a
given cancer, and to prove that it would be necessary to functionally identify complementary cancer genes. However, to date no genes have been identified in spontaneous tumors that complement the allegedly necessary oncogenes
to function as dominant cancer genes in diploid cells. Further, a consistent
correlation between an activated proto-onc gene, which is not sufficient to
transform a diploid cell, and a given cancer, has never been demonstrated.

For example, proto-ras mutations are only found in a minority of primary
tumors, but never in all tumors of a given type[1,25-28]. Moreover, transgenic mice have been generated that contain both an "activated" ras and a
reportedly complementary[40] myc gene in every cell[43]. Such mice all develop normally. However, they have a propensity to develop monoclonal tumors
later, although some do not develop tumors for 150 days[43]. Accordingly, it
was concluded that even a combination of "activated" ras and myc is not sufficient for carcinogenesis[43]. Further, it has been claimed, based on
studies of cultured cell lines, that there are consistent translocations of
proto-myc from chromosome 8 to either chromosome 2, 14 of 22 in all Burkitt's
lymphomas[49,50]. Yet, cytogenetic studies of primary Burkitt's lymphomas
have identified normal chromosomes 8, in up to 50% of the tumors which carried other clonal chromosome abnormalities instead[51,52]. It has been suggested that the consistent translocations observed in cultured cell lines
derived from such tumors are artifacts of selection in vitro[51]. Moreover,
many proto-myc translocations observed in cultured Burkitt's lymphoma cell
lines do not alter the germline structure of proto-myc, and others eliminate
the native promoter, yet do not provide an alternative promoter[1,2,49,50].
It is unlikely that promoter-less or normal proto-myc genes cause cancer.
Similarly, it has been claimed that retinoblastomas arise by a consistent
deletion or inactivation of the retinoblastoma-suppressor (Rb) gene[24].
However, a recent analysis found mutations or deletions of the Rb gene in
only 22% of 34 tumors analyzed[53].

Thus, there is neither a consistent correlation in support of, nor
functional proof for the hypothesis that the "activated" oncogenes known to

date are even necessary for carcinogenesis. Until there is functional evidence -or at least a consistent correlation- the hypothesis that certain "activated" oncogenes are necessary to initiate or maintain carcinogenesis remains unproven. Instead, aneuploidy and many quantitative changes are consistently found in tumor cells[2,35,46-48].

CANCER GENES GENERATED DE NOVO BY ILLEGITIMATE RECOMBINATION

Based on the fundamental differences between viral onc genes and proto-onc genes (Fig. 1), it is proposed here that cancer genes do not pre-exist in either retroviruses or cells, but are generated de novo. Such genes appear to be generated by rare illegitimate recombinations between cellular proto-onc genes and retroviral genes or other cellular genes. This proposal is based on the following proven or consistent examples:

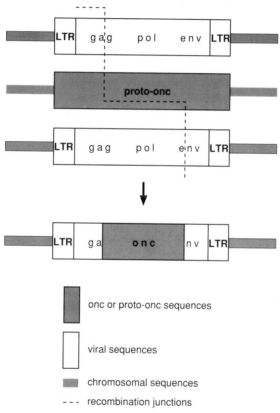

Fig. 3. Model for spontaneous transduction of oncogenic sequences from proto-onc genes by retroviruses via illegitimate recombination. Symbols are as in Figs. 1 and 2. We propose that a transforming retrovirus can be generated by two illegitimate DNA recombinations between an integrated retrovirus of about 9 kb and a cellular proto-onc gene (see Fig. 1).

1. The probability of a rearrangement per cell division,
    based on live-born babies with chromosome abnormalities
    is about[2,58]: 10^{-4}

2. The probability that a rearrangement affects a 10 kb
    proto-onc gene in a eukaryotic cell of 10^6 kb is: 10^{-5}

3. The probability that it rearranges an integrated
    provirus of 9 kb is also: 10^{-5}

Thus, the probability of generating an LTR-proto-onc hybrid
gene per mitosis is: 10^{-14}

And that of generating a virus involving two such
recombinations is: 10^{-28}

---

1. The retroviral onc genes are proved recombinant cancer genes (Fig.
1).

2. The recombinant retroviral LTR-proto-myc genes of the avian lymphomas (Fig. 2) are potential cancer genes. They function like viral onc genes
upon transfection into primary embryo cells and are probable onc genes in
viral lymphomas.

3. Further, the chromosome abnormalities found in nearly all virus-
negative tumors are consistent with this proposal. Some of the rearrangements that generated these abnormalities may have generated, by recombination, as yet unidentified cancer genes. The clonality of the chromosome
abnormalities of a given tumor, e.g., the marker chromosomes, further support
this view[2,35,46-48]. The clonality indicates that the chromosome rearrangement coincided with the origin of the tumor cell, and thus possibly
even caused it. Indeed, the retroviral onc genes may be viewed as oncogenic
chromosome rearrangements cloned in retrovirus vectors. Hence, tumors induced by such viruses are diploid -at least initially[2].

Clearly, most rearrangements based on illegitimate recombinations would
not generate cancer genes. Instead, they would inactivate genes or have no
effect. Very few combinations would be expected to generate "new" genes with
new functions such as viral onc genes (Fig. 1). Again, the origin of retroviral onc genes from cellular proto-onc genes and retroviruses may serve as a
model for how recombinant cancer genes are generated, because retroviruses,
once integrated into the chromosome of the host cell, are exactly like other
cellular genes. As illustrated in Fig. 3, two recombinations are necessary
to generate a retrovirus with an onc gene from a proto-onc gene and a retrovirus without an onc gene. The probability of generating an oncogenic retrovirus by this process based on illegitimate recombination between cellular
genes is estimated to be 10^{-28} per mitosis based on the assumptions listed
in Table 5. The low probability of such an event is quite consistent with
the reality of only 50 to 100 known isolates of such oncogenic recombinant
retroviruses in 80 years of retrovirus research[6,20].

However, it may be argued that the origin of oncogenic retroviruses is
not a model for the origin of cellular cancer genes by illegitimate chromosome recombination, because retroviruses were proposed to recombine with
cellular information via RNA-heterodimers[10,21,54,55]. This proposal is based

on the fact that all retroviral RNAs are diploid[56,57]. Hence, heterodimers may be formed that may recombine efficiently by copy choice transcription involving viral reverse transcriptase[54].

We have recently reexamined this proposal, using the same incomplete Harvey sarcoma provirus as the hypothetical recombination intermediate that

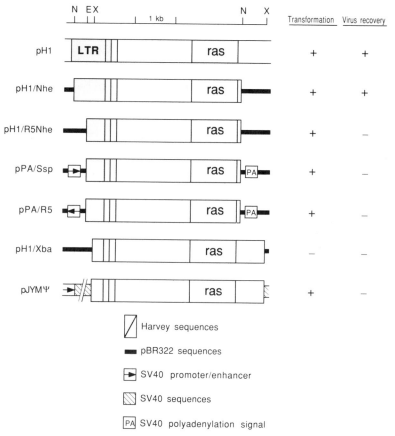

Fig. 4. Transformation and regeneration of Harvey sarcoma virus in cells transfected by incomplete Harvey proviruses with a single intact or partially deleted LTR. Harvey proviral constructions were transfected onto NIH 3T3 cells. Cells were scored for transformation 2 to 3 weeks later. It can be seen that only constructs with one complete (pH1) or nearly complete (pH1/Nhe) LTR regenerated infectious virus. However, nearly all constructs generated sufficient <u>ras</u> RNA to transform cells. Restriction enzyme sites that define the extent of the deletions of the LTR are N, Nhe I; E, EcoRV; X, Xba I. The large stippled and open boxes represent Harvey viral elements, and the solid black line indicates pBR322 sequences. The arrows and PA in the small boxes indicate SV40 promoter/enhancer and poly(A) signal sequences, respectively. The hatched boxes represent SV40 sequences. This figure is from reference 58.

was the original basis of the model[55]. In this model system, the incomplete 5' LTR-ras Harvey provirus (Fig. 4) was proposed to recover the missing 3'-LTR from a helper Moloney retrovirus by recombination between one LTR-ras RNA and one complete Moloney virus RNA linked in a heterodimer[55] (just as illustrated for the proposed chromosomal DNA recombinations shown in Fig. 3). However, this RNA-model did not exclude an alternate interpretation that proposes illegitimate DNA recombination between sequences flanking the incomplete LTR-ras provirus as the mechanism of virus regeneration[58].

In order to distinguish between RNA and DNA recombination, we have deleted the U3 element of the only LTR of the incomplete LTR-ras provirus and, in some constructs, substituted deleted retroviral promoter elements by a heterologous SV40 promoter to enhance transcription for heterodimer formation (Fig. 4). Indeed, the cells transfected with U3-deleted LTR-ras DNAs were transformed and the amount of LTR-ras RNA packaged by the helper virus was as much -or nearly as much- as that in the wild-type Harvey sarcoma virus[58]. However, virus recovery was observed only with LTR-ras proviruses that carried one complete LTR (Fig. 4). Since the RNAs transcribed from the intact and deleted templates are identical, heterodimer recombination should have occurred with all constructs capable of expressing RNA (Fig. 4). Because this was not observed, it follows that virus regeneration involved DNA recombination[58].

This model is consistent with the excision of SV40 or lambda viruses from host cells[59] or the spontaneous amplification of chromosomal regions observed in gene amplification[60]. An LTR-ras probe with one defective LTR can not, of course, regenerate a functional virus by this mechanism because an intact LTR is essential for a viable virus. Thus, there is no evidence against the view that the generation of retroviral onc genes proceeds via DNA recombination. However, there is independent evidence for DNA recombination, namely the occurrence of intron sequences in some retroviral onc genes[58] or the existence in avian lymphomas of partial recombinants like the LTR-proto-myc hybrid genes with oncogenic potential (Fig. 2).

It would appear then, that comparative analyses of viral onc genes and cellular proto-onc genes have not identified preexisting cellular oncogenes. Instead, these analyses have revealed fundamental genetic differences between viral onc genes and proto-onc genes, and that viral onc genes are the products of rare illegitimate recombinations. In view of the viral model and the clonal chromosome abnormalities in almost all cancer cells, it is proposed that cancer genes are generated de novo by rare illegitimate recombinations. To identify either autonomous cancer genes or combinations of mutually dependent ones, assays must be developed for transformation to malignancy of diploid cells.

REFERENCES

1. P. H. Duesberg, Activated oncogenes: Sufficient or necessary for cancer?, Science 228:669-677 (1985).
2. P. H. Duesberg, Cancer genes: Rare recombinants instead of activated oncogenes, Proc. Natl. Acad. Sci., USA 84:2117-2124 (1987).
3. P. H. Duesberg, Latent cellular oncogenes: The paradox dissolves, J. Cell Sci. Suppl. 7:169-187 (1987).
4. P. H. Duesberg, Cancer genes generated by rare chromosomal arrangements rather than activation of oncogenes, Med. Oncol. and Tumor Pharmacother. 4:163-175 (1987).
5. J. Tooze, "The Molecular Biology of Tumor Viruses," Cold Spring Harbor Laboratory, Cold Spring Harbor (1973).
6. R. Weiss, N. Teich, H. Varmus, and J. Coffin, "RNA Tumor Viruses: Molecular Biology of Tumor Viruses," 2nd ed., Cold Spring Harbor Laboratory, Cold Spring Harbor (1985).

7.   P. H. Duesberg, Retroviruses as carcinogens and pathogens: Expectations and reality, Cancer Res. 47:1199-1220 (1987).

8.   G. S. Martin and P. H. Duesberg, The a-subunit on the RNA of transforming avian tumor viruses: (I) Occurrence in different virus strains; (II) Spontaneous loss resulting in non-transforming variants, Virology 47:494-497 (1972).

9.   J. M. Coffin, P. N. Tsichlis, C. S. Barker, S. Voynow, and H. L. Robinson, Variation in avian retrovirus genomes, Ann. N.Y. Acad. Sci. 354:410-425 (1980).

10. H. M. Temin, Evolution of cancer genes as a mutation-driven process, Cancer Res. 48:1697-1701 (1988).

11. P. H. Duesberg and P. K. Vogt, Differences between the ribonucleic acids of transforming and nontransforming avian tumor viruses, Proc. Natl. Acad. Sci. USA 67:1673-1680 (1970).

12. P. H. Duesberg, Transforming genes of retroviruses, Cold Spring Harbor Symp. Quant. Biol. 44:12-27 (1979).

13. P. H. Duesberg, Retroviral transforming genes in normal cells? Nature 304:219-226 (1983).

14. N. D. Zinder, Infective heredity in bacteria, Cold Spring Harbor Symp. Quant. Biol. 18:261-269 (1953).

15. E. M. Scolnick, F. Rands, P. Williams, and W. P. Parks, Studies on the nucleic acid sequences of Kirsten sarcoma virus: A model for formation of a mammalian RNA-containing sarcoma virus, J. Virol. 12:458-463 (1973).

16. E. M. Scolnick and W. P. Parks, Harvey sarcoma virus: A second murine type C sarcoma virus with rat genetic information, J. Virol. 13:1211-1219 (1974).

17. P. Mellon, A. Pawson, K. Bister, G. S. Martin, and P. H. Duesberg, Specific RNA sequences and gene products of MC29 avian acute leukemia virus, Proc. Natl. Acad. Sci., USA 75:5874-5878 (1978).

18. D. K. Watson, E. P. Reddy, P. H. Duesberg, and T. S. Papas, Nucleotide sequence analysis of the chicken c-myc gene reveals homologous and unique regions by comparison with the transforming gene of avian myelocytomatosis virus MC29, delta-gag-myc, Proc. Natl. Acad. Sci., USA 80:2146-2150 (1983).

19. J. M. Bishop, Enemies within: The genesis of retrovirus oncogenes, Cell 23:5-6 (1981).

20. R. Weiss, N. Teich, H. Varmus, and J. Coffin, "RNA Tumor Viruses: Molecular Biology of Tumor Viruses," Cold Spring Harbor Laboratory, Cold Spring Harbor (1982).

21. J. D. Watson, N. H. Hopkins, J. W. Roberts, J. A. Steitz, and A. M. Weiner, in: "Molecular Biology of the Gene," Vol. II, Benjamin Publishing Co., New York (1987).

22. R. A. Weiss, The oncogene concept, Cancer Rev. 2:1-17 (1986).

23. A. G. Knudson, Jr., Hereditary cancer, oncogenes, and antioncogenes, Cancer Res., 45:1437-1443 (1985).

24. S. H. Friend, R. Bernards, S. Rogelj, R. A. Weinberg, J. M. Rapaport, D. M. Albert, and T. P. Dryja, A human DNA segment with properties of the gene that predisposes to retinoblastoma and osteosarcoma, Nature 323:643-646 (1986).

25. H. Varmus, The molecular genetics of cellular oncogenes, Ann. Rev. Genet. 18:553-612 (1984).

26. J. M. Bishop, The molecular genetics of cancer, Science 235:305-311 (1987).

27. C. Marshall, Human oncogenes, in: "RNA Tumor Viruses: Molecular Biology of Tumor Viruses," 2nd ed., R. Weiss, N. Teich, H. Varmus, and J. Coffin, eds., Cold Spring Harbor Laboratory, Cold Spring Harbor (1985).

28. M. Barbacid, Mutagens, oncogenes and cancer, Trends Genet. 2:188-192 (1986).

29. R. Nusse, The int genes in mammary tumorigenesis and in normal development, Trends in Gen. 4:291-295 (1988).

30. C. J. Tabin, S. M. Bradley, C. I. Bargmann, R. A. Weinberg, A. G. Papageorge, E. M. Scolnick, R. Dhar, D. R. Lowy, and E. H. Chang, Mechanism of activation of a human oncogene, Nature 300:143-149 (1982).

31. E. P. Reddy, R. K. Reynolds, E. Santos, and M. Barbacid, A point mutation is responsible for the acquisition of transforming properties by the T24 human bladder carcinoma oncogene, Nature 300:149-152 (1982).

32. W. S. Hayward, B. G. Neel, and S. M. Astrin, Activation of a cellular onc gene by promoter insertion in ALV-induced lymphoid leukosis, Nature 290:475-480 (1981).

33. G. S. Payne, J. M. Bishop, and H. E. Varmus, Multiple arrangements of viral DNA and an activated host oncogene in bursal lymphomas, Nature 295:209-214 (1982).

34. J. Cairns, "Cancer: Science and Society," W. H. Freeman and Company, San Francisco (1978).

35. S. Heim, N. Mandahl, and F. Mitelman, Genetic convergence and divergence in tumor progression, Cancer Res. 48:5911-5916 (1988).

36. K. Cichutek and P. H. Duesberg, Harvey ras genes transform without mutant codons, apparently activated by truncation of a 5' exon (exon-1), Proc. Natl. Acad. Sci., USA 83:2340-2344 (1986).

37. K. Cichutek and P. H. Duesberg, Recombinant Balb and Harvey sarcoma viruses with normal proto-ras coding regions transform embryo cells in culture and cause tumors in mice, J. Virol. (in press) (March, 1989).

38. R.-P. Zhou and P. H. Duesberg, myc proto-oncogene linked to retroviral promoter, but not to enhancer, transforms embryo cells, Proc. Natl. Acad. Sci., USA 85:2924-2928 (1988).

39. S. Pfaff and P. H. Duesberg, Two autonomous myc oncogenes in avian carcinoma virus OK10, J. Virol. 62:3703-3709 (1988).

40. H. Land, L. F. Parada, and R. A. Weinberg, Cellular oncogenes and multistep carcinogenesis, Science 222:771-778 (1983).

41. A. Balmain, M. Ramsden, G. T. Bowden, and J. Smith, Activation of the mouse cellular Harvey-ras gene in chemically induced benign skin papillomas, Nature 307:658-660 (1984).

42. S. H. Reynolds, S. J. Stowers, R. R. Maronpot, M. W. Anderson, and S. A. Aaronson, Detection and identification of activated oncogenes in spontaneously occurring benign and malignant hepatocellular tumors of the B6C3F1 mouse, Proc. Natl. Acad. Sci. USA 83:33-37 (1986).

43. E. Sinn, W. Muller, P. Pattengale, I. Tepler, R. Wallace, and P. Leder, Coexpression of MMTV/v-Ha-ras and MMTV/c-myc genes in transgenic mice: Synergistic action of oncogenes in vivo, Cell 49:465-475 (1987).

44. J. M. Adams, A. W. Harris, C. A. Pinkert, L. M. Corcoran, W. S. Alexander, R. Cory, D. Palmiter, and R. L. Brinster, The c-myc oncogene driven by immunoglobulin enhancers induces lymphoid malignancy in transgenic mice, Nature 318:533-538 (1985).

45. W. S. Alexander, J. W. Schrader, and J. Adams, Expression of the c-myc oncogene under control of an immunoglobulin enhancer in Eμ-myc transgenic mice, Mol. Cell Biol. 7:1436-1444 (1987).

46. S. R. Wolman, Karyotypic progression in human tumors, Canc. Met. Rev. 2:257-293 (1983).

47. J. M. Trent, Chromosomal alterations in human solid tumors: Implications of the stem cell model to cancer cytogenetics, Cancer Surv. 3:393-422 (1984).

48. A. Levan, Chromosomes in cancer tissue, Ann. N.Y. Acad. Sci. 63:774-792 (1956).

49. P. Leder, J. Battey, G. Lenoir, C. Moulding, W. Murphy, M. Potter, T. Stewart, and R. Taub, Translocations among antibody genes in human cancer, Science 227:765-771 (1983).

50. G. Klein, Specific chromosomal translocations and the genesis of B-cell derived tumors in mice and men, Cell 32:311-315 (1983).
51. R. J. Biggar, E. C. Lee, F. K. Nkrumah, and J. Whang-Peng, Direct cytogenetic studies by needle stick aspiration of Burkitt's lymphoma in Ghana, West Africa, J. Natl. Cancer Inst. 67:769-776 (1981).
52. R. Berger, A. Bernheim, F. Sigaux, F. Valensi, M.-T. Daniel, and G. Flandrin, Two Burkitt's lymphomas with chromosome 6 long arm deletions, Canc. Gen. & Cytogen. 15:159-167 (1985).
53. A. D. Goddard, H. Balakier, M. Canton, J. Dunn, J. Squire, E. Reyes, A. Becker, R. A. Phillips, and B. L. Gallie, Infrequent genomic rearrangement and normal expression of the putative Rb1 gene in retinoblastoma tumors, Mol. Cell. Biol. 8(5):2082-2088 (1988).
54. J. M. Coffin, Structure, replication, and recombination of retrovirus genomes: Some unifying hypotheses, J. Gen. Virol. 42:1-26 (1979).
55. M. P. Goldfarb and R. A. Weinberg, Structure of the proviruses within NIH 3T3 cells transfected with Harvey sarcoma virus DNA, J. Virol. 38:125-135 (1981).
56. P. H. Duesberg, Physical properties of Rous sarcoma RNA, Proc. Natl. Acad. Sci. USA 60:1511-1518 (1968).
57. W. F. Mangel, H. Delius, and P. H. Duesberg, Structure and molecular weight of the 60-70S RNA and the 30-40S RNA of the Rous sarcoma virus, Proc. Natl. Acad. Sci., USA 71:4541-4545 (1974).
58. D. W. Goodrich and P. H. Duesberg, Retroviral transduction of oncogenic sequences involves viral DNA instead of RNA, Proc. Natl. Acad. Sci., USA 85:3733-3737 (1988).
59. D. Hanahan, D. Lane, L. Lipsich, M. Wigler, and M. Botchan, Characteristics of an SV-40 plasmid recombinant and its movement into and out of the genome of a murine cells, Cell 21:127-140 (1980).
60. R. T. Schimke, S. W. Sherwood, A. B. Hill, and R. N. Johnston, Overreplication and recombination of DNA in higher eukaryotes: Potential consequences and biological implications, Proc. Natl. Acad. Sci., USA 83:2157-2161 (1986).

DISCUSSION

Moderator (Lance A. Liotta): We are going to hold questions for Peter to include them in the general discussion.

SUMMARY DISCUSSION

Moderator (Lance A. Liotta): We can summarize our presentations today as
follows: We have seen evidence that multiple genes are amplified, increased
in expression, or lost, and this correlates with tumorigenesis and
progression to metastases. Others may be are up-regulated genes that encode
for enzymes or cytoskeletal elements. Some of these may be so-called
suppressor genes. Their function is unknown. There may be even suppressor
genes for the metastatic phenotype, and I'll present evidence for a putative
metastases suppressor gene we've cloned tomorrow. But all of these
relationships are correlative in nature, that is, we're correlating the
changes of genetic expression with the biologic phenotype, and so the
unanswered question is of course the causal relationship between any of them
and progression of the tumor. In the future we need to develop means of
specifically inhibiting the genes of interest, so we can really study their
causal nature. If we could inhibit a ras gene product or inhibit any of the
other putative genes that play a role in progression with some antisense
mechanism, then maybe we could really see if they are necessary or just
correlated. I believe that there will be a revolution in pathology, in terms
of using combinations of genes and their relative amplification or expression
or loss as a new way of predicting the prognosis for an individual patient's
tumor. I foresee a situation in which the pathologist will not just look
under the microscope at the tumor but will analyze the relative pattern of
gene imbalance in the tumor DNA or in their pattern of expression. I think
that no matter what the role of these genes is, ultimately we will find out
that their importance for prognosis is unquestionable. I think another point
that should be raised in this regard is which genes are more important to
study and which have the better prognostic indication than others. We've
seen that a source of major controversy is the role of the erbB-2 gene, as
shown by Slamon, et al. originally. This correlates with breast cancer
aggressiveness, and I believe many pathology labs are poised to measure this
gene in their tumor samples. The difference of opinions demonstrate a major
gap in our resources. We do not have existing a very large tumor bank that
scientists throughout the country can use. What we need is a bank of
hundreds and hundreds of tumors with known clinical backgrounds, which differ
in low aggressiveness, high aggressiveness, high propensity for metastases,
and different states of histologic classification. I would hope that we will
be able to gather the resources of pathologists and physicians around the
country to create such a bank. We're not talking necessarily about viable
tumors, but just DNA, so that we can extend these correlative studies to much
larger studies. Maybe some of the members of the audience who've been
working in the field of tumor banking can add something to this question.
I'd like now to pose these questions to our discussion leaders and to invite
anyone else who wants to join in.

Boundaries between Promotion and Progression during Carcinogenesis
Edited by O. Sudilovsky *et al.*, Plenum Press, New York, 1991

Peter Duesberg: Do you think it is practical to use these as diagnostic
tools? It would be hard for pathologists to start hybridizing like Web
Cavenee with all his technicians.

Liotta: I think it will definitely be practical. We've only just begun to
develop astonishing new techniques for gene amplification that work on
samples obtained from parafin sections. That's just the beginning. I think
the technology will keep advancing and pathologists are just going to have to
keep up. Sure, I can see the development of centralized laboratories that
could do this as a reference situation.

Webster Cavenee: I believe that a lot of the genes which are being isolated
like those Tim Bowden was talking about, the association of that with fusion
protein technology and the making of monoclonal antibodies is perfectly
compatible with histopathology. I think that the level of technology is just
one step removed from the clinical lab, but it is not far coming.
Liotta: We are not going to see immunohistology given up in place of in situ
hybridization; immunohistology will become even more important because we
want to really look at the expression of these genes, particularly those that
are supposedly increased in relationship to cancer. That will best be done
by looking at the protein product with the immunohistology.

Cavenee: That's certainly true in heterogeneous tumors, as well.

Liotta: That's right, and you can look at a cell by cell basis.

Philip Frost: Once again, Lance, you're letting technology run ahead of prac-
ticality. The fact of the matter is that the prognosis of someone with adenocarcinoma of the lung is well established. I really don't think that whether
you can probe this or not is going to contribute to his prognosis because we
must face one ultimate reality: we don't have treatment for this disease. I
really don't know what basic fundamental advantage doing all this very sophisticated analysis is going to provide in terms of patient care.

Joseph Locker: I think it is a mistake to group all kinds of cancer together. Considering breast cancer, for example, about 60% of patients will be
cured by surgical intervention, and of those 40% that do not get cured, they
will eventually die of breast cancer, but therapies are very different in
managing their cancer. There are many decisions that have to be made about
whether they will respond, for example, to estrogen ablation therapy, or what
kind of chemotherapies should be given. Lung cancer is a disease that is
still a major challenge, but you cannot generalize over all of them. Regarding the sampling cancers to study, there are some assays that can be done
fairly easily. The quantitative hybridizations for gene amplification can be
done on paraffin blot material, and it is fairly easy to assemble a very comprehensive series of cases from that kind of a resource. On the other hand
one can get the specimens, but to do a meaningful study is not simply a
matter of getting specimens and doing assays, it is a matter of doing a good
patient followup and that takes years. In terms of therapy, it has to be
something beyond just the pathologist; it has to be an integrated study with
clinicians and with followup.

Liotta: I'd like to answer Dr. Frost's question in small part by asking him
a question. If one of your patients had breast cancer, would you say that
you don't need to look at the number of positive lymph nodes in that patient,
that you don't care because it doesn't do you any good to know the
prognosis? If somebody said that you could more accurately predict the
prognosis based on a Southern blot would you not use it?

214

Frost: If and when and how? You're talking fantasy. There's no evidence that that exists now. Just knowing how many nodes are involved is sufficient information. I don't have to do a Southern blot or measure the amount of message that's there. It doesn't at present help very much.

Liotta: Let's say then, that in a node negative patient (in which as you know 25% of them, -or even 35% depending on the study series- develop metastases and die), you knew that amplification of certain genes correlated with a poorer prognosis. Wouldn't you want that information?

Frost: I'm not sure, because I don't agree with Dr. Locker that 60% of breast cancer patients are cured by their surgery. I think that you'd have to show me some very good data to support that. In fact, if you really look at the data of the 7 year survivals, they're about the same over the last 20 years. It hasn't changed a great deal. I'm not saying that we're not going to ultimately develop something better. I hope we are, otherwise we are all wasting our time. What we have today is not very good. To encourage a lot more diagnostic procedures to give us more sophisticated ways of telling patients how ill they are doesn't make much sense to me at the moment.

Liotta: Once again I have a different point of view, and that is that the basic understanding which we are getting will lead to new strategies for therapeutics, and that drugs will be targeted against some of these key biological events.

Robert Callahan: I think it would be of value for a clinician to know or be able to identify those tumors that will or will not respond to a particular therapy. With respect to the breast tumor associations between prognosis and c-erbB amplification, there are about 7 laboratories who have examined panels of tumors ranging from 40 to approximately 200 tumors. No care was taken in their studies to characterize the nature of the panel with respect to the patient's history, or the geographical location from which the patients were drawn. Even more important than that, no association was made with the particular type of therapy that was given to the patients after surgery. That has turned out to be one of the major problems in follow up. I would submit that some of these mutations may be associated with a more aggressive tumor. It is also equally possible, if they got no therapy, that the tumor would relapse (specially is you assume that they are stage II, III, and IV patients. Some of these mutations may in fact be associated not necessarily with aggressive tumors but with patients which will not respond to a given therapy. In addition, I would like to restate that patients with stage I tumors, up until very recently, generally got no post-surgical therapy. 30% of these will relapse, and there is really no good marker for determining which one of them will be the ones that will relapse. The studies to date in fact have not focused on that select group of patients. I think they provide an opportunity for rather extensive retrospective studies using the vast warehouses of paraffin blocks that are available, in which we also know the post-surgery therapy, and what the outcome of the patient was.

Harry Rubin: Obviously, the underlying preconception for all of our discussion is that genetic changes are the cause of cancer. I must say Peter Duesberg qualified it a little by saying if the cancers are due to genetic events. I'd like to point out that the most common way that we turn genes on and off is during embryological development, aging as well. And there we make what is in fact hereditary changes in cells, where in the context of their organ or tissue they heritably perpetuate those changes in the offspring cells. Despite the fact that as far as we know their DNA is without any detectable mutation (although Peter points out that they probably do have

mutations). It is a remarkable thing, that they remain normally functioning. Now, Peter is engaged in numerology. I'd like for him to go one step
further. What we have to think about in cancer is really a question of
level, which may be hard to do. So, Peter, you get your little calculator
out and consider the mouse, as opposed to the human. We have about 1,000
times as many cells as a mouse. We get tumors in 50, 60, 70 years. Mice get
tumors in one or two years. If you go through the calculations, you would
have to say that if there are genetic events involved in producing tumors in
the mouse, they occur 100,000 times more frequently on a per cell basis or a
per time basis than those events occur in human beings. Yet, at least I know
of no genetic events like that. I don't know that mutations occur in mice
100,000 times more frequently than they occur in humans. In tissue culture
they occur at the same rate. I assume even illegitimate recombinations occur
at the same rate in mice. What I'm trying to point out is that we have to
consider different levels of organization in the genesis of tumors. Along
with Sir David Smithers, I consider tumorigenesis as basically a disease or
problem of organization, rather than a molecular problem per se. That does
not rule out a role for genetic changes, molecular changes ultimately, in the
final genesis of the tumor. But the question is what is the driving force.
And there is that dirty word that hasn't come up yet, but which I will bring
up again tomorrow: epigenetics.

Charles Boone: It should be mentioned that there is already a rapidly expanding use of hybridization technology by pathologists to diagnose the presence
of cancer-associated viruses in paraffin tissue sections, so that the next
step to analysis of gene amplification may not be such a big one. Human
Papilloma Virus types 16 and 18, for instance, have been very highly correlated with 80-90% of cervical carcinomas using hybridization technology (1,2).

Duesberg: I don't understand the import of that comment. Could you explain
it? I know the hybridizations have gone up a lot in sensitivity, but what
does that help in cancer? It's a great point, but what does it prove?

Boone: HPV type 16 and 18, I'm sure you must know, have been highly correlated with cervical carcinoma.

Duesberg: Like herpes 10 years earlier and Epstein Barr 20 years earlier...
and now with AIDS and HTLV-1 with leukemia. Where are the viruses? They
are in humans. They are human virus, so you have a good correlation...

Boone: No, but there are still problems. A lot of the flat condylomata
contain the virus and you can see these lesions in the women who later
develop...but that doesn't...

Moderator: Let's hear some other points of view on some different topics.

Rubin: I want Peter to answer my numerology.

Moderator: I think he has to get his calculator.

Duesberg: Yes, I have to get the calculator, of course, but I didn't see why
you said 100,000 times more.

Rubin: A thousand times less cells and get cancer 50 times more frequently.

Duesberg: Is that known for wild mice, or is that for AKR mice? AKR mice
have been bred for years to get these cancers. What I don't know is what the
breast cancer risk is for wild mouse, or the colon cancer or lung cancer. In
humans we are looking at a rather outbred population.

Rubin: Maybe the difference is 5 orders of magnitude.

George Michalopoulos: It might be the wrong time for calculations, but if you look at a rat liver which is 6-8 grams and has about 10^9 cells, then the human, at 10^{14} cells would weigh 1,000 kilograms. That's I think off by a factor of at least about 10. Another thing about mouse tumors: we have to remember that a mouse is a much smaller animal compared to humans. A pea size tumor in the brain of a mouse would have tremendous implications, caus- ing cerebral edema, herniation of the cerebellum and all that; but in the human a pea size tumor of the brain would probably be undetected for years and years. These kinds of frequencies in small animals may be due to the fact that there's not much space for tumors to expand.

Jose Russo: At the Michigan Cancer Foundation we have followed up 1200 patients with breast cancer for almost 8 years. The interesting aspect of this study is that the survival rate and the recurrence of these patients depends upon how the subsets are done, meaning by histological, nuclear, mitotic grade, estrogen receptor, lymph node status and so on. According to how they are divided, the survival rate and the effect of treatment is different. Therefore, the idea to use genetic probes to assess prognosis is good, but they must be applied to a specific subset of patients in which the survival or recurrence rate are known. The problem with all the markers that have been published in the literature is that the number of patients is very small, and most of them, without too much clinical background. Therefore, ge- netic probes must be studied in a large number of cases and in specific sub- set of patients in which the clinical history is known. When a study of this nature is done, all the data must be analyzed under stringent statistical analysis in order to see if there is a real correlation or not. Until some- thing like this is done, we must be extremely careful how to interpret the data.

Cavenee: I think that's exactly right. That's a clinical trial. It just hasn't been done because the number of cases are not large enough. It's the same thing that Dr. Liotta was pointing out. The sort of analyses that can be done are clear, whether they're causal or just markers, it's a different issue. But those need to be analyzed in a large, well-documented series.

Liotta: And undoubtedly it will be a different series of markers for each different kind of tumor.

Russo: Exactly; one of the problems, and I agree with Philip Frost, is that the oncologist needs to treat the patient and needs to do something with the patient. But maybe the most important question that we need to ask is who is the patient that needs to be treated? I strongly believe that if we divide the patients according to specific subsets the treatment could be more efficient.

Liotta: You could consider a scenario where a node negative patient has a very aggressive appearing tumor from the genetic analysis. You may want to treat that patient more aggressively with known therapy and follow him up more carefully. Sure, it's not a cure for cancer, but it might give us some edge on management, in actual fact.

Callahan: As a clinician and a surgical oncologist, it's a very important issue now, since De Vita's recent promulgation of suggesting chemotherapy for basically all patients. It is a particular problem for us in the very early stage of breast carcinoma. We've always felt that patients who are estrogen receptor positive and lymph node negative with small tumors should have an extremely favorable prognosis. The fact that the NSABP have lumped everybody together and have treated them suggests that perhaps everyone should be get- ting chemotherapy. I think it is very important to make distinctions and pick out those patients who would benefit from adjuvant types of chemotherapy or adjuvant treatments versus those who are not. Dr. Duesberg, it seems to me that your calculations presume 100% efficiency of expression of cancer, when in actual fact, for example, yesterday Dr. Nowell suggested clearly that there are other factors that will suppress the expression of cancer. We know from immunologic data that there probably are various things that can prevent the expression of cancers. How do you factor into your equation that perhaps there is 10^7 or a 10^9 order of suppression which can be generated by a good immune system?

Duesberg: What I was calculating what is the net result of cancer. I did not consider tumors that might arise and disappear as a result of a suppressor or because of an immune surveillance, or any other type of suppressive mechanism. What I was calculating was the bottom line of cancer.

Callahan: If you use the figure, the incidence of one cancer in every five human beings...

Duesberg: That's correct. That is what is happening.

Callahan: ...that is the net result of all these various factors, so that it would seem to me that your numbers may be totally irrelevant other than as a theoretical calculation of the cancer importance that the human being would have to...

Duesberg: It seems to me that these are the most relevant numbers. Those are the ones that hit you or me. Those are the only relevant numbers in my opinion.

Callahan: There is a huge denominator which is orders of magnitude...

Duesberg: I don't care about the 10^5 ras mutations that I have every morning and I eliminate every evening, according to Dr. Weinberg. I worry about the one in three that three of us will get in our lifetime. Those are serious. Weinberg's I can live with.

Moderator: I don't think we are going to be changing people's opinions here.

Duesberg: That is not an opinion. This is a clarification of fact. I don't think most of us have cancer every morning and get over it during the day, or the evening and get over it during the night. Maybe Dr. Weinberg and you do, but I don't know about that.

Moderator: We didn't cover our point that was raised earlier by Dr. Smith, and that is what is the most relevant model, the human breast cancer model or the mouse breast cancer, and is there different mechanisms of carcinogenesis in these two models. It was based on a question asked to Dr. Callahan. Helene, do you want to say anything in regard to your question?

Smith: Yesterday we talked about the tyranny of definitions and the tyranny of words, and it seems to me that in the original Todaro-Huebner concept of oncogenes, elegantly described by Peter, these authors were talking about a single cause by a single gene that did the whole job. But from what we're learning about all different kinds of epithelial cancers, both from various model systems and from studying humans, perhaps our problem is that we're still using an old word that has ambiguous meanings when we really need a new word. A possible mistake would be to throw out the baby with the bath water because what we used to call oncogenes, are irrelevant since they don't do the whole job in every cancer. If you think about the possibility that if it were random and one in 10^7 colon cancers would be expected to have a mutated ras, and instead we find 30% of colon cancers having a mutated ras...

Unidentified speaker: 67% in some.

218

Smith: Now that doesn't seem to be random to me. It is hard to ignore the possibility that that has something to do with the development of that cancer. Maybe our problem is in trying to think that it is only one cause and only one step is needed because of the original definitions of those models.

Moderator: Does anyone want to answer the question about the mouse versus human. I think they both have numerous things to offer. Both model systems give you a lot of information. You do experiments on one that you can't do in others. We are going to benefit from using both.

References

1. Beckmann, A., et al, "Detection and localization of Human Papilloma Virus DNA in human genital condylomata by in situ hybridization with biotinylated probes. J. Med. Virol. 16:265-273, 1985.
2. Review: Grody, W.W., et al., "In situ viral DNA hybridization in diagnostic surgical pathology." Human Path. 18:535-543, 1987.

GENETIC INSTABILITY AND TUMOR DEVELOPMENT

Peter C. Nowell

Department of Pathology and Laboratory Medicine
University of Pennsylvania
Philadelphia PA 19104

This paper will discuss in very general terms, with relatively few data, the issue of genetic instability and what role it may play in the process of tumor progression. It will include both comments on tumor progression in general and on instability from a cytogenetic approach.

Tumors usually begin with a single altered cell and then expand as a clonal neoplasm through multiple divisions. They also seem often to have an increased propensity to generate variations that result in clinical malignancy. In particular, we see it in the common epithelial tumors which show considerable heterogeneity and multiple subpopulations with diverse properties. Those include increased aggressiveness, capacity for invasion and metastasis, and other aspects that we associate with tumor progression[1-5]. The time frame for this, of course, can be enormously variable. One can think in terms of a benign tumor first, with subsequent acquisition of the characteristics that we associate with malignancy. Clearly there are situations in which this evolution occurs earlier, so that by the time we can see a macroscopic tumor, the growth is already heterogeneous, and aggressive subpopulations have come to predominate. Chromosome studies, and related molecular genetic findings, have begun to provide evidence on the sequence of somatic genetic events that may underlie the biological and clinical progression just outlined. One of the best documented is in chronic granulocytic or chronic myelogenous leukemia (CML) in which one has a clone that expands, with a translocation between chromosomes 9 and 22. The C-abl oncogene is moved to chromosome 22, producing a "hybrid gene," that generates a tyrosine kinase with altered function from the parental proto-oncogene[24,25]. Since this translocation is in every cell of the tumor, it presumably contributes importantly to the expansion of the clone that we recognize clinically as CML. There are some isoenzyme data that suggest that a clone may expand in some patients before this translocation takes place[5]. In the context of tumor progression, however, one typically sees a clone with this translocation during the chronic phase of the disease, and then in the vast majority of cases (sometime within five years) there is a rather abrupt change to a much more aggressive phenotype. At that point, the leukemic cell population contains, in addition to the Philadelphia chromosome, one or more additional cytogenetic changes representing a subclone change. It is simply a way of providing signposts, but in this circumstance we typically see a subclone that is expanding and overgrowing the original population. The additional change is often an extra Philadelphia chromosome (Ph), an extra chromosome 8, or an isochromosome for the long arm of 17, replacing a chromosome 17[27]. Except for the second Ph, we do not yet have information on the specific genes involved in these additional changes, however, they seem to result in more aggressive

Boundaries between Promotion and Progression during Carcinogenesis
Edited by O. Sudilovsky *et al.*, Plenum Press, New York, 1991

221

growth and failure to differentiate. It is interesting that in acute myeloid leukemia, arising de novo, the most common cytogenetic finding is an extra dose of chromosome 8 without any other change[22,23]. Another frequent abnormality in various stages of myeloid leukemia and some other neoplasms, is an isochromosome 17. It does suggest that there are one or more genes on chromosome 8, that in extra dosage, play a role in leukemogenesis. There may be several genes on chromosome 17, perhaps a suppressor gene on the short arm and one or more stimulatory genes (oncogenes) on the long arm, similarly involved.

This type of study is now being extended to lymphomas in a number of laboratories, including our own collaborations with Dr. Carlo Croce, and certain patterns have begun to emerge. In many of the relatively low-grade, indolent follicular lymphomas in this country, there is a t(15;18) chromosome translocation, often as the only karyotypic abnormality in which a small piece of chromosome 18 is moved to the long arm of chromosome 14. Molecular studies have indicated that this involves a previously unrecognized gene, which we have called bcl-v1, that is brought into juxtaposition with the immunoglobulin heavy chain locus on chromosome 14[31]. As a result, the bcl-v gene is deregulated. Although the function of the gene is not known, it has many of the characteristics of the c-myc proto-oncogene in terms of the time course of its expression in normal lymphocyte growth. Also, the gene has been shown very recently to function as a weak oncogene in certain transfection assays.

This is an example of an indolent lymphoma in which translocation of the bcl-v gene (the first of a number of new oncogenes that are now being recognized through the study of chromosome translocations in B-cell and T-cell tumors) apparently plays an important role in the initial stages of the development of the neoplasm. Later on, as the disease progresses either with or without therapy, a number of cases are being recorded in which a second translocation or rearrangement occurs in the neoplastic cells. In some instances, it is the t(8;14) translocation that is seen in the Burkitt lymphoma[31-33] (Fig. 1). In other cases, it is a different rearrangement, leading to altered function of the c-myc gene or other genes. The result is that the disease changes from a low-grade slowly growing lymphoma to an aggressive, high-grade process. Thus, these kinds of chromosomal studies are beginning to provide clues to the locations of some of the genes involved in various stages of lymphoid neoplasia.

In non-hemic tumors, similar sequences of events have been described in a few instances, either in experimental neoplasms or in patients. The limited molecular data involve a few known oncogenes as well as putative tumor suppressor genes. For example, we have been involved in studies of malignant melanomas and related precursor lesions. There are very few cytogenetic data on the latter (nevi, dysplastic nevi), but they do show non-random involvement of the distal portion of the long arm of chromosome 10[44]. These and other findings suggest that there may be a locus in this region which plays a role in the development of neoplasms of neural crest origin, including both CNS tumors[48,49] and tumors derived from melanocytes.

In fully developed malignant melanomas, one typically sees a variety of cytogenetic changes, with non-random involvement of the short arm of chromosome 1 (1p), both arms of chromosome 6, and extra dosage of chromosome 7[44] (Fig. 2). There are several possible sites for suppressor genes on 1p and probably also a gene that plays a role in familial melanoma[46,47], but none has yet been isolated or defined. The most specific association in advanced melanomas is between extra dosage of chromosome 7 and expression of the receptor for epidermal growth factor. Only tumors with extra number 7s express the receptor[10]. Initial molecular studies suggest that the epidermal

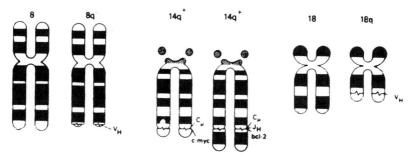

Fig. 1. Diagram of the t(8;14) and t(14;18) chromosome translocations observed in B cell lymphomas that bring the c-myc and bcl-2 genes into
association with the immunoglobulin heavy chain locus on chromosome
#14, resulting in deregulation of the proto-oncogenes. Low-grade
follicular lymphomas often have only the t(14;18) rearrangement, but
may progress to a more aggressive stage when the second translocation
is acquired[32,33].

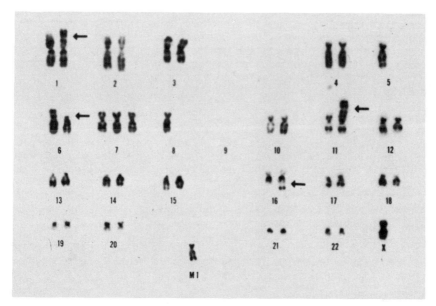

Fig. 2. Karyotype of a cell line derived from a metastatic malignant melanoma, illustrating the numerous chromosome abnormalities typically
observed in such tumors. Nonrandom alterations include translocations and deletions involving the short arm of chromosome #1, similar
abnormalities of chromosome #6, and extra copies of chromosome #7,
indicating the multiple steps and various genetic mechanisms involved
in generating the fully developed malignant phenotype[44].

growth factor receptor gene which is located on 7p so it may be a single extra dose of the gene. This may be enough to have the receptor expressed on the cell surface, giving an additional growth advantage to an already neoplastic cell. Interestingly, this is also often associated with loss of all or a part of a chromosome 10, suggesting that perhaps the gene we mentioned earlier has some regulatory role with respect to the epidermal growth factor receptor.

These, then, are the kinds of cytogenetic clues that are beginning to indicate a sequence of events in tumor progression in different neoplasms, helping to focus on some of the genes involved, and also on sites in the genome where there may well be other genes, as yet undefined, that are important contributors to the process. There are a variety of different mechanisms by which one or another of these genes may be altered. We have to remember that host regulatory factors, immunological and otherwise, can have a major effect on how a specific tumor evolves. We simply do not know enough about the details of normal and local growth regulation to make all-encompassing, doctrinaire statements about neoplastic growth. We have identified only a few of the genes and gene products involved in human tumors, and much remains to be learned at all stages of tumor progression.

Now let us turn to genetic instability -another aspect of the problem. There are some data indicating that tumor cell populations in general are more susceptible to mutagenesis, broadly defined, than are comparable normal tissues[56-59]. These come primarily from studies of cell lines, enumerating mutations or chromosomal rearrangements, spontaneous or induced. However, the information is limited, and there is always the question of what is the appropriate control. In vivo, once a tumor has progressed extensively to the point where it is extremely aneuploid, mitotic abnormalities at almost every cell division are common, indicating considerable instability in the system, and providing the basis for continuing evolution at this gross level of somatic genetic change.

There are a number of possible mechanisms that may underlie this apparent instability at different stages of tumor development. There are acquired mutations that alter the stability of the genome and increase the probability that further genetic changes will occur. There is evidence in Drosophila and lower organisms for such "mutator" genes and various types can be envisioned in tumors[20,56,60,61]. There could be an alteration in a gene that regulates DNA repair; or it could be a gene involved in maintaining the fidelity of DNA synthesis. It could be even a gene that regulates the stability of the mitotic apparatus itself, influencing the probability of gross chromosomal errors at mitosis. Once a chromosomal alteration is present, the occurrence of other mitotic abnormalities is increased when a cell divides. Thus, a number of different acquired alterations could contribute to genetic instability, even within the same tumor as it evolves.

There are also individuals who are born with an unstable genome, "chromosomal fragility" syndrome, such as ataxia telangiectasia, Fanconi's anemia, or Bloom's syndrome[56,64]. These individuals have a constitutional defect of one sort or another in DNA "housekeeping", resulting in an increased probability of chromosome breakage and in the generation of aneuploid clones that eventually progress to frank malignancy. An interesting consideration is how many of us have similar but "subclinical" defects in genetic stability. There is some evidence from studies of familial bowel cancer and melanoma, that although there may not be spontaneous chromosome breakage in lymphocyte cultures, the cells of family members may show sensitivity to mutagenic agents[47,56,59]. It will be important to determine how often an inherited gene, that plays a role in human neoplasia, acts by increasing the lability of the genome rather than through a direct growth regulatory mechanism.

Finally, we must also recognize that what may appear to be genetic instability in a tumor may in some cases reflect external factors. Someone occupationally exposed to radiation or to certain chemicals may be continually bombarding an early neoplasm with mutagenic agents. Even nutritional factors, such as deficiencies in certain amino acids, may contribute to increased mutability. Furthermore, although we normally think of "promoting" agents as providing a nonspecific stimulus for proliferation, it now seems clear that under some circumstances cell injury and their resultant proliferation can induce increased levels of mutagenic oxygen radicals[71].

Thus, apparent genetic instability in tumors can be mediated through a number of different mechanisms; these are not necessarily mutually exclusive and may vary in different individuals. Much remains to be learned, and this aspect of tumor progression has been relatively neglected; more research is needed on the mechanisms that underlie the sequential somatic genetic changes involved.

Recent findings in B-cell chronic lymphocytic leukemia provide one final example of the questions that remain to be answered. In a proportion of these patients, the neoplastic B cells constitute a chromosomally abnormal clone. An extra chromosome 12 is the most common alteration. A number of these patients have now been followed with serial chromosome studies for up to 10 years. In interesting contrast to CML, the B-CLL cells are remarkably stable karyotypically, whether or not they have an initial chromosomal change[28]. One patient, for example, has been followed for ten years with a karyotype containing 49 chromosomes, including trisomy for 12, 18 and 19. He has received radiotherapy and chemotherapy for exacerbations of his leukemia and for lung cancer, and yet the karyotype has remained unchanged. Clearly, in this disease one can have clinical progression without <u>visible</u> additional genetic changes in the neoplastic cells; despite exposure to genotoxic agents, the karyotype may be quite stable. This seems to represent one end of the spectrum of instability in tumor cells populations, and indicates the need to avoid overly-simplistic generalizations as we attempt to understand better the complex phenomenon of tumor progression in different neoplasms and different patients.

REFERENCES

1. L. Foulds, Tumor progression, <u>Cancer Res</u>. 17:355-356 (1957).
2. J. Cairns, Mutation, selection and the natural history of cancer, <u>Nature</u> 255:197-200 (1975).
3. P. Nowell, The clonal evolution of tumor cell populations, <u>Science</u> 194:23-28 (1976).
4. G. Klein, Lymphoma development in mice and humans: diversity of initiation is followed by convergent cytogenetic evolution, <u>Proc</u>. <u>Natl</u>. <u>Acad</u>. <u>Sci</u>. <u>USA</u> 76:2442-2446 (1979).
5. P. Fialkow, Clonal origin of human tumors, <u>Annu</u>. <u>Rev</u>. <u>Med</u>. 30:135-143 (1979).
6. A. Arnold, J. Cossman, A. Bakhshi, et al., Immunoglobulin gene rearrangements as unique clonal markers in human lymphoid neoplasms, <u>N</u>. <u>Engl</u>. <u>J</u>. <u>Med</u>. 309:1593-1599 (1983).
7. P. C. Nowell, C. M. Croce, Chromosomes, genes, and cancer, <u>Am</u>. <u>J</u>. <u>Pathol</u>. 125:8-15 (1986).
8. J. L. Biedler, P. W. Malera, and B. A. Spengler, Chromosome abnormalities and gene amplification: comparison of antifolate-resistant and human neuroblastoma cell systems, <u>in</u>: J. D. Rowley and J. B. Ultmann, eds., "Chromosomes and Cancer: From Molecules to Man", Academic Press, New York (1983).

9. K. Alitalo and M. Schwab, Oncogene amplification in tumor cells, Adv. Cancer Res. 47:235-259 (1986).
10. H. Koprowski, M. Herlyn, G. Balaban, et al., Expression of the receptor for epidermal growth factor correlates with increased dosage of chromosome 7 in malignant melanoma, Somatic Cell Molec. Genet. 1:297-302 (1985).
11. S. H. Friend, T. P. Dryja, and R. A. Weinberg, Oncogenes and tumor-suppressing genes, N. Engl. J. Med. 318:618-622 (1988).
12. M. F. Hansen and W. K. Cavenee, Tumor suppressors: recessive mutations that lead to cancer, Cell 53:172-173 (1988).
13. G. Klein and E. Klein, Evolution of tumors and the impact of molecular oncology, Nature 315:190-195 (1985).
14. J. M. Bishop, The molecular genetics of cancer, Science 235:305-311 (1987).
15. L. E. Babiss, S. G. Zimmer, and P. D. Fisher, Reversibility of progression of the transformed phenotype in Ad5-transformed rat embryo cells, Science 228:1099-1101 (1985).
16. G. Nicolson, and L. Milas, eds., "Cancer Invasion and Metastasis: Biologic and Therapeutic Aspects," Raven Press, New York (1984).
17. K. Lapis, L. Liotta, and A. Rabson, eds., "Biochemistry and Molecular Genetics of Cancer Metastasis," Martinus Nijhoff, The Hague (1986).
18. R. Muschel and L. A. Liotta, Role of oncogenes in metastases, Carcinogenesis 9:705-710 (1988).
19. G. L. Nicolson, Tumor cell instability, diversification, and progression to the metastatic phenotype: From oncogene to oncofetal expression, Cancer Res. 47:1473-1487 (1987).
20. J. P. G. Volpe, Genetic instability of cancer: Why a metastatic tumor is unstable and a benign tumor is stable, Cancer Genet. Cytogenet. 34:125-134 (1988).
21. S. Heim, N. Mandahl, and F. Mitelman, Genetic convergence and divergence in tumor progression, Cancer Res. 48:5911-5916 (1988).
22. J. J. Yunis, The chromosomal basis of human neoplasia, Science 221:227-236 (1983).
23. J. D. Rowley, Biological implications of consistent chromosome rearrangements in leukemia and lymphoma, Cancer Res. 44:3159-3168 (1984).
24. J. Groffen, J. R. Stephenson, N. Heistercamp, et al., Philadelphia chromosomal breakpoints are clustered within a limited region, bcr, on chromosome 22, Cell 36:93-99 (1984).
25. E. Shtivelman, B. Lifshitz, R. P. Gale, et al., Fused transcript of abl and bcr genes in chronic myelogenous leukaemia, Nature 315:550-554 (1985).
26. S. S. Clark, J. McLaughlin, M. Timmons, et al., Expression of a distinctive BCR-ABL oncogene in Ph¹-positive acute lymphocytic leukemia (ALL), Science 239:775-777 (1988).
27. J. D. Rowley, Ph-positive leukaemia, including chronic myelogenous leukaemia, Clin. Haematol. 9:55-86 (1980).
28. P. C. Nowell, L. Moreau, P. Growney, et al., Karyotypic stability in chronic B cell leukemia, Cancer Genet. Cytogenet. 33:155-160 (1988).
29. P. Nowell, J. Finan, D. Glover, et al., Cytogenetic evidence for the clonal nature of Richter's syndrome, Blood 58:183-186 (1981).
30. P. C. Nowell, L. Jackson, A. Weiss, et al., Ph-positive chronic myelogenous leukemia followed for 27 years, Cancer Genet. Cytogenet. 34:57-61 (1988).
31. C. M. Croce and P. C. Nowell, Molecular basis of human B cell neoplasia, Blood 65:1-7 (1985).
32. C. E. Gauwerky, J. Hoxie, P. C. Nowell, et al., Pre-B-cell leukemia with a t(8;14) and a t(14;18) translocation is preceded by follicular lymphoma, Oncogene 2:431-435 (1988).

33. D. de Jong, B. M. H. Voetdijk, G. C. Beverstock, et al., Activation of the c-myc oncogene in a precursor-B-cell blast crisis of follicular lymphoma, presenting as composite lymphoma, N. Engl. J. Med. 318:1373-1378 (1988).

34. P. Leder, J. Battey, G. Lenoir, et al., Translocations among antibody genes in human cancer, Science 222:765-770 (1983).

35. J. J. Yunis, G. Frizzera, M. M. Oken, et al., Multiple recurrent genomic defects in follicular lymphoma: A possible model for cancer, N. Engl. J. Med. 316:79-84 (1987).

36. M. Isobe, G. Russo, F. G. Haluska, et al., Cloning of the gene encoding the delta subunit of the human T-cell receptor reveals its physical organization within the alpha-subunit locus and its involvement in chromosome translocations in T-cell malignancy, Proc. Natl. Acad. Sci. USA 85:3933-3937 (1988).

37. F. Hecht, R. Morgan, B. K.-M. Hecht, et al., Common region on chromosome 14 in T-cell leukemia and lymphoma, Science 226:1445-1446 (1984).

38. G. Russo, M. Isobe, L. Pegoraro, et al., Molecular analysis of a t(7;14) (q35;q32) chromosome translocation in a T cell leukemia of a patient with ataxia telangiectasia, Cell 53:137-144 (1988).

39. N. Sadmori, K. Miyuki, K. Nishino, et al., Abnormalities of chromosome 14 at band 14q11 in Japanese patients with chronic T-cell leukemia/ lymphoma, Cancer Genet. Cytogenet. 17:279-282 (1985).

40. B. Vogelstein, E. R. Fearon, S. R. Hamilton, et al., Genetic alterations during colorectal-tumor development, N. Engl. J. Med. 319:525-532 (1988).

41. E. Solomon, R. Voss, V. Hall, et al., Chromosome 5 allele loss in human colorectal carcinomas, Nature 328:616-619 (1987).

42. L. A. Cannon-Albright, M. H. Skolnick, D. T. Bishop, et al., Common inheritance of susceptibility to colonic adenomatous polyps and associated colorectal cancers, N. Engl. J. Med. 319:533-537 (1988).

43. B. S. Danes, E. J. Gardner, and M. Lipkin, Studies on the identification of genetic risk for heritable colon cancer, Cancer Detect. Prev. 8:349-365 (1985).

44. A. H. Parmiter and P. C. Nowell, The cytogenetics of human malignant melanoma and premalignant lesions, in: L. Nathanson, ed., "Malignant Melanoma: Biology, Diagnosis, and Therapy," Kluwer Academic Publishers, Boston (1988).

45. J. M. Cowan, R. Halaban, and U. Francke, Cytogenetic analysis of melanocytes from premalignant nevi and melanomas, J. Natl. Cancer Inst. 80:1159-1164 (1988).

46. N. C. Dracopoli, B. Alhadeff, A. N. Houghton, et al., Loss of heterozygosity at autosomal and X-linked loci during tumor progression in a patient with melanoma, Cancer Res. 47:3995-4000 (1987).

47. S. J. Bale, A. Chalravarti, and M. H. Greene, Cutaneous malignant melanoma and familial dysplastic nevi: Evidence for autosomal dominance and pleiotropy, Am. J. Hum. Genet. 38:188-196 (1986).

48. S. H. Bigner, J. Mark, P. C. Burger, et al., Specific chromosomal abnormalities in malignant human gliomas, Cancer Res. 88:405-411 (1988).

49. C. D. James, E. Carlbom, J. P. Dumanski, et al., Clonal genomic alterations in glioma malignancy stages, Cancer Res. 48:5546-5551 (1988).

50. M. J. Birrer and J. D. Minna, Molecular genetic of lung cancer, Semin. Oncol. 15:226-235 (1988).

51. I. U. Ali, R. Lifereau, C. Theillet, et al., Reduction to homozygosity of genes on chromosome 11 in human breast neoplasia, Science 238:185-188 (1987).

52. M. J. van de Vijver, J. L. Peterse, W. J. Mooi, et al., Neu-protein overexpression in breast cancer: Association with comedo-type ductal carcinoma in situ and limited prognostic value in stage II breast cancer, N. Engl. J. Med. 319:1239-1245 (1988).

53. E. P. Gelmann and M. E. Lippman, Understanding the role of oncogenes in human breast cancer, in: M. Sluyser, ed., "Growth Factors and Oncogenes in Breast Cancer," Ellis Horwood Ltd., Chichester (England) (1987).

54. T. Boveri, Zur Frage der Entstehung maligner Tumoren, Gustave Fischer Verlag, Jena (Germany) (1914).

55. D. Von Hansemann, Uber asymmetrische Zellteilung in Epithelkrebsen und der biologische Bedeutung, Virchows Arch. Pathol. Anat. Physiol. 119:298-307 (1890).

56. J. German, ed., "Chromosome Mutation and Neoplasia, Alan R. Liss, New York (1983).

57. R. Sager, Genetic instability, suppression, and human cancer, in: L. Sachs, ed., "Gene Regulation in the Expression of Malignancy," Oxford University Press, London (1985).

58. V. Ling, A. F. Chambers, J. F. Harris, et al., Quantitative genetic analysis of tumor progression, Cancer Metastasis Rev. 4:173-194 (1985).

59. R. Gantt, K. K. Sanford, R. Parshad, et al., Enhanced G_2 chromatid radiosensitivity, an early stage in the neoplastic transformation of human epidermal keratinocytes in culture, Cancer Res. 47:1390-1397 (1987).

60. J. Cairns, The origin of human cancers, Nature 289:353-357 (1981).

61. L. Loeb and L. Kunkel, Fidelity of DNA synthesis, Annu. Rev. Biochem. 52:429-457 (1982).

62. G. Poste, J. Doll, and I. J. Fidler, Interactions between clonal subpopulations affect the stability of the metastatic phenotype in polyclonal populations of B16 melanoma cells, Proc. Natl. Acad. Sci. USA 78:6226-6230 (1981).

63. W. C. Lambert and M. W. Lambert, DNA repair deficiency and cancer in xeroderma pigmentosum, Cancer Rev. 7:56-81 (1987).

64. F. Hecht, DNA ligase I, Bloom's syndrome, and cancer, Cancer Genet. Cytogenet. 30:181-182 (1988).

65. J. J. Yunis and A. L. Soreng, Constitutive fragile sites and cancer, Science 226:1199-1204 (1984).

66. G. R. Sutherland and R. N. Simmers, No statistical association between common fragile sites and nonrandom chromosomal breakpoints in cancer cells, Cancer Genet. Cytogenet. 31:9-15 (1988).

67. M. B. Sporn and G. J. Todaro, Autocrine secretion and malignant transformation of cells, N. Engl. J. Med. 303:878-880 (1980).

68. M. Sluyser, ed., "Growth Factors and Oncogenes in Breast Cancer," Ellis Horwood Ltd., Chichester (England) (1987).

69. E. Farber and C. Cameron, The sequential analysis of cancer development, Adv. Cancer Res. 31:125-225 (1980).

70. P. Nowell, Preleukemia: Cytogenetic clues in some confusing disorders, Am. J. Pathol. 89:459-476 (1977).

71. B. N. Ames, R. Magaw, and L. S. Gold, Ranking possible carcinogenic hazards, Science 236:271-280 (1987).

72. K. Yamashina, B. E. Miller, and G. H. Heppner, Macrophage-mediated induction of drug-resistant variants in a mouse mammary tumor cell line, Cancer Res. 46:2396-2401 (1986).

DISCUSSION

Joe W. Grisham: Peter, I think it is generally accepted that the rate of mutation in single genes, either spontaneous or induced, in mouse and human are similar. Yet, ease of transformation in vitro among them differs dramatically; there are some skeptics who think that human cells have not yet been transformed in vitro by chemicals. Also, there is the observation that in rodent cells chromosomes are very unstable and aneuploidy develops readily in culture. Do you have any insight into the mechanistic aspects of chromosomal instability in normal rodent cells, or on the regulation of chromosomal stability in human cells?

Peter C. Nowell: Let me give a short answer to a long question, Joe. No, I don't. I think there may be people in the audience who can respond to that, but I think that as far as I know nobody has worked on this apparent species difference or began to define it.

Harry Rubin: I'd like to come back to a point that you mentioned in passing when you started. You said that there are some cases of chronic myelogenous leukemia where the pathological symptoms or the standards of the disease show up before there is evidence of the Philadelphia chromosome. The Philadelphia chromosome appears later. Is that right?

Nowell: I think there are several problems, actually. There is the situation that you've described, but that is extremely rare. I only know of one or two cases, and then there is always the issue of what label are you putting on the particular syndrome. What I was just talking about was the work of Fialcow, who showed that even though myeloid elements might show the Philadelphia chromosome, some of the B cells in the clone did not. (I think that there is general agreement that the neoplastic clone arises from a totipotential hematopoietic cell, or multipotential, if you prefer).

Rubin: That is the second part of my question. So, chromosomal lesions can occur in some cells while others remain normal, i.e., the lymphocyte series. Isn't that what Fialcow found?

Nowell: That's another part of the story. I think that it is clear that this particular translocation we described and which has now been described at the molecular level, arises in a multipotential stem cell in the bone marrow; but the majority of the cells that expand are in the myeloid lineage, and so that's the label we put on the disease. This suggests that this particularly altered tyrosine kinase is more important for myeloid expansion than it is for lymphoid. Conversely, when you see the same translocation at the cytogenetic level in a patient with an acute lymphocytic leukemia, the breakpoint on chromosome 22 is slightly more proximal, and as a result you get a slightly different gene product. The supposition is that this altered tyrosine kinase is more active in allowing cells of the lymphoid lineage to expand. So, I don't want to use the word normal, because I think we're talking quantitatively here. It is also true that in most cases of typical CML, the circulating lymphocytes are not part of the neoplastic clone and have no chromosomal abnormality.

Rubin: The question really was, in those few cases where some symptoms or hematological evidence appeared before the chromosomal event occurred, do you have any thought about what was the driving force (since there were no chromosomal abnormalities)? Is it possible that we're missing cases like that, that they are not so rare?

Nowell: I don't have any explanation for what the initial event could be. I think that in that kind of situation there is a clonal expansion within the marrow which precedes the appearance of the Philadelphia chromosome. But in fact, we don't have any evidence that that occurs, except for the indirect evidence from tissue culture of B cell lines, indicating that they came out from some Ph positive patients. It could be any of the mechanisms that we have been talking about here earlier on: point mutation, viral induced clonal expansion, any phenomenon that you can envision that could initiate a clonal process.

George Yoakum: With regard to the B-CLL karyotype you showed, I think it was clear that that was not really a normal karyotype.

Nowell: I didn't say that it was. I should have emphasized that there was trisomy 12 in there, as well as an extra 18 and 19. One of the more remarkable facts was that this case had stayed this way unchanged for 10 years. From the first time we saw this patient, he had that abnormal karyotype, and it didn't change through 10 years of exacerbations and extensive therapy.

Yoakum: The stability of that karyotype to his status of disease is one that is a single point in time. We don't really have a panel of patients in whom you can show us a similar phenomenon, so that we can attribute it to something. It may simply be an anomaly of his particular constitution making this particular clone of cells stable for some reason that isn't clear.

Nowell: I didn't put the data up, but we just published a new long term series, since the techniques to induce these cells to divide became available recently. We have 21 CLL patients now that we have followed with sequential chromosome changes, all of them for more than 2 years, most of them for 5 years, and a few for 10 years. Half of those had chromosomally abnormal clones, half of them did not. In none of them did we see any karyotypic evolution, even though clinical progression did occur in half the patients. We did have several other patients who showed karyotypic evolution and died within the first two years, and so they didn't fall into that group. With karyotypic evolution, when it does occur in CLL, you get aggressive disease and poor prognosis. My point was simply that, it certainly suggests that the karyotype, even the abnormal karyotype in this disease, is very stable. Now, whether that means that it is not making variants or whether the variants are being more effectively eliminated, or whether tumor progression involves submicroscopic genetic changes, we don't know.

Unidentified Speaker: Where does the term promotion fit in your concept of tumor evolution in humans?

Nowell: I don't want to get into definitions, as we were requested not to, but operationally, I use promotion to indicate non-specific stimulation of proliferation. It could occur before a critical mutagenic event, as in the case of certain forms of hyperplasia. It could occur after mutagenesis, as in the model we've heard about involving radiation followed by subtotal hepatectomy in rodents. I think even that simplistic kind of concept blurs, because it is now clear that what we are calling non-specific stimuli of proliferation may, in fact, have some mutagenic component.

Sudilovsky: I had to fight for the microphone, so I hope that my question is appropriate, Peter. I believe one of the problems is how to measure genetic instability. It is obvious that if one finds chromosomal alterations then one knows that there is genetic instability; but are there any other parameters to measure genetic instability? The second question is that chromosomal changes occur before malignant changes, like in the melanotic dysplasias, and in Down's syndrome. What is their significance in regard to progression? Aren't those cells "progressed" already, and are some individuals born with progressive changes?

Moderator (Russel G. Grieg): You have ten seconds, Peter.

Nowell: Yes, there are other mechanisms for looking at instability. They include measuring mutation rates, either induced or spontaneous. You can also look at the capacity for DNA repair enzymes or at the fidelity of DNA polymerases, and so on. Yes, chromosome changes do occur before gross malignancy. You can get clones that expand and are clinically benign, which

means they haven't acquired the additional capacity to invade and to metastasize. Even in the hematopoietic system you can have expanded clones as in
the ataxia telangiectasia patient, of T cells that are chromosomally abnormal. They expand to a size that you can see easily in the peripheral blood
lymphocytes, but they are regulated and don't, therefore, come under the
rubric of neoplasm. <u>I see this as a quantitative continuum</u>, and the same
applies. Now, when you get to constitutional chromosome changes, as in
Down's syndrome, that's an entirely different story and I would prefer to
save it for a later discussion.

EDITOR'S NOTE

An interesting interpretation of the meaning of structural chromosomal
changes, such as aneuploidy in Down's syndrome, is expressed in this volume
by G. Barry Pierce (in the Discussion to his presentation).

UNKNOWN PRIMARY TUMORS: AN EXAMPLE OF ACCELERATED

(TYPE 2) TUMOR PROGRESSION*

Philip Frost

Departments of Cell Biology and Medicine
The University of Texas M. D. Anderson Cancer Center
Houston, TX

The acquisition of the malignant phenotype (invasiveness and metastasis) by tumor cells has been attributed to tumor progression, a term used by Foulds to describe the acquisition of permanent irreversible changes in a neoplasm[1]. Progression in turn has generally been presumed to result from "genetic instability" that results in the emergence of neoplastic cells that have lost control of the mechanisms governing or regulating gene expression[2,3]. From these ideas evolved the proposal that tumor progression is generally unidirectional, since genetic or genomic changes favoring malignant cells with some form of a presumed growth advantage generally come to dominate tumor growth. The advantages acquired by these cells were hypothesized to be due to an increasing rate of genomic instability that accompanied the increasingly malignant phenotype[4,5]. Thus, it is believed that only cells with an increased rate of genomic instability become malignant (metastatic). In short the transformation of normal cells into tumor cells is thought to be accompanied by the destabilization of the genome, leading to tumor heterogeneity which in turn accompanies tumor progression to a malignant phenotype. One difficulty with this paradigm is that we have only a minimal understanding of the events responsible for the transformation of a normal to a malignant cell. It has therefore been accepted a priori that any analysis of tumor initiation, promotion and progression accept this gap in knowledge and proceed from there.

We would like to propose a change in thinking regarding tumor progression. Instead of assuming that all tumors progress to a malignant phenotype by well defined steps, we would suggest that while this is true for some tumors (colon[6,7], cervix, melanoma[8]) it is not true for other tumors such as lung, renal and prostatic carcinomas.

DO ALL TUMORS PROGRESS?

It has been shown that the various steps in the malignant cascade can be histologically defined in colon carcinoma, melanoma, cervical carcinoma, etc. The most detailed analysis of this form of progression (Type 1) is colon carcinoma in which histologic, karyotypic, oncogene and DNA methylation changes can be shown to occur as a benign colonic polyp becomes a liver

*Supported in part by Grant number 38953 from the PHS and Cancer Development
 Funds of The University of Texas M. D. Anderson Cancer Center

Boundaries between Promotion and Progression during Carcinogenesis
Edited by O. Sudilovsky *et al.*, Plenum Press, New York, 1991

metastasis[7,9,10]. However, it is difficult to visualize such a cascade of events in other human tumors such as pancreatic, renal cell, oat cell, lung or prostate carcinoma, for no clearly (histologically) defined pre-malignant lesions have been described for these tumors. Thus, it must be assumed that this latter group of tumors acquire a malignant phenotype directly by unrelated mechanisms or soon after the transformation event; they undergo accelerated progression (Type 2). The most dramatic group of tumors of this type are the unknown primary tumors which will be discussed below.

TYPE 1 TUMOR PROGRESSION

There are several essential features to this classic form of tumor progression. These were defined by Foulds[1] who emphasized that as tumors progress they acquire permanent irreversible changes. From this principle, that was based on histopathologic observations, has evolved the view that the changes observed during progression are due to genomic instability, a term used to explain the marked variability and heterogeneity seen in neoplastic cells. In conjunction with this view it was proposed that as tumors progress from the benign to the malignant phenotype, their rate of genomic instability increases[1]. In contrast, we have argued that the rate of genomic instability is constant, i.e., as a cell becomes more malignant, the rate of generation of karyotypic or phenotypic abnormalities need not increase[4]. The accumulation of genomic alterations at a constant rate achieves the same diversity as would an accelerated rate of instability. It is therefore not surprising that the rate of spontaneous mutation and the generation of karyotypic abnormalities are no different in benign and malignant (metastatic) cells[12-14]. In fact, treatment with mutagens appears, in some cases, to produce more mutations in benign as opposed to metastatic cells[15]. In addition, teleological arguments support the view that a constant rate of "genomic instability" is more likely to produce a malignant cell with a greater survival potential. This view is based on the following reasoning. Genomic instability is in part unpredictible and as such the phenotypic changes resulting from genotypic alterations may or may not be of benefit to the cell. Mutations, rearrangements or altered regulation of the expression of essential genes could be lethal. Indeed, the occurrence of non-beneficial or even lethal genetic changes may be quite high and some tumors do regress[16]. However, at a constant rate of change, non-lethal "hits" are cumulative and some cells acquire the ability to apparently absorb non-lethal changes, for they do survive. Malignant cells are therefore those selected for their ability to absorb numerous non-lethal genetic "hits". We would propose that the chance of these cells surviving, if the rate of instability increased with time, would be very low -for as the rate of change increased, the risk of a lethal genetic alteration would increase and cells would ultimately lose their ability to absorb and tolerate such changes.

The concept of genomic instability has also fostered a common view that tumors are extensively heterogeneous. While one cannot disagree with the obvious heterogeneity of neoplastic cells, I would take issue with the relevance of all the changes observed and would emphasize that despite considerable heterogeneity, tumors from particular organs do have basic features in common. This view is based on the hypothesis that not all genes are regulated in an identical manner[17]. Thus, genes determining organ specificity are inviolate, i.e., colon tumors express colon specific genes and despite the disorganization brought on by transformation, colon tumors do not express genes of an unrelated organ phenotype. In fact, on a molecular level evidence for reproducible chromosomal deletions and k-ras mutations during the progression of colon carcinoma have already been reported[7]. In addition, the clinical behavior of most tumors is remarkably reproducible. Similarly, Imm+ variant clones obtained from a heterogeneous parent population can protect against a challenge with viable parent cells. This must mean that

234

despite their heterogeneity, all cells in the parent population must share a common antigen with the Imm⁺ clone. These three points argue for limits in heterogeneity, a view which supports the contention that genomic instability does not produce endless phenotypic variations[17], but rather allows for the selection of cells with specific changes likely related to the original genetic background of the transformed cell, that provide for a growth advantage.

The work of Vogelstein et al., particularly strengthens the idea that while multiple genomic changes may occur ([7] and Vogelstein, personal communications) there are only some that are relevant to the development of neoplasia. It is the identification of those specific [most likely organ specific] changes that represents one of the major challenges in tumor biology. However, there may well be early genomic changes [seen after transformation] that are common to all tumors but are not readily detectable at the time neoplasia is clinically recognizable and thus available for analysis. This hypothesis has led us to attempt to identify such early changes using a unique human tumor model.

TYPE 2 TUMOR PROGRESSION

While many different tumors fit this category -pancreas, kidney, prostate are but a few examples- we have chosen Unknown Primary Tumors (UPT) as the best example of tumors that do not detectably progress through the steps outlined for Type 1 progressing tumors. By definition, UPT present as metastases in sites where they could not have originated (adenocarcinoma in bone or lymph node)[18]. Furthermore, UPT are unique in that the primary tumor cannot be identified. It has been proposed that the primary tumor is either too small to detect, or has involuted[18]. Either of these possibilities infer that subsequent to transformation the primary tumor has changed its phenotype and presumably its genotype. It has either been deleted by numerous lethal changes or affected by alterations that severely limited its growth rate. These events had to have occurred after the fully malignant cells had already metastasized, though some would argue that UPT metastases develop soon after transformation and no primary tumor ever forms. These cells can therefore be presumed to have undergone a more rapid series of changes, possibly because they were "more genomically unstable" and were able to dramatically shift their phenotypes because of numerous genotypic changes. UPT are presumed, therefore, to have migrated out of the primary tumor before it had reached the minimally detectable size of 1 cm (10^9 cells, 30 doublings). Thus, UPT may bring us somewhat closer to transformation -albeit likely still many steps removed, but closer than metastases from large tumors of known origin, and the observation of consistent cytogenetic changes in these cells could reflect on earlier events after transformation. This is an important issue since many human tumors have numerous karyotypic abnormalities, the relative importance of which cannot be ascertained[19]. In UPT, we may have the opportunity to identify cytogenetic abnormalities which may be specific for early changes in specific tumors and could reflect on the early effects of genomic instability. If such changes can be seen, then a more focused cytogenetic search of tumors of human origin can be undertaken.

Our own studies of the cytogenetics of four UPT lend support to the view that these tumors, while they have a heterogeneous origin, have cytogenetic features in common. We have identified a high frequency of structural alterations in chromosomes 1, and 7 in these tumors[2]. For example, simple deletions of the long arm of chromosome 7 involving band region q11-q21 was seen in three of the four cases and deletions of the short arm of chromosome 1 was seen in all cases. While a more detailed description of these cases is reported elsewhere, we would propose that the analysis of additional UPT patients will show similar alterations consistent with the view that UPT are a unique type of neoplasia. As mentioned earlier, it is also possible that the observed chromosomal changes may represent early cytogenetic alterations occurring soon after transformation.

In summary, we suggest that the commonly accepted views regarding tumor progression be adjusted to take into account that:

1. The rate of genomic instability need not increase as Type 1 tumors progress from the benign to the malignant phenotype. A constant rate of genomic instability can result in the same degree of (limited) diversity and malignancy because the genotypic changes are cumulative.

2. Genomic instability does not result in only random phenotypic changes. There are limits to tumor heterogeneity likely dictated by the organ specific gene background of the transformed cell.

3. UPT may provide a means for studying cytogenetic and molecular changes that occur closer to the transformation event.

4. Tumors do not all progress by the classic mechanisms (Type 1) described for colon carcinomas. Most tumors likely progress by an accelerated mechanism (Type 2) that is yet to be defined.

REFERENCES

1. L. Foulds, The experimental study of tumor progression: a review, Cancer Res. 14:327-339 (1954).
2. P. C. Nowell, The clonal evolution of tumor cell populations, Science 194:23-28 (1976).
3. P. C. Nowell, Mechanisms of tumor progression, Cancer Res. 46:2203-2207 (1986).
4. W. A. Kendal and P. Frost, Constancy of genomic instability in tumor progression, J. Theor. Biol. 126-369-371 (1987).
5. W. A. Kendal and P. Frost, Genetic instability and tumor progression, Pathol. Immunopathol. Res. 5:455-467 (1986).
6. B. C. Morson and I. M. P. Dawson, "Gastrointestinal Pathology," 2nd ed., Blackwell Scientific, London (1979).
7. B. Vogelstein, E. R. Fearon, S. R. Hamilton, et al., Genetic alterations during colorectal-tumor development, NEJM 319:535-532 (1988).
8. Y. H. Pilch, Malignant Melanoma, in: "Surgical Oncology", Y. H. Pilch, McGraw Hill, New York (1984).
9. S. Goelz, B. Vogelstein, S. R. Hamilton, and A. P. Feinberg, Hypomethylation of DNA from benign and malignant human colon neoplasms, Science 228:187-190 (1985).
10. J. Bos, E. Fearon, S. Hamilton, M. Verlaan de Vries, J. van deBoom, A. van der Eb, and B. Vogelstein, Prevalence of ras gene mutations in human colorectal cancers, Nature 327:293-297 (1987).
11. M. Cifone and I. J. Fidler, Increasing metastatic potential is associated with increasing genetic instability for clones isolated from murine neoplasms, Proc. Natl. Acad. Sci. 78:6949-6952 (1981).
12. W. S. Kendal and P. Frost, Metastatic potential and spontaneous mutation rates: Studies with two murine cell lines and their recently induced metastatic variants, Cancer Res. 46:6131-6135 (1986).
13. W. S. Kendal, R. U. Wang, T. C. Hsu, and P. Frost, Rate of generation of major karyotypic abnormalities in relationship to the metastatic potential of B16 murine melanoma, Cancer Res. 47:3835-3841 (1987).
14. W. S. Kendal, R. Y. Wang, and P. Frost, Spontaneous mutation rates in cloned murine tumors do not correlate with metastatic potential, whereas the prevalence of karyotypic abnormalities in the parental tumor does, Intl. J. Cancer 40:408-413 (1987).

15. P. Frost, W. S. Kendal, B. Hunt, and M. Ellis, The prevalence of ouabain resistant variants after mutagen treatment: failure to correlate the frequency of variant expression with the metastatic phenotype, Invasion and Metastasis 8:73-86 (1988).
16. T. C. Everson and W. H. Cole, "Spontaneous Regression of Cancer," Saunders Pub., Philadelphia (1966).
17. P. Frost, W. S. Kendal, and R. S. Kerbel, The limits of tumor hetero-geneity, in: "Neo-adjuvant Chemotherapy," Colloque INSERM/John Libbey Eurotext Ltd. 137-57-60 (1986).
18. J. L. Abbruzzese, M. N. Raber, and P. Frost, An effective strategy for the evaluation of unknown primary tumors, Cancer Bulletin (1988).
19. C. Bell and P. Frost, Characterization of two cell lines derived from human metastatic adenocarcinomas of unknown primary origin (UPT), Proc AACR 29:28 (1988).

DISCUSSION

Moderator (Russel G. Greig): First question: before your presentation, you mentioned to some of us a War Games scenario. What were you going to say about it?

Philip Frost: Well, the War Games scenario. There was a movie called "War Games," and in the movie there was a computer that had gone berserk and was searching for the sequence number for launching missiles. There was a sequence of about 12 numbers and it went along searching for them; every time it found a number that it somehow intuitively knew was critical to the launch sequence, it locked it into position. Finally the programmers found a way of getting around the computer program before the missile was launched by making the computer understand that the game it wanted to play was a failed game. Nobody wins.

I've tried to make an analogy to cancer in terms of the same thing. What happens in a cancer cell is that transformation occurs. Again, we are stuck with this very, very serious problem in not understanding what that is. But, once it occurs, I would propose that the tumor cell does certain key things. It tries to lock in essential genes for survival, so that those genes are not altered. Those genes may be locked in by mechanisms that are similar to the mechanisms that lock in organ specificity. Organ specificity is locked in very early on, otherwise the embryo would be a tower of Babel. Once that happens, once the tumor cell locks in those genes, then it says anything else you want to change is alright with me. I don't care. I've got what I need. So that once you lock in what is essential for growth and for division, any other chromosomal changes, karyotypic changes, molecular changes that occur may be not relevant to the development of the tumor. That's what the War Games scenario is. Now, if we could teach this cancer cell that this is a no win situation, then we'd be in very good shape.

Harry Rubin: You said that it's very unusual for tumors in a particular tissue to express genes from another tissue. At least one case that I can think of is in Rous transformation, when Rudine and Weintraub got expression of 1,000 new genes. One of them was a transcription of the hemoglobin RNA. I don't think hemoglobin was actually produced, but messenger RNA.

Frost: This was in a Rous sarcoma system?

Rubin: That's right.

Frost: I'm talking about human tumors.

Rubin: Well, that's an example of a possibility. We hear a lot about oncogenes which were silent in a particular tissue and then are produced later. There must be genes that are expressed in another tissue somewhere, whether they are identified as a specific differentiated phenotype is beside the point. That's number one. Number two is, really I have a question. When you mention that there are such things as premalignant lesions, I'm excluding adenomas and things like that.

Frost: I didn't say there are no such things. I said they haven't been defined as far as I know.

Rubin: Well, I wanted to hear your comments about claims that have been made at various times that look intriguing to me but never seem to get followed up, of behavior of supposedly normal cells in either patients with cancers or patients closely related to individuals with cancer or from families of cancer history, such as their fibroblasts have an embryonic mode of invading collagen or, as Catherine Sanford says, that they have a poor DNA repair period if you x-ray them during the G2 period, or, as Aroni and Macieira-Coello say, that there's something funny about an extended S period in normal fibroblasts from patients, let's say, with mammary cancer. Or Kopelovich's claim which was put up here by an earlier speaker that in people who had a tendency to get colon cancer or had colon cancer their normal fibroblasts had a disrupted or an abnormal cytoskeleton development. So these are all, I don't know if you want to call them lesions, but characteristics of supposedly normal cells in some way associated with malignancy. Do you believe them? Is there any credence to them? What are you thinking about them?

Frost: I don't know. There are some areas there that are controversial. Some people argue that they've seen changes in fibroblasts. Others have not been able to reproduce them. I don't have the answer. I think that if there is such a thing as a cancer diathesis -diathesis is a tendency towards something; you can have a diathesis for kidney stones, for instance- I don't know of any absolute evidence that you can document it in the normal cells of those patients. There are people who claim they can, but I'm no expert in that area, so I wouldn't comment.

George Yoakum: Philip, I think there's one point I'd like to clarify with you. That is the question of rate of karyotypic abnormality. I think it's perfectly reasonable to contend that cells may be able to take up an increased rate of accumulation of abnormal karyotypes without having to maintain a constant rate of increase. In other words, they may be able to be higher than normal without having to go on to commit suicide as you tended to suggest in your presentation. Secondly, I think that the ouabain experiment is an interesting one, but could have been done better with chemotherapeutic drugs. I think you may have been addressing the phenotypic niche of the cells more accurately, possibly, than ouabain would. I also noticed that on your list there are tumors which metastasize immediately and that we really have very poor ways of diagnosing cancer without terribly invasive techniques. It appears to me that in the case of breast cancer, for instance, when methods came along that were less invasive and when more screening was done, better treatment was offered. In those cases I think that there are statistics that would suggest just clearly that there are premalignant conditions that occur in those tumors. And that may very well be the case for all of the others; it's just that we can't look at the pancreas very well every day.

Frost: I think that you'd have to show me those statistics to convince me. I don't believe that. Secondly, the procedures used are really not all that invasive. A CAT scan will pick up a 1 cm tumor. There is no other technique for measuring anything smaller than that.

Yoakum: Who gets a CAT scan every year?

Frost: Well, if you'd come to M. D. Anderson you'd get one every year. The point is that there is a procedure that's relatively straightforward. I'm not advocating doing CAT scans on everybody every year, but in colon tumors you can do occult blood and get an easy assessment that something is wrong. And I agree that it's possible that some of these tumors are picked up earlier simply because of that. But, you know, many colon patients present with metastasis. They never have any occult blood. They never have any symptoms. They present with pain in the right upper quadrant because they have liver metastases and then you find the primary tumor. I'm trying to remember your other two questions. Oh, yes, on the rate business, we can talk about it privately but I don't see how that can happen. I mean, if you keep increasing the rate as the tumor gets more malignant, the rate of change keeps increasing at the same time.

Daniel Longnecker: I receive part of your presentation very warmly, in that I think you're doing occupational psychotherapy. As an autopsy pathologist, I always regard the unknown primary as a failure on my part, and I assume that a number of oncologists feel the same way, so maybe we can feel a little better. I would like to side with Dr. Yoakum on the issue of Type 1, Type 2, and unclassified, and lobby that you should move pancreas to Type 1 or at least unclassified. I've been involved in an autopsy series that says that in aged individuals, more than 50% of us have either focal hyperplastic or clonal abnormalities that are similar to things that are seen in carcinogen treated animals.

Frost: That point is well taken, and I'll talk to you about that. The list is arbitrary, in a sense. Obviously, I'm not familiar with everything that is going on in the world, so I appreciate your telling me that.

Oscar Sudilovsky: I will need your help, Phil, to ask this question. If you can show us again the slide stained with Feulgen, I want to draw your attention to the alterations in the "normal epithelium" showing karyotypic changes in chromosome 5 before there is a hyperplastic epithelium. That is to me an announcement of genetic instability in that chromosome. And now, I would like you to go backward again, to the slide in which you presented the scheme of initiation, progression, etc.

Frost: That's all the way to number one!

Sudilovsky: Is that too much? The reason I want to show these two slides is because chromosome alterations in the colon occur before progression. That means that the initiated cells do have a change in their genomic stability, and I wonder whether progression in that slide should have been placed rather to the left, in the part that you mentioned as transformation, so that progression occurs during promotion or immediately after initiation. That is the problem that I want to bring up in regard to the boundaries. Progression may occur simultaneously with or instead of promotion, immediately after initiation. I would like to have your comments on that.

Frost: You mean you want to change where the...I'm not...

Sudilovsky: Progression occurs before a benign tumor happens.

Frost: The problem with all of that is that we don't know what happens during transformation. So what we're really doing is guessing, and you can draw the schema any way you like.

Sudilovsky: I refer you to your slide with the definition of progression by Foulds, saying that there is the need of an irreversible change in the cell.

239

Isn't that what the karyotypic alterations in chromosome imply? Because there is indeed a modification in the amount of genomic information in those chromosomes.

Frost: That's fair enough.

SUMMARY BY MODERATOR

Moderator (Russel G. Grieg): Let me just summarize very briefly my
impressions. I think both speakers have demonstrated that karyotypic
changes, some random, some not so random, accompany tumor progression; but
the mechanism of these changes remains somewhat cloudy. We have not enough
evidence that these changes are going to have prognostic or therapeutic
implications; however I think everyone involved in the area would agree that
they are worthy of systematic and methodical pursuit in order to establish
whether they will be relevant to the patients. I think one of the catalysts
for the future studies will be Dr. Nowell's comment that we have to avoid
doctrinaire statements if we are to make progress in what is a very cloudy
area, confused by some very awkward semantics.

TUMOR HETEROGENEITY AND INTRINSICALLY CHEMORESISTANT SUBPOPULATIONS

IN FRESHLY RESECTED HUMAN MALIGNANT GLIOMAS*

Joan R. Shapiro*, Bipin M. Mehta, Salah A. D. Ebrahim,
Adrienne C. Scheck, Paul L. Moots, and Martin R. Fiola

George C. Cotzias Laboratory of Neuro-Oncology
    and Department of Neurology
Memorial Sloan-Kettering Cancer Center,
Department of Neurology,
Cornell University Medical College [J.R.S.]
New York, NY 10021

INTRODUCTION

Malignant gliomas continue to defy clinical treatment despite aggressive therapy that includes a combination of surgery, radiotherapy and chemotherapy. In the last 5 years, our success rate has not improved from 50% mortality within six months and 90% mortality within 1 1/2 years[1]. Furthermore, the design of new treatment strategies has been limited by our lack of understanding of the fundamental processes which occur in this tumor. Therefore, if new drugs are to be developed and rational therapeutic approaches devised we must understand the complex biology of human malignant gliomas that contribute to cellular resistance.

To meet these objectives, major research efforts continue to define characteristics of this tumor system and phenotypes associated with tumor progression[2,3]. These investigations have begun to yield data that define genotypic changes that may serve as hallmarks of the disease. Such changes include amplification[4-8] or loss of specific genes[9] that seem to occur during tumor progression.

Work in our laboratory has focused on the problem of chemoresistance. For chemotherapy of brain tumors to advance beyond its current empiric approach, it must be made specific, both in terms of its role as a cellular poison and in terms of methods to define individual cellular susceptibility. With respect to the latter problem we have begun to define the characteristics of intrinsically chemoresistant malignant glioma cell(s). This type of investigation required that we define (i) what cells exist in the tumor population at the time of resection and, (ii) which of these cells, if any, survives to repopulate the tumor mass following treatment with irradiation and/or chemotherapy.

Initially we reported that malignant human gliomas are karyotypically

*This work is supported by Grants CA 25956 and CA 07848 of the National
 Institutes of Health, Bethesda, MD and the Preuss Foundation for Brain Tumor
 Research and Moulin Foundation.

Boundaries between Promotion and Progression during Carcinogenesis
Edited by O. Sudilovsky *et al.*, Plenum Press, New York, 1991

243

heterogeneous tumors with numerous cell populations and isolated cell types[10]. Subpopulations within the heterogeneous tumor had variable stability patterns in culture[11] and often responded differently to chemotherapeutic agents, thus making tumor heterogeneity a therapeutically relevant issue[12]. Along with other investigators[11,13-19], we examined freshly resected gliomas for non-random chromosomal changes. These karyotypic studies were extended to include an examination of regional heterogeneity within the same tumor[20,21] in an attempt to identify phenotypic changes that might be associated with these karyotypic changes. Karyotypic analyses performed in the first 72-105 hours following tumor resection[11] and cytogenetic analyses on early passage cells maintained in vitro, demonstrated rapid cell selection[22]. Further, as in vitro evolution occurred in both the parental populations and clones isolated from them, one could demonstrate altered karyotypes and drug sensitivity. In the ensuing sections we will briefly review some of these findings and conclude with our latest observation in which we will describe the identification of a minor subset of cells in primary tumors that are chemoresistant and capable of repopulating the tumor mass. This chemoresistant cell has a specific chromosomal constitution and several aberrant phenotypes that may provide such cells with a growth and/or survival advantage. Additional evidence for this is the fact that identical cells are found in recurrent tumor samples from these same patients.

Karyotypic Heterogeneity in Freshly Resected Human Gliomas

Freshly resected human gliomas are relatively easy to grow in vitro and numerous lines have been established for analysis. However, it is crucial that the cytogenetic analysis be performed using freshly resected cells to ensure that the data obtained is free of artifacts that arise due to the influence of a long-term tissue culture environment. Such an environment permits adaptations and/or genetic drift to select the cell types that populate the heterogeneous mixture of cells. For this reason we established the following protocol that allows us to examine first division cells following tumor resection (Fig. 1). Two techniques are utilized to obtain as many metaphase cells as possible. The Reference Set of karyotypes establishes the "fingerprint" pattern of that tumor. This permits examination of the chromosomal constitution of the dividing tumor cell and of the changes that occur as the tumors' cellular populations evolve[23,24].

The histogram depicted in Fig. 2 illustrates some of the differences in the distribution of chromosome numbers in freshly resected tumor samples. Each of the patients represented in the histogram has the same pathological grade of tumor yet the distribution of cells in each tumor is quite different. For example, the distribution of cells may be primarily hypodiploid (JC), near-diploid (MK), bimodel (MS), trimodel (MA), or extensively hyperdiploid (HFA and EI) in which the majority of cells have 3n and 4n chromosome numbers.

Our protocols also permit us to clone cells directly from the freshly dissociated tumor population. Clonal analysis of the tumor coded MA (Fig. 2) further supports the morphologic as well as karyotypic heterogeneity described above (Fig. 3-5). Clone MAC-19 (Fig. 3) is typical of many of the near-diploid clones in which chromosome 22 is frequently monosomic in a 2n cell. In addition, this karyotype depicts the loss of several other chromosomes that are unique to this clone. MAC-32 is representative of the tetraploid clones (Fig. 4). Most of the chromosome numbers are represented by 4 homologues, suggesting that this cell was derived from a 2n cell that had undergone endopolyploidization by any one of a number of mechanisms[25]. The karyotypic analysis of this clone also demonstrated the under-representation of chromosome 22; the loss of chromosome 4 is unique to this karyotype. Unlike the previous two clones, clone MAC-24 is not representative of the

244

majority of cells observed in the primary cytogenetic analysis of the MA tumor (Fig. 5). While many of the karyotypes in the primary analysis demonstrated the gain and loss of specific chromosomes (gain of chromosome 7 and loss of chromosome 10 and 13) only a few karyotypes demonstrated the gain of chromosome 22 as depicted in this karyotype (Fig. 5). In clone MAC-24, chromosomes are represented by a single homologue (chromosomes 10, 13 and 21) or multiple homologues (chromosomes 7, 12, 15, 17, and 22). When one analyzes 15-25 karyotypes from such a clone several chromosomes are consistently over-represented (chromosomes 7 and 22) while others are consistently under-represented (chromosome 10 and 13). Additionally, this clone contains the only marker chromosome observed in the primary analysis, del(3)(p21).

Regional Heterogeneity in Freshly Resected Human Gliomas

One question that is raised by such analyses is why tumors with the same histopathology should appear so divergent in chromosome number. As depicted in Fig. 2, the variability in chromosome number among the tumors contrasted to the patients' clinical courses, in that 5 of the 6 patients died approximately 1 year after tumor resection; only the patient coded MK is 4+ years. One possible explanation for the tumors' cellular variability was that the biopsies used for these studies represented only one part of these tumors, and that other sections might have been more representative.

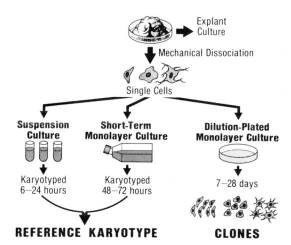

Fig. 1. Cell culture protocol. The chart depicts by flow diagram the protocol used to dissociate glioma tumors into single cells. First the tissue is minced to a fine pulp, then aspirated twice through an 18 gauge needle. The cell mixture is passed over a mesh screen to separate single cells from tissue clumps. Single cells are divided for suspension and short-term culture (cytogenetics), dilution-plating (cloning), monolayer cultures and drug testing (not included in this diagram). Reprinted with permission, Shapiro, JR et al, 1981).

To investigate regional heterogeneity, biopsies removed from different regions of a high grade tumor were examined for karyotypic differences. As an example, the following data were derived from a freshly resected human glioma coded HFA. Fig. 6 illustrates the distribution of tumor cells, by chromosome number, in two different regions of the same tumor. Figs. 7-10 illustrate the karyotypic differences and evolution of cell types in these two regions. The tumor region designated HF contained a stem-line karyotype

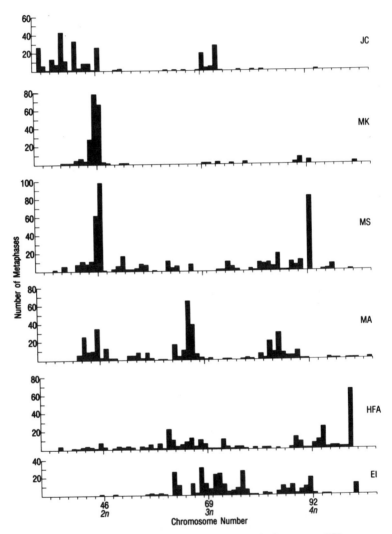

Fig. 2. The distribution of cells containing specific chromosome numbers in 6 human glioblastomas multiforme. The chromosome preparations were made from cells illustrated in our flow diagram. Two hundred to 250 cells are counted and karyotyped to determine the chromosomal constitutions of subpopulations present in the tumor sample. Reprinted with permission, Shapiro, JR and Shapiro, WR, 1985.

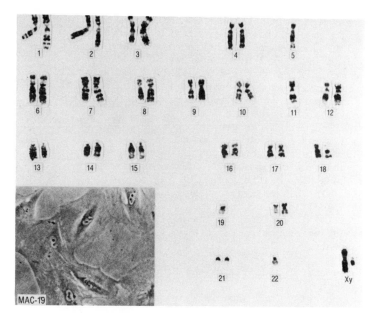

Fig. 3. Representative stem-line karyotype of the near-
diploid clone, MAC-19. The karyotypic deviation
is: (2n±) 42,XY,-5,-11,-19,-22. Giemsa banding,
x1,100.

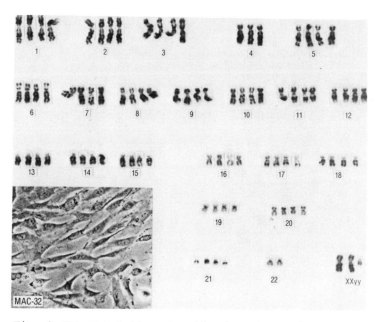

Fig. 4. Representative stem-line karyotype of the tetraploid clone, MAC-32. Each homologue is represented by 4 homologous chromosomes (the loss of
chromosome 4 is unique to this karyotype) except
chromosome 22. Eight karyotypes prepared from
this clone were similarly missing 2 copies of
chromosome 22. Giemsa banding, x1,100.

(most frequent karyotype) in which the chromosomes had normal Giemsa-bands and only a single sex chromosome was missing. Karyotypes prepared from cells that contained 46 chromosomes demonstrated two karyotypic patterns, one with 46,XX chromosomes and normal Giemsa-bands and a second one illustrated in Fig. 7. In this cell the single X was missing but an additional change had occurred; chromosome 7 was over-represented. Fig. 8 illustrates the karyotypic evolution observed in this region of the tumor. This cell has the over-representation of chromosome 7 and the loss of one sex chromosome. However, additional numerical changes, the most frequent of which was the loss of chromosome 22, and a marker chromosome, del(6)(q21) were also present. The tetraploid populations were a continuum of these findings in which the over-representation of chromosome 7 and the under-representation of chromosomes 22 and the X were maintained. The region designated HFA was markedly more aneuploid. In this analysis no normal 46,XX Giemsa-banded karyotypes were observed. Additional rearrangements included several translocations t(3;7) and t(15;18) in addition to the del(6)(q21) (Figs. 9 and 10). Thus, karyotypic evolution is most notable when two distal regions of a tumor are observed separately.

Chemosensitivity of Parental Tumors and Clones from the Same Tumor

That heterogeneity (and especially regional heterogeneity) is important in the chemotherapy of brain tumors and can be demonstrated readily by the differences in chemosensitivity of cellular populations isolated by early

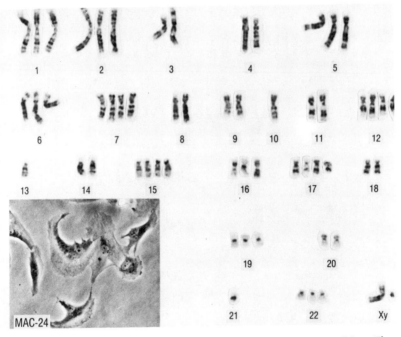

Fig. 5. Representative stem-line karyotype of clone MAC-24. The karyotypic deviation is: (2n±) 58,XY,+1,+2,+5,+6,+7,+7, +12,+12,+15,+15,+16,+17,+17,+19,+22,-10,-13,-21,. Twenty-one karyotypes were prepared from this clone and all chromosomal constitutions of these karyotypes demonstrated an over-representation of chromosomes 1,7,17 and 22 and an under-representation of chromosomes 10 and 13. Giemsa banding, x1,100.

Fig. 6. The two histograms HF and HFA illustrate the distribution of metaphase cells by chromosome numbers. Freshly resected glioblastoma tumor was removed from a female patient and was dissected into two regions at the interface of what appeared grossly to be normal cortex (HF) adjacent to tumor (HFA). Each region was dissociated into a pool of single cells and subjected to cytogenetic analysis within 72 hours post-plating. Over two hundred cells were counted for each region. Region HF was predominantly near-diploid (2n) in chromosome number while region HFA was primarily hyperdiploid with the majority of cells having triploid (3n) and tetraploid (4n) chromosomal complements. (Reprinted with permission, Shapiro, JR, 1986).

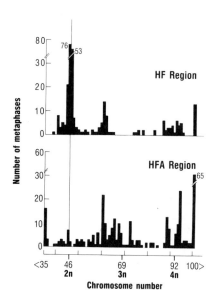

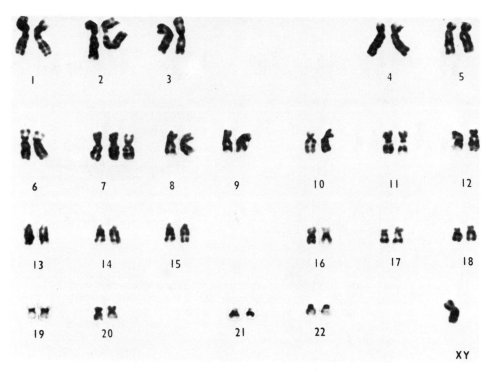

Fig. 7. A metaphase cell prepared from the suspension cultures of the HF region. This karyotype contains 46 chromosomes and is representative of tumors that are primarily near-diploid in chromosome number. It is monosomic for a sex chromosome and trisomic for chromosome 7. The Giemsa bands (G-bands) appear normal for all chromosomes. x1,100. (Reprinted with permission, Shapiro, JR, 1986).

passage techniques[12]. Dose response curves of colony forming assays (CFA) in clones grown from a freshly resected human malignant glioma and treated with BCNU differed significantly, although the differences were not large. Similarly, dose-response curves of clones treated with cisplatin differed, the differences also being small. In our study of regional heterogeneity in gliomas, we found that mixed cell populations from different regions had different chemosensitivities, reflecting intrinsic cellular resistance[20]. In Figure 11, the distribution of cells by chromosome number is illustrated for 2 separate regions of tumor RM. The chromosome patterns within the two regions (RM and RM1) demonstrated karyotypic heterogeneity similar to that described for HF and HFA. The chemosensitivity profile of the two regions reflect these karyotypic differences (Fig. 12). The more extensively aneuploid region RM1 demonstrated greater sensitivity to BCNU than region RM in which the majority of cells were near-diploid. Our analysis of clones from region RM[22] illustrates more acutely the relationship between chromosome number and chemosensitivity (Fig. 13). All the clones with near-diploid chromosome number were resistant to BCNU while clones that were hyperdiploid (60 or more chromosomes/metaphase) were sensitive. We have continued to

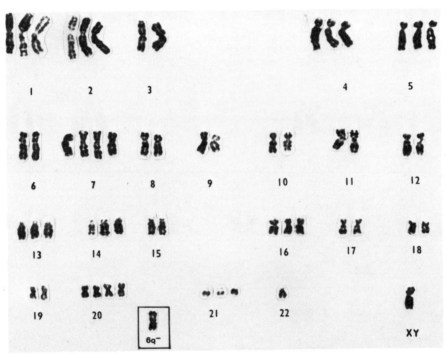

Fig. 8. A metaphase cell prepared from the suspension cultures of the HF region. This karyotype contains 56 chromosomes with normal G-bands and one marker chromosome designated del(6)(q21), in which the lower portion of the q arm is deleted (enclosed area, 6q-). The del(6)(q21) was the only marker chromosome seen more than once and therefore represents a clonal marker for this region. Cells carrying marker chromosomes represented less than 2% of the total cells analyzed in this region. In addition to the marker chromosome this cell has undergone numerous segregational errors; chromosomes 7 and 20 are tetrasomic, chromosomes 1, 2, 4, 5, 13, 14, 16, and 21 are trisomic and chromosomes 22 and the sex chromosome are monosomic. (Reprinted with permission, Shapiro, JR, 1986).

study intrinsic and acquired resistance to chemotherapeutic agents among glioma cell populations and clones[26-29]. Such studies have continued to indicate that the most resistant cell in the freshly resected glioma is near-diploid in chromosome number, while hyperdiploid cells tend to be sensitive.

Karyotypes of Intrinsically BCNU-Resistant Cells

The above studies permitted us to gather a large number of resistant clones and cell types. The karyotypic analysis of these cells began to reveal a pattern of chromosome retention that was consistent between different tumors, this being the over-representation of chromosomes 7 and 22[28,29]. This consistent finding suggested that something on these two chromosomes might confer a selective and/or survival advantage to these cells.

Information about the possible significance of chromosome 7 has been the focus of several investigations. These studies demonstrated that the ampli-

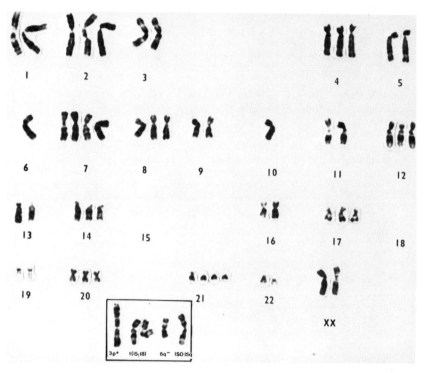

Fig. 9. This karyotype was prepared from the suspension cultures of the HFA region. It is a cell with 56 chromosomes, five of which are marker chromosomes (enclosure). This cell is representative of this region in that several markers were found in most of the metaphases analyzed. One marker (6q-) is identical to the marker in Fig. 8 and suggests that some of the cells from this clonal sub-population have undergone additional karyotypic evolution. Chromosomes 15 and 18 are now represented as a translocation chromosome, t(15;18) and chromosome 3 has additional chromatin added to the p arm (3p+), and chromosome 6 has split horizontally at the centromere to form an isochromosome, i(6p). (Reprinted with permission, Shapiro, JR, 1986).

251

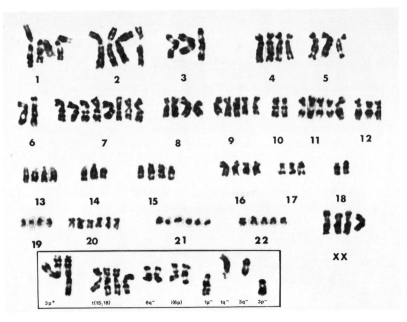

Fig. 10. A karyotype of a metaphase cell with 106 chromosomes, which
was obtained from the short-term cultures of the HFA
region. This cell may have arisen by endoreduplication
because four of the eight marker chromosomes are duplicated. In addition to the four markers seen in Fig. 4,
chromosomes 1 and 3 have other rearrangements involving the
loss of either the p or q arm of the chromosomes. One
chromosome is notably over-represented (chromosome 7),
which is typical of cells in advanced stages of karyotypic
evolution. (Reprinted with permission, Shapiro, JR, 1988).

Fig. 11. The distribution of cells in two
separate regions of tumor RM, designated RM and RM1. The cells in
region RM are primarily near-diploid
in chromosome number. The second
major peak of cells contains chromosomal complements of 92 in which the
majority of chromosome numbers are
represented by 4 homologous chromosomes. Such cells are believed to
develop by endopolyploidization[25].
In region RM1 the major cell type is
also near-diploid but there are two
other peaks of cells with 3n and 4n
chromosome numbers. The 4n cells of
region RM1, while tetraploid in chromosome number, contain chromosomal
complements that have chromosomes
represented by a single homologue or
3-5 homologues. The evolution of
this population depicts numerous
segregational errors.

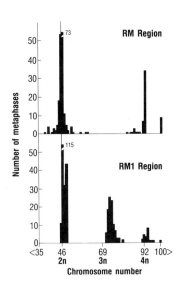

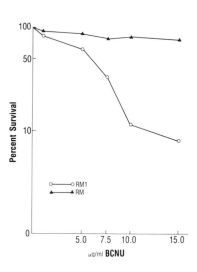

Fig. 12. BCNU semilog dose-response curve
of tumor regions RM and RM1
using a colony forming assay.
The cells were tested in early
passage (passage 2). Region RM
is predominantly near-diploid in
chromosome number and resistant
to BCNU while region RM1 (almost
an equal mixture of near-diploid
and hyperdiploid cells) demonstrates a modest sensitivity to
BCNU. (Reprinted with permission, Shapiro, WR, and Shapiro,
JR, 1986).

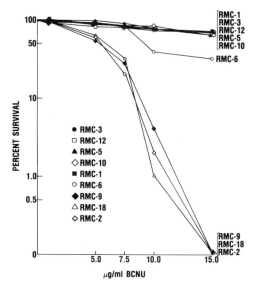

Fig. 13. BCNU semilog dose-response curve of
clones isolated from tumor MK. The
clones selected for testing in the
colony forming assay were near-diploid
in chromosome number (RMC-1, RMC-3,
RMC-12, RMC-5, RMC-10, RMC-6) or
hyperdiploid (RMC-9, RMC-18, RMC-2).
As the curves demonstrate all the
clones with near-diploid chromosome
numbers were resistant to doses of
15 µg/ml BCNU. The hyperdiploid
clones with chromosome numbers greater
than 60 chromosomes/ metaphase were
moderately sensitive having ED_{50} of
5.5, 5.5, and 5.8 µg/ml BCNU. (Reprinted with permission, Shapiro, JR
and Shapiro, WR, 1985).

fication of the EGFR gene was related to the extra copies of chromosome 7 or rearrangement of this gene in human freshly resected gliomas and glioma cell lines[4-6,30-34]. Other genes mapped to this chromosome that may have significance in human gliomas is the A chain polypeptide gene of platelet-derived growth factor (PDGF)[35], the p-glycoprotein multiple drug resistance gene (MDR)[36], and the ornithine decarboxylase gene (ODC)[37] which will be discussed below. In all, some 65 genes have been assigned to chromosome 7[38].

Our interest in chromosome 22 was spawned by the contrast we observed in the karyotypes of the BCNU-resistant cells (Fig. 14) when compared to the majority of tumor cells in the freshly resected tumor. Chromosome 22 was first noted to be under-represented in the karyotypes of many human gliomas[14,15]. This loss of chromosome 22 was observed in all grades of human gliomas, the astrocytoma, the anaplastic astrocytoma and the glioblastoma multiforme[17,22]. In our analyses of freshly resected human gliomas where we could do extensive karyotyping (200-400 cells) a large proportion of karyotypes (sometimes as great as 65% of the cells) demonstrated the loss of this chromosome[10]. Only a small proportion of the cells were diploid and an even smaller proportion were polysomic for this chromosome (1-8%) especially in the glioblastomas. In contrast, the BCNU-resistant clones or parental population subjected to repeated exposure of BCNU always had two or more copies of chromosome 22 or a rearrangement involving chromosome 22[29]. Was this chromosomal constitution selected as a result of the drug pressure or was it induced by the presence of the drug? In tumors like EI where stable rearrangements are identified in the majority of cells and where extensive karyotyping was undertaken we were able to identify karyotypes in the

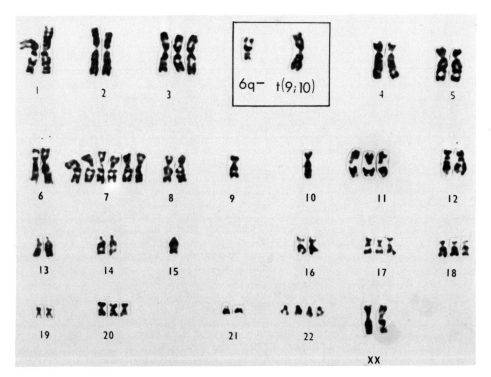

Fig. 14. The stemline karyotype from the EI tumor prepared from a monolayer culture treated with repeated and increasing concentrations of BCNU (1.0, 5.0, 7.5, and 10.0 μg/ml BCNU). The karyotypic deviation is: (2n±) 56,XX,+3,+7,+7,+7,+7,+11,+17,+18,+20,+22,+22,-9,-10,-15, del(6)(q14), t(9;10)(q12;p11). Giemsa-banded, x1,100.

freshly resected untreated cells that were similar to the BCNU-resistant cells (Fig. 15).

The importance of the over-representation of chromosomes 7 and 22 might reside with genes that are assigned to this chromosome. The c-sis oncogene that encodes the B-chain of PDGF is mapped to chromosome 22[39,40]. Tumor cell lines from several different sarcomas and human gliomas were known to secrete a PDGF or PDGF-like molecule and these tumors appeared to exhibit autocrine regulation of this growth factor[41-47].

The chromosomal constitution of the BCNU-resistant cells suggested that the growth factor PDGF may be functional in these freshly resected cells and thus we sought to confirm this hypothesis by biochemical and molecular techniques.

Characterization of BCNU-Resistant Cells

Our work on freshly resected human gliomas suggests that individual cells and/or small foci of cells within the tumor are capable of secreting a growth stimulatory factor[48]. This was first visualized by immunohistochemical staining of frozen sections prepared from the freshly resected tumors. We then demonstrated that the conditioned medium obtained from primary cells in early passage would stimulate NIH 3T3 cells (measured by the uptake of ^{3}H-thymidine) but this stimulation would increase 4-10 fold if such cells were challenged with repeated and increasing concentrations

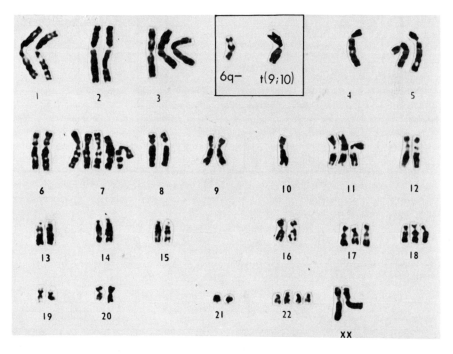

Fig. 15. A karyotype prepared from the primary cytogenetic analysis (72 h monolayer preparation) of tumor EI. The karyotypic deviation is: (2n+) 55,XX,+3,+7,+7,+7,+11,+17,+18,+22,+22,-4,-10, del(6)(q14), t(9;10)(q12;p11). Giemsa-banded, x1,100.

255

of BCNU (0.1 µg/ml to 10 µg/ml) prior to use in this assay[29]. This stimulatory activity could be inhibited by pre-treating the conditioned medium with anti-PDGF or 2-mercaptoethanol (BME) suggesting that this stimulatory activity might be PDGF or PDGF-like. To confirm the specificity of this factor we did competitive binding studies utilizing highly purified iodinated PDGF[49]. Phosphorylation analyses also confirmed that the material secreted by the BCNU-resistant cells was capable of stimulating the PDGF receptor, thus confirming the autocrine nature of this system[50]. Southern blot hybridization analyses have confirmed that the BCNU-resistant cells have additional copies of the PDGF A- and B-chain genes when compared to the primary untreated cells. The amplification is approximately that expected for the 2-3 extra copies of chromosome 7 and 22 that we visualize in the karyotype[51].

All these observations were made on our in vitro system[28,29], and while the karyotypes suggested that these cells were in the tumor at the time

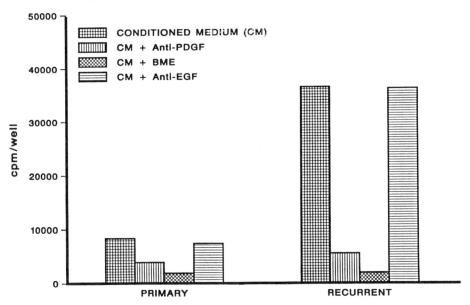

Fig. 16. The ³H-thymidine uptake stimulated by conditioned medium (CM) obtained from freshly resected samples of primary and recurrent tumor samples from the same patient coded GJ (gliosarcoma). The cells from each tumor sample were plated in 24-well Linbro dishes (5x10⁴), incubated 24 h, then fed growth medium without serum for 72 h. The media were collected, brought to 0.2% with human serum albumin and clarified by centrifugation at 5000g. The media was lyophilized, reconstituted to 1/4 volume with distilled water, dialyzed at 4°C against phosphate buffered saline and aliquoted into 4 samples. One sample was pre-treated with anti-PDGF, a second with 2-mercaptoethanol (BME), the third with anti-EGF and the fourth received no pretreatment. The untreated samples from both the primary and recurrent tumor demonstrated a stimulatory effect. This effect was inhibited with both anti-PDGF and BME but not anti-EGF. The CM obtained from the recurrent tumor produced a 2-fold increase in stimulation over the primary tumor.

of resection (Fig. 15), they did not demonstrate the potential of this subset of cells to be important in the regrowth of tumor. To determine if this cell has an in vivo correlate, we analyzed recurrent tumor samples from 5 patients whose primary tumor samples were analyzed previously. These patients were all randomized to BCNU treatment protocols. When tumor progression was noted in these patients, they underwent a second surgical procedure. The time span between the first and second surgeries was 4 months for tumor DI and approximately one year to eighteen months for tumors JD, EM, BD, and GJ.

Each of these five recurrent tumors have been analyzed cytogenetically and in all cases the dominant cell observed was identical to a subset of the cells observed in the primary analysis. In all 5 tumors this subset of cells was a minor population in the primary untreated tumor (representing no more than 4-36% of the cells). In the recurrent samples, this BCNU-resistant cell now represented 63-96% of the cell population. As shown for cells treated in vitro, conditioned medium collected from these recurrent tumors had a marked stimulatory effect which could be inhibited by pre-treatment of the conditioned medium with anti-PDGF or 2-mercaptoethanol (Fig. 16). While the receptor and molecular analyses are in progress on several of these tumors, those that have been analyzed do secrete a PDGF-like peptide that binds to the PDGF receptor and stimulates phosphotyrosine activity[50]; these tumors also have an increased PDGF A and B gene copy number and expression[51].

In addition to the gene for the PDGF A-chain, the multidrug resistance gene locus (MDR) has also been mapped to chromosome 7. While the MDR gene has never been associated with drug resistance for any alkylating agent such as BCNU, the BCNU-resistant populations analyzed to date have all demonstrated an increased copy number for the MDR gene consistent with the increase in polysomy of chromosome 7[52]. We are currently investigating the expression of these genes using northern and western blot analyses.

Elevated concentrations of polyamines have also been described in malignant cells that exhibited chemoresistance to alkylating agents. To determine if our BCNU-resistant cells had similar increased biosynthesis, we analyzed cells from freshly resected primary tumors, recurrent tumors, and cells treated in vitro with BCNU. BCNU-resistant cells obtained from recurrent tumors demonstrated a 5-7 fold increase over primary tumor cells. A similar increase in polyamine biosynthesis (5 fold) was demonstrated in the cells treated in vitro with 10 µg/ml BCNU (Heston et al, unpublished data). These findings suggest that the polysomy of chromosome 7 may cause an amplification of ornithine decarboxylase (ODC). This enzyme is the rate limiting enzyme for polyamine synthesis and the genes encoding it have been mapped to chromosomes 2 and 7[37].

To explore the possibility that other mechanisms may be operating to cause BCNU resistance in these cells, we are analyzing the activity of O^6-methyltransferase, glutathione and glutathione related enzymes in collaboration with Dr. Daniel Yarosh. Preliminary data has demonstrated that the activities of these enzymes in all tumor samples were normal.

Thus, all the tumors that have the specific chromosomal constitution in which the over-representation of chromosomes 7 and 22 were observed have also demonstrated the ability to secrete PDGF peptides. The aberrant expression of PDGF in these cells seems to be a relevant biomarker that delineates a subset of cells that can carry additional lesions (aberrant expression of the MDR gene, and elevated polyamine biosynthesis). We believe that this combination of lesions associated with the over-representation of chromosomes 7 and 22 confer a selective and/or survival advantage to these cells. When the selective pressures acting on the tumor populations are altered (irradiation and chemotherapy) cells with a selective advantage such as those described above ultimately have the ability to repopulate the tumor mass.

In summary, our data suggest that when many patients become symptomatic, their malignant tumors (anaplastic astrocytoma and glioblastoma multiforme) have so progressed that, contained within the cellular heterogeneity of that tumor, are minor subsets of intrinsically chemoresistant cells. Further characterization of such cells may provide new clues for drug development and/or new therapeutic strategies.

REFERENCES

1. W. R. Shapiro, Therapy of adult malignant brain tumors: what have the clinical trials taught us?, Sem. Onc. 13:38-45 (1986).
2. M. L. Rosenblum, and C. B. Wilson, "Brain Tumor Biology," S. Karger, Basel (1984).
3. M. D. Walker and D. G. T. Thomas, eds., "Biology of Brain Tumours," Martinus Niijhoff Publishers, Boston (1986).
4. T. A. Libermann, N. Razon, A. D. Bartal, Y. Yarden, J. Schlessinger, and H. Soreq, Expression of epidermal growth factor receptors in human brain tumors, Cancer Res. 44:753-760 (1984).
5. T. A. Libermann, H. R. Nusbaum, N. Razon, R. Kris, I. Lax, H. Soreq, N. Whittle, M. D. Waterfield, and J. Schlessinger, Amplification and overexpression of the EGF receptor gene in primary human glioblastomas, J. Cell Sci. 9 suppl. 3:161-172 (1985).
6. T. A. Libermann, H. R. Nusbaum, N. Razon, R. Kris, I. Lax, H. Soreq, N. Whittle, M. D. Waterfield, A. Ullrich, and J. Schlessinger, Amplification, enhanced expression and possible rearrangement of EGF receptor gene in primary human brain tumours of glial origin, Nature 313:144-147 (1985).
7. A. J. Wong, S. H. Bigner, D. D. Bigner, K. W. Kinzler, S. R. Hamilton, and B. Vogelstein, Increased expression of the epidermal growth factor receptor gene in malignant gliomas is invariably associated with gene amplification, Proc. Natl. Acad. Sci. U.S.A. 84:6899-6903 (1987).
8. K. W. Kinzler, S. H. Bigner, D. D. Bigner, J. M. Trent, M. L. Law, S. J. O'Brien, A. J. Wong, and B. Vogelstein, Identification of an amplified, highly expressed gene in a human glioma, Science 236:70-73 (1987).
9. C. D. James, E. Carlbom, J. P. Dumanski, M. Hansen, M. Nordenskjold, V. P. Collins, and W. K. Cavenee, Clonal genomic alterations in glioma malignancy stages, Cancer Res. 48:5546-5551 (1988).
10. J. R. Shapiro, W.-K. A. Yung, and W. R. Shapiro, Isolation, karyotype, and clonal growth of heterogeneous subpopulations of human malignant gliomas, Cancer Res. 41:2349-2359 (1981).
11. J. R. Shapiro and W. R. Shapiro, Clonal tumor cell heterogeneity, in: M. L. Rosenblum and C. B. Wilson, eds., "Progress in Experimental Tumor Research: Brain Tumor Biology," S. Karger, Basel (1984).
12. W.-K. A. Yung, J. R. Shapiro, and W. R. Shapiro, Heterogeneous chemosensitivities of subpopulations of human glioma cells in culture, Cancer Res. 42:992-998 (1982).
13. J. Mark, Chromosomal characteristics of neurogenic tumours in adults, Hereditas 68:61-100 (1971).
14. A. A. Al Saadi and F. L. Latimer, Nonrandom chromosomal abnormalities in human brain tumors, Am. J. Hum. Genet. 32:61A (1980).
15. K. Yamada, T. Kondo, M. Yosioka, and H. Oami, Cytogenetic studies in twenty human brain tumors: association of no. 22 chromosome abnormalities with tumors of the brain, Cancer Genet. Cytogenet. 2:293-307 (1980).
16. J. A. Rey, M. J. Bello, J. M. de Campos, J. Benitez, M. C. Ayoso, and E. Valcarcel, Chromosome studies in two human brain tumors, Cancer Genet. Cytogenet. 10:159-165 (1983).

17. S. H. Bigner, J. Mark, M. S. Mahaley, and D. D. Bigner, Patterns of the early, gross chromosomal changes in malignant human gliomas, Hereditas 101:103-113 (1984).

18. S. H. Bigner, J. Mark, D. E. Bullard, M. S. Mahaley, and D. D. Bigner, Chromosomal evolution in malignant human gliomas starts with specific and usually numerical deviations, Cancer Genet. Cytogenet. 22:121-135 (1986).

19. S. H. Bigner, J. Mark, P. C. Burger, M. S. Mahaley, D. E. Bullard, L. H. Muhlbaier, and D. D. Bigner, Specific chromosomal abnormalities in malignant human gliomas, Cancer Res. 88:405-411 (1988).

20. J. R. Shapiro, P.-Y. Pu, A. N. Mohamed, S. L. Neilsen, N. Sundaresan, and W. R. Shapiro, Regional heterogeneity in high grade gliomas, Proc. Amer. Assn. Cancer Res. 25:375 (1984).

21. J. R. Shapiro, Biology of gliomas: heterogeneity, oncogenes, growth factors, Sem. Onc. 8:4-15 (1986).

22. J. R. Shapiro and W. R. Shapiro, The subpopulations and isolated cell types of freshly resected high grade human gliomas: their influence on the tumors's evolution in vivo and behavior and therapy in vitro, Cancer Metast. Rev. 4:107-124 (1985).

23. S. Wolman, Karyotypic progression in human tumors, Cancer Metast. Rev. 2:257-293 (1983).

24. J. D. Rowley, Chromosome abnormalities in cancer, Cancer Genet. Cytogenet. 2:175-198 (1980).

25. T. Oksala, and E. Therman, Mitotic abnormalities and cancer, in: "Chromosomes and Cancer," J. German, ed., John Wiley Sons, New York (1974).

26. J. R. Shapiro, P.-Y. Pu, A. N. Mohamed, W. R. Shapiro, J. H. Galicich, and S. A. D. Ebrahim, Correlation of chromosome number and BCNU sensitivity in freshly resected human astrocytoma, anaplastic astrocytoma, and glioblastoma multiforme, Cancer (Submitted) 1990.

27. J. R. Shapiro, W. R. Shapiro, A. N. Mohamed, and P.-Y. Pu, BCNU-sensitivity in parental cells and clones from four freshly resected near-diploid human gliomas: an astrocytoma, and anaplastic astrocytoma and two glioblastoma multiforme, J. Neuro. Oncol. (Submitted) (1990).

28. J. R. Shapiro, W. R. Shapiro, and P.-Y. Pu, BCNU-resistant cell types in the heterogeneous populations of two hyperdiploid human malignant gliomas, J. Neuro. Oncol. (Submitted) (1990).

29. P.-Y. Pu, W. R. Shapiro, and J. R. Shapiro, Effect of repeated BCNU exposure on freshly resected human malignant gliomas in vitro: enrichment of BCNU-resistant near-diploid cell populations, J. Neuro. Oncol. (Submitted) (1990).

30. C. Bell, G. Harsh, M. Rosenblum, P. Meltzer, and J. Trent, Numeric and structural alterations of chromosome 7 in human brain tumors; correlation with expression of epidermal growth factor receptor (EGFR), Proc. Am. Assoc. Cancer Res. 27:37 (1986).

31. W. Henn, N. Blin, and K. D. Zang, Polysomy of chromosome 7 correlated with overexpression of the erb-B oncogene in human glioblastoma cell line, Hum. Genet. 74:104-106 (1986).

32. J. A. Rey, M. J. Bello, J. M. de Campos, M. E. Kusak, and S. Moreno, On trisomy of chromosome 7 in human gliomas, Cancer Genet. Cytogenet. 29:323-326 (1987).

33. S. H. Bigner, A. J. Wong, J. Mark, L. H. Muhlbaier, K. W. Kinzler, B. Vogelstein, and D. D. Bigner, Relationship between gene amplification and chromosomal deviations in malignant human gliomas, Cancer Genet. Cytogenet. 29:165-170 (1987).

34. P. A. Humphrey, A. J. Wong, B. Vogelstein, H. S. Griedman, M. H. Werner, D. D. Bigner, and S. H. Bigner, Amplification and expression of the epidermal growth factor receptor gene in human glioma xenografts, Cancer Res. 48:2231-2238 (1988).

35. C. Betsholtz, A. Johnsson, C.-H. Heldin, B. Westermark, P. Lind, M. S. Urdea, R. Eddy, T. B. Shows, K. Philpott, A. L. Mellor, T. J. Knott, and J. Scott, cDNA sequence and chromosomal localization of human platelet-derived growth factor A-chain and its expression in tumor cell lines, Nature 320:695-699 (1986).

36. D. R. Bell, J. M. Trent, H. F. Willard, J. R. Riodan, and V. Ling, Chromosomal location of human p-glycoprotein gene sequences, Cancer Genet. Cytogenet. 25:141-148 (1987).

37. R. Winqvist, T. P. Makela, P. Seppanen, O. A. Janne, L. Alhonen-Hongisto, J. Janne, K. H. Grzeschik, and K. Alitalo, Human ornithine decarboxylase sequences map to chromosome regions 2pter-p23 and 7cen-qter but are not coamplified with the NMYC oncogene, Cytogenet. Cell Genet. 42:133-140 (1986).

38. Human Gene Mapping 9: Ninth International Workshop on Human Gene Mapping, Cytogenet. Cell Genet., Vol. 46, Nos. 1-4 (1987).

39. R. C. Dalla-Favera, R. C. Gallo, A. Giallongo, and C. M. Croce, Chromosomal localization of the human homolog (c-sis) of the simian sarcoma virus onc gene, Science 218:686-688 (1982).

40. D. C. Swan, O. W. McBride, K. C. Robbins, D. A. Keithley, E. P. Reddy, and S. A. Aaronson, Chromosomal mapping of the simian sarcoma virus onc gene analogue in human cells, Proc. Natl. Acad. Sci. U.S.A. 79:4691-4695 (1982).

41. C.-H. Heldin, B. Westermark, and A. Wasteson, Chemical and biological properties of a growth factor from human-cultured osteosarcoma cells: resemblance with platelet-derived growth factor, J. Cell. Physiol. 105:235-246 (1980).

42. M. Nister, C.-H. Heldin, A. Wasteson, B. Westermark, A platelet-derived growth factor analog produced by a human clonal glioma cell line, Ann. N.Y. Acad. Sci. 397:25-33 (1982).

43. T. D. Graves, A. J. Owen, and N. N. Antoniades, Evidence that a human osteosarcoma cell line which secretes a mitogen similar to platelet-derived growth factor require growth factors present in platelet-poor plasma, Cancer Res. 43:83-87 (1983).

44. C. Betsholtz, C.-H. Heldin, M. Nister, B. Ek, A. Wasteson, and B. Westermark, Synthesis of a PDGF-like growth factor in human glioma and sarcoma cells suggests the expression of the cellular homologue to the transforming protein of simian sarcoma virus, Biochem. Biophys. Res. Commun. 117:176-182 (1983).

45. P. Pantazis, P. G. Pelicci, R. Dalla-Favera, and H. N. Antoniades, Synthesis and secretion of proteins resembling platelet-derived growth factor human glioblastoma and fibrosarcoma cells in culture, Cell Biol. 82:2404-2408 (1985).

46. A. N. Mohamed, P.-Y. Pu, W. R. Shapiro, and J. R. Shapiro, Correlation of BCNU resistance in human glioma cells with over-representation of chromosome 22 and production of a factor resembling platelet-derived growth factor, Proc. Am. Assoc. Cancer Res. 26:32 (1985).

47. B. Westermark, A. Johnsson, Y. Paulsson, C. Betsholtz, C.-H. Heldin, M. Herlyn, U. Rodek, and H. Koprowski, Human melanoma cell lines of primary and metastatic origin express the genes encoding the chains of platelet-derived growth factor (PDGF) and produce a PDGF-like growth factor, Proc. Natl. Acad. Sci. U.S.A. 83:7197-7200 (1986).

48. P. L. Moots, M. K. Rosenblum, and J. R. Shapiro, Demonstration of platelet-derived growth factor immunoreactivity in glioblastomas and gliosarcomas, Neurol. 38:358 (1988).

49. D. W. Kimmel, J. M. Cunningham, D. B. Donner, and J. R. Shapiro, Binding studies of platelet-derived growth factor (PDGF) provide evidence for autocrine stimulation in a glioma cell line, Proc. Am. Assoc. Cancer Res. 27:216 (1986).

50. B. M. Mehta, S. A. D. Ebrahim, D. Andrews, and J. R. Shapiro, Selection of BCNU-resistant cells in primary and recurrent human gliomas; a correlation between BCNU clinical resistance and a PDGF autocrine growth pathway, Proc. Am. Assoc. Cancer Res. 29:49 (1988).
51. A. C. Scheck, P. L. Moots, B. M. Mehta, S. A. D. Ebrahim, and J. R. Shapiro, BCNU-resistant human glioma cells exhibit autocrine regulation of platelet-derived growth factor, J. Cell. Biochem. Suppl. 13D (In Press) (1989).
52. A. C. Scheck, W. D. W. Heston, and J. R. Shapiro, Expression of platelet-derived growth factor, ornithine decarboxylase, and the multi-drug resistance gene locus in BCNU-resistant human glioma cells, Proc. Am. Assoc. Cancer Res. 30:64 (1989).

DISCUSSION

Moderator (Russel G. Greig): Given your extensive experience with this type of tumor, could you address the title of this colloquium and your opinion as to at which point in time does genetic instability occur in the natural history of this particular neoplasm?

Joan R. Shapiro: The regional study I presented on tumor AI may be the most informative study in attempting to answer this question. Tumor AI was an astrocytoma. Seven different regions from this tumor were examined cytogenetically. Six of the 7 regions demonstrated a large proportion of normal karyotypes and karyotypes missing only a single sex chromosome. Only one region, AI7, contained karyotypes suggesting greater karyotypic evolution. Thus, tumors like AI demonstrate that while most of the cells in a low grade tumor are similar, genetic instability already exists at the time of clinical presentation, and mutant cells are being generated that differ in both their chromosomal complement and their phenotypic expression. There are a number of mechanisms known that cause the loss or gain of chromosomes in cancer cells[1].

Greig: Would you also like to comment on the non-metastatic behavior of this type of neoplasm? We've had a lot of discussions this morning on tumor progression and how quickly metastatic potential develops in particular tumors, but pathologists tell me this is not a particularly metastatic neoplasm.

Shapiro: Extracranial sites metastases are relatively common in childhood medulloblastoma, but there are a number of reports of gliomas metastasizing outside of the brain. This is more likely to occur with multiple operations. The relative rarity of this event may be related to the size of the tumor at presentation, as Alvord[2] has suggested. Metastases within the cerebrospinal fluid are more common, occurring in younger patients and with longer survival[3]. Perhaps if our patient population survived longer we would see an increase in extracranial metastatic lesions[4].

Harry Rubin: In your talk you noted that patients have focal seizures 10-15 years before a tumor was resolved. Is that a late occurrence in this tumor and are the symptoms caused by some local state of disorganization in that tissue? Does it have something to do with aging or loss of cells? Have autopsy studies of such patients revealed cellular lesions that would account for these symptoms?

Shapiro: Seizures can be caused by several different abnormalities. A variety of pre-existing lesions have been documented pathologically. These occur in inherited disorders, i.e., neurofibromatosis and tuberous sclerosis and in association with acquired lesions, i.e., multiple sclerosis or trauma[4]. The most likely cause of seizures in patients who later are found to have a tumor, is the tumor itself, which may induce the fit before the tumor is visible on imaging techniques. The seizure is presumably caused by some kind of stimulation -perhaps biochemical- induced by the tumor cells. The seizure arises in the cortex of the brain, and this induction is likely to occur earlier at a smaller tumor volume if the tumor starts in the cortex rather than in the white matter where it must grow to a larger size to affect the cortical neurons. There is no evidence that both tumors and seizures arise from a common cause.

Rubin: So there is possibly some kind of premalignant lesion which is not even a cellular lesion in a way; in other words, it's just a local area of disturbance?

Shapiro: That's right. Edema of the brain could produce this kind of symptoms.

Sidney Crawl: I think Dr. Rubin's question is answerable. There are a large number of patients who've had seizure surgery for focal seizures, and about 30% of them have histologically demonstrable tumors at the excision site that were not demonstrable without histopathology.

Shapiro: Such operations were performed before modern imaging techniques were available. With MRI, the incidence of "unknown" causes of focal seizures has markedly declined.

Crawl: The literature on that is fairly well established[5].

REFERENCES

1. T. Oskala and E. Therman, Mitotic abnormalities and cancer, in: "Chromosomes and Cancer," J. German, ed., John Wiley Sons, New York (1974).
2. E. C. Alvord, Why do gliomas not metastasize?, Arch. Neurol. 33:73-75 (1976).
3. W.-K. A. Yung, B. Horton, and W. R. Shapiro, Meningeal gliomatosis: a review of 12 cases, Ann. Neurol. 8:605-608 (1980).
4. D. S. Russell and L. J. Rubinstein, "Pathology of Tumours of the Nervous System," The Williams and Wilkins Co., Baltimore (1977).
5. R. J. Gumnit and I. E. Leppik, The epilepsies, in: "The Clinical Neurosciences," Vol. 1, R. N. Rosenberg, ed., Churchill Livingstone, New York (1983).

GENETIC INSTABILITY OCCURS SOONER THAN EXPECTED: PROMOTION,

PROGRESSION AND CLONALITY DURING HEPATOCARCINOGENESIS IN THE RAT

Oscar Sudilovsky, Lucila I. Hinrichsen, Tom K. Hei, Cecilia M.
Whitacre, Jian H. Wang, Sriram Kasturi, Shi H. Jiang, Ronald
Cechner, Stella Miron and Fadi Abdul-Karim

Institute of Pathology
Case Western Reserve University
Cleveland, OH 44106

INTRODUCTION

The basic question of "at which point in time does genetic instability
occur in the natural history of cancer" can be answered only with another
question: in which organ or model? Dr. Shapiro has just given us an
impressive account of her work in human malignant gliomas. We also attempted
to explore the problem but in a different organ, species and model system. We

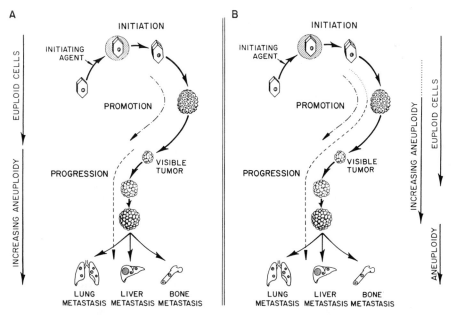

Fig. 1.   The natural history of neoplasia. (A) Common concept of promotion
           and progression. (B) Postulation showing overlapping of promotion
           and progression, based on the DEN/CD and DEN/CD+PHB model.

Boundaries between Promotion and Progression during Carcinogenesis
Edited by O. Sudilovsky *et al.*, Plenum Press, New York, 1991

found that in the rat liver treated with diethylnitrosamine (DEN) and a choline deficient (CD) diet, genomic instability expressed by aneuploidy takes place during the promotion treatment, long before hepato-carcinomas can be diagnosed (1,2). The presence of aneuploidy implies that irreversible genetic changes, characteristic of progression, occur during dietary promotion (Fig. 1). Since increasing evidence points out that most malignant hepatomas are monoclonal in origin, this fact and the studies previously mentioned have led us to propose now a scheme of cell renewal which explains the overlapping of promotion with progression arising in the clonally replicating foci of preneoplastic populations.

PROMOTION AND PROGRESSION IN THE RAT LIVER

A number of experimental models (3-8), in which non-genotoxic or genotoxic substances act as initiators and a variety of non-carcinogenic compounds are used as promoters, have attempted to reproduce some of the stages of the multistep process of carcinogenesis in the rat liver. The development of hepatocarcinomas in these models is predated by the appearance of localized epithelial proliferations of non-tumoral hepatic cells, called enzyme-altered foci (EAF) or altered hepatic foci (AHF). These lesions are recognized by morphologic changes which serve either as positive or negative "markers" for new cell populations. The most widely used among the positive markers is γ-glutamyl transpeptidase (GGTP), a membrane-bound enzyme not seen in the adult rat hepatocyte (5,6,9-11). Other positive markers include the presence of α-fetoprotein (12), glucose-6-phosphate dehydrogenase (13) and placental glutathione-S-transferase (14). Of the negative markers, the most "popular" are canalicular ATPase (11), glucose-6-phosphatase (11,13) and resistance to iron accumulation (15). Histochemical and immunological techniques have recorded the phenotypic diversity of EAF in treated livers. None of the markers, however, have been shown to be critical for the early or late phases of carcinogenesis (16). A putative first cellular step during initiation has been identified using the positive marker, placental glutathione-S-transferase (17-20), but unequivocal proof that this marker is essential for neoplastic transformation is still lacking. Studies from many laboratories (21-25) have shown that following removal of the promoting agent, most of the EAF fade out by gradual controlled death (apoptosis) of the foci (26) or by "redifferentiation" and blending with normal liver parenchyma (23) in a fashion similar to that of skin papillomas induced by the "two stage" procedure (see article by Slaga, this volume, Table 2). Those EAF that persist after the promoter is withheld are believed to give origin to malignant neoplasms at a later time of the life of the animal (23). The number of tumors observed under those circumstances is considerably smaller than the number of "persistent nodules," suggesting that only a minority of them can become hepatomas. Of considerable interest, however, is the fact that if withdrawal of the promoting agent is followed after some time by further promoting treatment, the potential for renewal of the same number of EAF is preserved (27). This finding is in agreement with the postulation that each individual focus originates from a single initiated cell.

THE CLONAL NATURE OF ENZYME-ALTERED FOCI

The concept that each focus originates from the clonal expansion of a single initiated cell is important in any attempt to unravel, as we shall do later, genetic changes emerging during the preneoplastic stages of carcinogen-esis. Although in some studies of liver carcinogenesis a polyclonal origin could be postulated (28), increasing information indicates that both preneoplastic foci and hepatocarcinomas are the progeny of a single initiated cell (29,30).

264

Integration of hepatitis B virus genome in the cellular DNA of hepatocellular carcinomas has been used in humans, in an attempt to demonstrate the clonal origin of the tumors (31). Number of restriction fragments and molecular sizes of different parts of the primary neoplasia, in the same liver and in its metastasis, were analyzed by Southern blotting. Of 14 cases of hepatocarcinomas, 13 were monoclonal in origin. In one case, however, one of the samples demonstrated a clone dissimilar from that in other fragments of the tumor. The possibility of superinfection or chromosomal DNA rearrangement of an initial monoclonal tumor could explain this result, although the question of polyclonality cannot be completely ruled out with this technology.

Using heterozygous females of PGK-1 mosaic mice, Rabes et al. (29) analyzed the X-linked isozymes phosphoglycerate kinase-1A (PGK-1A) and phosphoglycerate kinase-1B (PGK-1B) of hyperplastic nodules and normal surrounding liver. As expected for heterozygous females, the normal adjacent liver demonstrated a typical mosaic pattern (50-70% PGK-1A and 50-30% PGK-1B), while the nodules consisted of pure populations of either one of the enzymes. This is as it should be according to the Lyon hypothesis, if the population originates from a single initiated cell. A more rigorous study has been performed recently in rats by Iannaccone and associates (30), who observed simultaneously the histology and the enzyme mosaicism of GGTP in the PVG-RT1^a and PVG-RT1^c lineages. Of 499 lesions, 151 were derived from the PVG-RT1^a cells and 323 originated from the PVG-RT1^c lineage. The remaining 25 represented mixed lesions which were of the same small size as many of the normal genetically marked patches in the liver tissue.

These experiments with chimera and X-linked mosaic animals seem fairly conclusive in demonstrating the clonal nature of enzyme-altered foci. Similar results were obtained in mice using ornithine carbamoyltransferase isozymes in chemically induced hepatocarcinomas in mosaic females (32), thus confirming the monclonality of malignant liver tumors.

CHROMOSOMAL ALTERATIONS IN MALIGNANT NEOPLASIAS

It is known, in addition, that most neoplasias consist of cells with abnormal DNA complement (33-35). Specific changes in chromosomes exhibited by the majority of malignant tumors (36) support the clonal origin of neoplasias. With primary hepatomas, however, there have been no real efforts in methodologically analyzing their DNA content. Although some flow cytometry experiments have detected the presence of predominantly diploid cell populations (37,38), other studies, both in humans (39,40) and in animals (41-46) have substantiated the existence of aneuploidy. The development of aneuploidy is one of the most reliable indicators of progression, which some have called the third or final stage of carcinogenesis (47,48). It has been assumed that in solid tumors aneuploidy occurs as a late event in the evolution of malignancy (49). However, the time at which such genomic changes take place is not known. Multistage models of liver carcinogenesis offered us the possibility of examining ploidy alterations in each of the numerous EAF before and after the acquisition of malignancy. In the remainder of this article, therefore, we will discuss the results of experiments done while attempting to obtain some answers to the basic question posed at the beginning of this essay.

NUCLEAR DNA CONTENT OF ENZYME-ALTERED FOCI

Changes in the ploidy content of cells can be evaluated by analyzing the DNA complement during both mitosis and interphase or during mitosis alone. The first can be done by either cytospectrophotometry (or image analysis) or flow cytometry and the latter by the use of karyotyping methodology. The

265

parameters investigated by those means are different and should not be confused. Karyotyping and flow cytometry require the preparation of cell suspensions, with all the restrictions entailed in such techniques. When assessing the ploidy of hepatomas one can excise the tumors and, after dissociation, examine the cells by either method. But these procedures are not practical with preneoplastic foci of less than 1 mm in diameter, since individual dissections are very difficult to achieve (38) and small foci dissociate when collagenase is utilized (50). One could use, alternatively, the mass technique of flow cytometry of the entire liver (37,38) but that would make virtually impossible the clonal analysis of any individual focus. Cytospectrophotometry, on the other hand, can be performed in tissue sections without the need of cell suspensions. This method allows in addition an exact outline of each individual focus and permits the identification of more than one type of cell in the fixed, paraffin embedded sections. As a matter of fact, when preserving the clonal growth pattern is a primary consideration, it is the only methodology currently available for the determination of nuclear DNA content in EAF.

Consequently, we used cytospectrophotometry to assess if ploidy changes or other alterations denoting genomic instability, occur before cancer is established (51). Male Fischer 344 rats were partially hepatectomized and 18 hours after the operation were given a single i.p. dose of DEN (50 mg/kg body weight). In order to rapidly obtain EAF, they were fed a CD diet containing 0.05% phenobarbital (PHB). Controls were administered saline i.p. and a choline sufficient (CS) diet. The animals were sacrificed at 10, 16 and 29 wks after the beginning of promotion and sections of liver were stained with a combined GGTP-Feulgen stain (52). Cytospectrophotometric measurements of nuclear DNA values, in control hepatocytes and most EAF from treated rats, had a trimodal distribution with peaks corresponding to the 2N, 4N and 8N ploidy. At each treatment time, however, a minority of EAF depicted an aneuploid pattern. Fifteen out of 50 EAF in 10 rats were aneuploid, with hypertetraploid or hyperdiploid and hypertetraploid peaks. The remaining parenchymal cell population of those foci was euploid.

Although these initial studies indicated that genomic instability could emerge early in the process of carcinogenesis, the experimental design used did not allow us either to distinguish a possible association of aneuploidy with various components of the treatment, or to see if there were responses related to sex differences. Furthermore, we needed to demonstrate in long term tests that the hepatomas we obtained using our model were for the most part aneuploid. We thus decided to undertake -as part of the project- more extensive experiments using Sprague Dawley rats of both sexes in order to see if, in addition to the above parameters, there were differences in strain responses. These, in fact, were found and will be reported at a later date.

In short term studies, we hepatectomized both male and female Sprague Dawley rats, inoculated them with DEN or saline and after 1 wk on a CS diet we formed 4 treatment groups/sex. Each group was then fed CS, CS+PHB, CD and CD+PHB, sacrificed 16 wks later, and analyzed as before. Following hepatectomy, an additional female group received saline instead of DEN and was given a CD diet subsequent to 1 wk on CS. The small size of the foci in this strain was most evident in the females given DEN+PHB which developed, at the end of the study, significantly smaller foci than the males. Because the foci in such rats did not reach the minimum area required to count 75 to 100 suitable nuclei, it was not possible to evaluate the isolated effects of DEN or CD. The final number of animals included in the study was reduced to 14 males and 14 females which, nevertheless, yielded 79 foci among the various arrays.

Cytospectrophotometry demonstrated that control hepatocytes and hepatocytes adjacent to foci in treated animals had the typical trimodal nuclear DNA pattern (Fig. 2). The hepatocytes in the EAF of treated rats were mostly euploid and had a multimodal distribution. Because of variations in the number of cells in each ploidy range, a given focus was classified as diploid if more than 70% of the hepatocytes were in the 2N range, tetraploid when better than 70% were in the 4N range, and heterogeneous if both ploidy types were about the same (as displayed in Fig. 2).

Fig. 2. Representative histograms of DNA distribution in treated rats. All histograms were from the same section. Broken lines represent the mean value of DNA content in $\log_{10}$ AU (arbitrary unit of nuclear DNA content) of diploid (2C), tetraploid (4C), and octoploid (8C) hepatocytes. The histograms of control rats, which did not have foci, consisted only of panels similar to the upper and middle ones in this figure (not shown). (Reproduced from reference 1, Wang, J.H. et al, Cancer Res. 50: 7571-7576, 1990, with permission of the authors and publisher.)

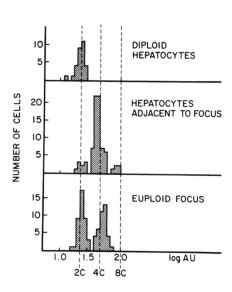

Table 1 demonstrates that when both sexes were considered, approximately 9% of the foci subjected to DEN/CD+PHB were diploid while the ratio increased to about 55% in those who had DEN/PHB. In contrast, the percentage of foci classified as tetraploid in the same animals was 48.5 and 10, respectively. On the other hand, those who fared DEN+CD had a diploid:tetraploid ratio of 23:35. The proportion of heterogeneous foci remained relatively constant at 30-35%, after the different treatments.

These data, therefore, indicate that dietary manipulations affect the ploidy of hepatocytes in the EAF without changing the nuclear DNA content of the surrounding liver parenchyma. Bannasch (53) has defined "preneoplasia" as a lesion consisting of phenotypically altered cell populations which, although not neoplastic in nature, have a high probability of progressing to the tumoral stage. Within this context, ploidy changes in foci induced by some dietary manipulations may be precursors or manifestations of genomic instability, to be revealed later by the appearance of aneuploidy or some other form of chromosomal alteration. Indeed, although expected because of our previous studies, we found aneuploidy in a minority of the foci (Fig. 3, Tables 1 and 2).

TABLE 1   Ploidy distribution in rat liver EAF

All rats were treated as indicated in text.   Controls were given saline i.p.
injections and fed CS diet.

| TREATMENT | DIPLOID | | TETRAPLOID | | HETEROGENEOUS | | ANEUPLOID | |
	M	F	M	F	M	F	M	F
DEN/PHB	8[a]	3	2	0	7	0	0	0
DEN/CD	3	3	5	4	4	5	1	1
DEN/CD+PHB	2	1	5	11	2	8	2	2

[a]Number of foci.   All focal intersections were ~450 μm in diameter.

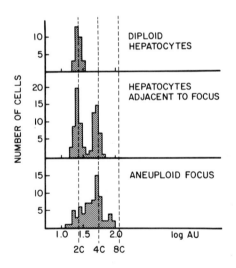

Fig. 3.   Representative histogram of DNA distribution in treated rats. Histograms were obtained under conditions similar to those for Fig. 1, except that the lower panel shows an aneuploid focus. (Reproduced from reference 1, Wang, J.H. et al., Cancer Res. 50:7571-7576, 1990, with permission of the publisher and authors)

Such ploidy changes were evident only in rats administered DEN and fed either the CD diet alone (7.7% in either males or females) or the CD diet supplemented with PHB (18.2% in males; 9.1% in females). Aneuploid cells were not seen in animals given DEN+PHB, suggesting that CD (regarded as a weak initiator as well as a strong promoter) may be the factor responsible for the appearance of genetic instability. Due to its clastogenic properties, administration of DEN could be a precondition for this event. Such possibility has not been completely ruled out because of the small size of the DEN+CS foci in our experiments.

TABLE 2. Distribution of modal nuclear DNA in aneuploid foci

Conditions of the experiment as stated in Table 1.

SEX	TREATMENT	MODAL DNA CONTENT				
		2N	4N	8N	HYPERDIPLOID	HYPERTETRAPLOID
Male	DEN/CD		42.0[a]	22.0	36.0	
	DEN/CD+PHB		21.5±3.5		78.5±3.5	
Female	DEN/CD				67.0	33.0
	DEN/CD+PHB		67.0±24			33.0±24

[a]Mean ± S.D.

Having substantiated the modifications of ploidy in EAF, confirmed the presence of aneuploidy in a small number of them and investigated the importance of strain and sex in the carcinogenesis model used, we set out to prove that aneuploidy is a common feature in malignant rat hepatomas attained by dietary manipulations.

In long term studies we utilized 6 wk old female Sprague Dawley and Fischer 344 rats. The protocol was the same as in the short term experiments, with the exception that promotion was done only with a CD+PHB diet. The control group received a saline injection 18 hrs after hepatectomy. The animals were then placed on a CS diet until the end of the investigation, which was 18 months after starting promotion in the treated rats. Samples of the quickly excised liver were fixed in 10% buffered formaldehyde and embedded in paraffin. Serial sections cut consecutively at a thickness of 6 and 14 μm, were stained with hematoxylin-eosin to locate tumor areas, and with Feulgen to measure nuclear DNA content. The microscopic image of the Feulgen stained liver sections were screened in a monitor and DNA quantitation was done with a Sperry System 1800 computer with a Pulnix TCM50 color camera, a PCVISION digitizer, a Mitsubishi color monitor and a Microsoft mouse. The system was provided with an Image Measure Software (Microsciences Inc.) such that, once the nucleus was outlined, the density of the image was integrated and corrected for background shading. The individual logarithmic values and the mean and standard deviations of all determinations were then calculated automatically. A description of the system has been published (54,55).

Approximately 100 nuclei from each hepatocarcinoma were randomly selected and measured. Age-matched rat liver sections were placed adjacent to tumor tissues and 75 to 80 nuclei, stained simultaneously, were used to establish the reference ploidy peaks. Nuclear DNA content of normal and hepatoma cells was expressed, as before, in logarithmic arbitrary units and the results were plotted in a histogram. Ploidy was considered to be in the diploid range when the mean of a peak in the tumor was within one standard deviation of the mean of the diploid peak in control hepatocytes. An abnormal or aneuploid value was declared when the mean of any given hepatoma peak was at more than one standard deviation of the corresponding ploidy in the control nuclei, or when a broadly based histogram without intervening peaks was noted. A peak encompassing 70% or more nuclei was regarded as the ploidy of the tumor cells stemline; when none of the peaks examined reached the 70% boundary, the hepatocarcinoma was considered heterogeneous. Ergo, it is possible that tumors classified as euploid have aneuploid cells; conversely, aneuploid carcinomas may contain a number of euploid nuclei.

TABLE 3. Ploidy distribution in rat hepatocarcinomas

Conditions of the experiment as stated in the text.

STRAIN	DI PLOID	TETRA PLOID	HETEROG EUPLOID	HYPO DIPL	HYPER DIPL	HYPO TETR	HYPER TETR	HETEROG ANEUPLOID
Sp Dwley	1	2	2	8	1	0	0	5
F344	1	2	1	2	6	2	2	3

Although the experiment above has not yet been completely analyzed, the data described in Table 3 indicate that in round numbers 75% of the malignant tumors examined in either strain were aneuploid. This is at variance with the outcome of the short term experiments, in which only a minority of GGTP positive foci had an aneuploid modal DNA content. A representation of the results achieved thus far is also depicted in Figs. 4 and 5.

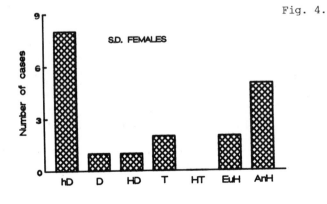

Fig. 4. Hepatocarcinomas in female Sprague Dawley rats: Modal nuclear DNA content distribution. See text for protocol of treatments. hD: hypodiploid; D: diploid; HD: hyperdiploid; T: tetraploid; HT: hypertetraploid; EuH: euploid, heterogeneous; AnH: aneuploid, heterogeneous.

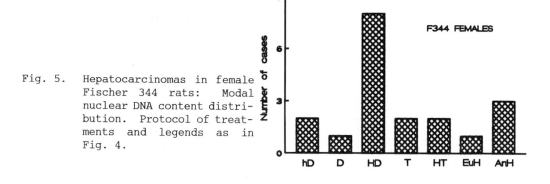

Fig. 5. Hepatocarcinomas in female Fischer 344 rats: Modal nuclear DNA content distribution. Protocol of treatments and legends as in Fig. 4.

Hence, these long term studies have ascertained that both euploid and aneuploid hepatocarcinomas can be detected after dietary manipulations in the rat. Using flow cytometry some investigators reported (56,57), that malignant rat hepatomas are predominantly diploid. This discrepancy may be more apparent than real since we also obtained a small number of tumors consisting mostly of diploid cells. The difference resides probably with the effects of the CD treatment, which seemingly was never used by these authors. In an effort to explain the genomic changes we observed both during preneoplasia and in liver carcinomas, and to account also for the existence of euploid foci and some predominantly diploid or tetraploid hepatocarcinomas, we propose a general model for cell development throughout the preneoplastic stage. This model, originating from ideas first integrated by Nowell (28), is based on the concept of the clonality of each individual focus and is supported by results reported here and by other investigators (1,28,41,45,46,58-60). The hypothesis, which should apply to organs other than the liver, is summarized in Figs. 6 and 7.

Any of the normal euploid cell types could be initiated and each of those cells can originate an individual focus when exposed to promoters. If the initiated cell is diploid, the two-hit event postulated by Knudson (61) has a higher probability to occur in recessive mutations (if a suppressor gene mechanism is operating). The promoted cells have growth advantage over the normal cells and clonal expansion and focus formation begins. The legend to Fig. 6 explains the salient features of the expanded population. Because of lack of space, the case of aneuploid cells that do not undergo the additonal event(s) required for them to become malignant are not included.

Notice that an important property in the model is the fact that the ploidy distribution in a focus changes with time, i.e., the foci are asynchronous, at least with respect to their nuclear DNA content. A predominantly diploid focus could become heterogeneous or aneuploid at a later date, due to their microenvironment or because the production of more variants (28) makes certain cells grow faster than others. In the arbitrary focus shown in Fig. 6, for example, a cross-section through the third generation following initiation will have 20% diploid, 40% tetraploid, 20% hypodiploid and 20% hypertetraploid. Two generations later, the proportion would be 17% diploid, 33% octoploid, 17% hyperdiploid, 17% hypotetraploid and 17% hypertetraploid. On the right side of the drawing (9th hypothetical generation), when cancer can be diagnosed histologically, 22% of the hepatoma cells are diploid, 11% are tetraploid and 66% are aneuploid. The situation could have been reversed,

271

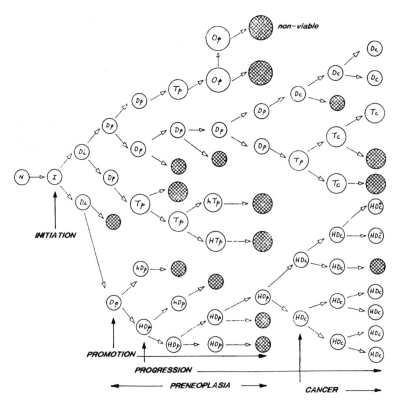

Fig. 6. Clonal model of hepatic foci: progression-linked or persistent
focus. The normal cell (N) from which every focus originated can be
of any of the euploid types. A carcinogen initiated changes in a
normal cell (I), which then replicates and "fixes" those changes.
For convenience, this has been diagrammed as diploid initiated (Di).
These cells have growth advantage when subjected to promoting
treatment (Dp), allowing for their clonal expansion and the
beginning of focus formation. A variable proportion of promoted
cells become (or are already) tetraploid (Tp) and octoploid (Op),
and proliferate differentially and independently of the age of the
liver of the animal. This is so because the rigorously controlled
polyploidization in normal hepatocytes in altered in the cells
comprising a focus, indicating a distorted kinetic or genetic
stability. Such genomic instability may consist of point mutations
or chromosomal breakage and rearrangements that would not be
detected by flow cytometry, cytospectrophotometry or Feulgen image
analysis. However, these alterations increase with every cell
replication and, eventually, become manifested by the emergence of a
new population with higher or low number of chromosomes. Changes
occurring during the promoting treatment or after progression, are
denoted in the drawing by HDp (hyperdiploid, promoted/progressed),
hDp (hypodiploid, promoted/progressed), HTp (hypertetraploid,
promoted/progressed), etc. A diverse amount of cells is not viable,
as depicted, because of toxic effects, immune responses, insufficient or deficient chromosomal complement, mitotic changes, etc.
After malignancy has been established, the carcinomatous variants
are specified by the subscript c. In any given tumor it may be
possible to observe a predominantly diploid population or a mixture
of diverse euploid cells. On the other hand, aneuploid cells could

constitute most of the cell population as pointed out in the picture (Hd_c and hD_c). Additions and subtractions in the number of chromosomes which occur at a later stage are signalled by the superscripts "+" or "-". These stepwise changes in ploidy are determined by a number of known (selection pressures), suspected (microenvironment), or unknown factors. They result in two basic characteristics of the cells in a focus: asynchronism and heterogeneity.

---

with the majority of the cancer cells being diploid, had the assumption been made that the aneuploid cells in previous generations were nonviable.

It should be noted that changes in ploidy may create variant cells in the focus, with different properties. This translates into the production of heterogeneous cell populations within that focus which, together with asynchrony, is an important biological characteristic of those cells.

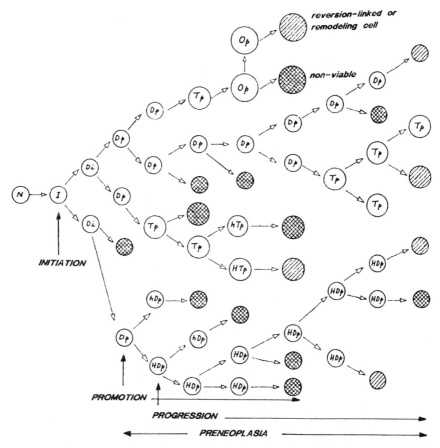

Fig. 7. Clonal model of hepatic foci: reversion-linked or remodelling focus. The vast majority of foci generated by the initiation-promotion model will eventually reverse their cytochemical or cytomorphological alterations, particularly after suspension of the promoting treatment. The diagram portrays changes found in most foci, with their cells destined to disappear or to become "differentiated" and remodelled within the non-affected parenchyma. Legends are the same as in Fig. 6.

Figure 7 represents the maturation or reversion changes that occur in the majority of the foci (53), with their viable cells developing probably partial differentiation in their morphological and cytochemical features. In the end they cannot be distinguished from the surrounding normal liver parenchyma, as the focus disappears as such.

This darwinian postulation of the development of a focus within the concept of multistage carcinogenesis, allows us to understand the variety of ploidy distributions between similar lesions, the possibility of a two-hit hypothesis and the existence of diploid and aneuploid hepatomas. Furthermore, it points out an aspect of the preneoplastic foci that has not been stressed in the past: the asynchronism and the heterogeneity of their cells.

CONCLUDING REMARKS

Much still has to be done in order to resolve the specific question of at which point in time does genetic instability occur. Our rat liver model with a diet deficient in choline indicates that it can happen early in the carcinogenesis process. But the juncture at which the genome becomes unstable may change in different systems or when using parameters other than aneuploidy to estimate stability. Timing genetic instability has more to it than just academic interest. The span in which it takes place determines when heterogeneity in a cellular population will commence. This, in turn, has enormous implications in the premature appearance of metastatic variants and problems related to them in the treatment of malignant tumors. It also presents an important challenge, because diagnosing preneoplastic lesions before the emergence of instability would, undoubtedly, benefit their therapy.

ACKNOWLEDGEMENTS

This work was supported by grants from the USPHS CA25164, CA35362 and CA45716. We would like to thank Eulalia C. Sudilovsky, Oscar R. Garcia, Richard Banozic, Ronald Sobecks and Venkatesh Krishnamurti for invaluable assistance. We also thank Jeannie Nagy for exceptional secretarial help.

REFERENCES

1. O. Sudilovsky, and T.K. Hei, Aneuploidy and progression in promoted preneoplastic foci during hepatocarcinogenesis in the rat, Cancer Lett. 56:131-135, 1991.
2. J.H. Wang, L.I. Hinrichsen, C.M. Whitacre, R.L. Cechner, and O. Sudilovsky, Nuclear DNA content of altered hepatic foci in a rat liver carcinogenesis model, Cancer Res. 50:7571-7576, 1990.
3. C. Peraino, R.J.M. Fry, and E. Staffeldt, Reduction and enhancement by phenobarbital of hepatocarcinogenesis induced in the rat by 2-acetylaminofluorene, Cancer Res. 31:1506-1512, 1971.
4. E. Scherer, and P. Emmelot, Foci of altered liver cells induced by a single dose of diethylnitrosamine and partial hepatectomy: their contribution to hepatocarcinogenesis in the rat, Eur. J. Cancer 11:145-154, 1975.
5. D. Solt, and E. Farber, A new principle of the analysis of chemical carcinogenesis, Nature 263:701-703, 1976.
6. H.C. Pitot, L. Barsness, T. Goldsworthy, and T. Kitagawa, Biochemical characterization of stages of hepatocarcinogenesis after a single dose of diethylnitrosamine, Nature 271:456-458, 1978.

7. H. Shinozuka, B. Lombardi, S. Sell, and R.M. Iammarino, Enhancement of DL-methionine-induced carcinogenesis in rats fed a choline-devoid diet, J. Natl. Cancer Inst. 61:813-817, 1978.

8. J.K. Reddy, M.S. Rao, and D.E. Moody, Hepatocellular carcinomas in acatalasemic mice treated with nafenopin, a hypolipidemic peroxisome proliferator, Cancer Res. 36:1211-1217, 1976.

9. S. Fiala, and A.E. Fiala, Acquisition of an embryonal biochemical feature by rat hepatomas. Experientia 26:889-890, 1970.

10. M.H. Hanigan, and H.C. Pitot, Gamma-glutamyltranspeptidase -its role in hepatocarcinogenesis, Carcinogenesis 6:165-172, 1985.

11. T.D. Pugh, and S. Goldfarb, Quantitative histochemical and autoradiograph studies of hepatocarcinogenesis in rats fed 2-acetylaminofluorene followed by phenobarbital, Cancer Res. 38:4450-4457, 1978.

12. S. Sell, Distribution of α-fetoprotein and albumin-containing cells in the livers of Fischer rats fed four cycles of N-2-fluorenylacetamide, Cancer Res. 38:3107-3113, 1978.

13. P. Bannasch, Dose-dependence of early cellular changes during liver carcinogenesis, Arch. Toxicol. Suppl 3:111-128, 1980.

14. K. Sato, A. Kitahara, K. Satoh, T. Ishikawa, M. Tatematsu, and N. Ito, The placental form of glutathione S-transferase as a new marker protein for preneoplasia in rat chemical hepatocarcinogenesis, Gann 75:199-202, 1984.

15. G.M. Williams, The pathogenesis of rat liver cancer caused by chemical carcinogens, Biochem. Biophys. Acta 605:167-189, 1980.

16. H.C. Pitot, H.C. Glauert, and M. Hanigan, The significance of selected biochemical markers in the characterization of putative initiated cell populations in rodent liver, Cancer Lett. 29:1-14, 1985.

17. R.G. Cameron, Identification of the putative first cellular step of chemical hepatocarcinogenesis, Cancer Lett. 47:163-167, 1989.

18. R.G. Cameron, Comparison of GST-P versus GGT as markers of hepatocellular lineage during analysis of initiation of carcinogenesis, Cancer Invest. 6:725-734, 1988.

19. M.A. Moore, K. Nakagawa, K. Satoh, T. Ishikawa, and K. Sato, Single GST-P positive liver cells--putative initiated hepatocytes, Carcinogenesis 8:483-486,1987.

20. K. Yokota, U. Singh, and H. Shinosuka, Effects of a choline-deficient diet and a hypolipidemic agent on single glutathione S-transferase placental form-positive hepatocytes in rat liver, Jpn. J. Cancer Res. 81:129-134, 1990.

21. G.W. Teebor, and F.F. Becker, Regression and persistance of hyperplastic hepatic nodules induced by N-2-fluorenylacetamide and their relationship to hepatocarcinogenesis, Cancer Res. 31:1-3, 1971.

22. G.M. Williams, and K. Watanabe, Quantitative kinetics of development of N-2-fluorenylacetamide-induced, altered (hyperplastic) hepatocellular foci resistant to iron accumulation and their reversion or persistance following removal of carcinogen. J Natl. Cancer Inst. 61:113-121, 1978.

23. K. Enomoto, and E. Farber, Kinetics of phenotypic maturation of remodelling of hyperplastic nodules during liver carcinogenesis, Cancer Res. 42:2330-2335, 1982.

24. M. A. Moore, H.J. Hacker, and P. Bannasch, Phenotypic instability in foci and nodular lesions induced in a short term system in the rat liver, Carcinogenesis 4:595-603, 1983.

25. S. Takahashi, B. Lombardi, and H. Shinozuka, Progression of carcinogen-induced foci of γ-glutamyltranspeptidase-positive hepatocytes to hepatomas in rats fed a choline-deficient diet, Int. J. Cancer 29:445-450, 1982.

26. A. Columbano, G.M. Ledda-Columbano, P.M. Rao, S. Rajalakshmi, and D.S.R. Sarma, Occurrence of cell death (apoptosis) in preneoplastic and neoplastic liver cells. A sequential study. Am. J. Pathol. 116:441-446, 1984.

27. S. Hendrich, H.P. Glauert, and H.C. Pitot, The phenotypic stability of altered hepatic foci: effects of withdrawal and subsequent readministration of phenobarbital, Carcinogenesis 7:2041-2045, 1986.

28. P.C. Nowell, The clonal nature of neoplasia, Cancer Cells 1:29-30, 1989.

29. H.M. Rabes, Th. Bücher, A. Hartmann, I. Linke, and M. Dunnwald, Clonal growth of carcinogen-induced enzyme-deficient preneoplastic cell populations in mouse liver, Cancer Res. 42:3220-3227, 1982.

30. W.C. Weinberg, L. Berkwits, and P.M. Iannaccone, The clonal nature of carcinogen-induced altered foci of γ-glutamyl transpeptidase expression in rat liver, Carcinogenesis 8:565-570, 1987.

31. M. Esumi, T. Aritaka, M. Arii, K. Suzuki, H. Mizuo, T. Mima, and T. Shikata. Clonal origin of human hepatoma determined by integration of hepatitis B virus DNA. Cancer Res. 16:5767-5771, 1986.

32. S. Howell, K.A. Wareham, and E.D. Williams, Clonal origin of mouse liver cell tumors, Am. J. Pathol. 121:426-432, 1985.

33. J.J. Yunis, Specific fine chromosomal defects in cancer: an overview. Human Pathol. 12:503-515, 1981.

34. N. Böhm, and W. Sandritter, DNA in human tumors: a cytophotometric study, Curr. Top. Pathol. 60:151-219, 1975.

35. W. Sandritter, Quantitative pathology in theory and practice, Pathol. Res. Pract. 171:2-21, 1981.

36. T.O. Caspersson, Quantitative tumor cytochemistry, Cancer Res. 39:2341-2355, 1979.

37. V. Digernes, Chemical liver carcinogenesis: monitoring the process by flow cytometric DNA measurements, Environ. Health Perspect. 50:195-200, 1983.

38. G. Saeter, P.E. Schwarze, J.M. Nesland, N. Juul, E.O. Pettersen, and P.O. Seglen, The polyploidizing growth pattern of normal rat liver is replaced by divisional diploid growth in hepatocellular nodules and hepatocarcinomas. Carcinogenesis 9:939-945, 1988.

39. Y. Koike, Y. Suzuki, A. Nagata, S. Furuta, and T. Nagata, T., Studies on DNA content of hepatocytes in cirrhosis and hepatomas by means of microspectrophotometry and radioautography. Histochem. 73:549-562, 1982.

40. T. Ezaki, T. Kanematsu, T. Okamura, T. Sonoda, and K. Sugimachi, DNA analysis of hepatocellular carcinoma and clinicopathologic implications. Cancer 61:106-109, 1988.

41. H.F. Stich, The DNA content of tumor cells. II. Alterations during the formation of hepatomas in rats, J. Natl. Cancer Inst. 24:1283-1297, 1960.

42. F.F. Becker, R.A. Fox, K.M. Klein, and S.R. Wolman, Chromosome patterns in rat hepatocytes during N-2-fluorenylacetamide carcinogenesis, J. Natl. Cancer Inst. 46:1261-1269, 1971.

43. F.F. Becker, K.M. Klein, S.R. Wolman, R. Asofsky, and S. Sell, Characterization of primary hepatocellular carcinomas and initial transplant generations, Cancer Res. 33:3330-3338, 1973.

44. H. Mori, T. Tanaka, S. Sugie, M. Takahashi, and G.M. Williams, DNA content of liver cell nuclei of N-2-fluorenylacetamide-induced altered foci and neoplasms in rats and human hyperplastic foci, J. Natl. Cancer Inst. 69:1277-1281, 1982.

45. M. Sarafoff, H.M. Rabes, and P. Dörmer, Correlations between ploidy and initiation probability determined by DNA cytophotometry in individual altered hepatic foci, Carcinogenesis 7:1191-1196, 1986.

46. L. Hinrichsen, S. Miron, R. Cechner, and O. Sudilovsky. Hepatocarcinogenesis in the rat: nuclear DNA content in hepatocarcinomas, Proc. Am. Assoc. Cancer Res. 31:155, 1990.

47. H.C. Pitot, and H.A. Campbell, Quantitative studies on multistage carcinogenesis in the rat. In Tumor Promoters: Biological approaches for mechanistics studies and assay system (Progr. Cancer Res. Ther. V.

34), R. Langenbach, E. Elmore, and J. Carl Barrett, eds., Raven Press, New York, 1988, pp 79-95.

48. C. Farber, and D.S.R. Sarma, Biology of disease. Hepatocarcinogenesis: a dynamic cellular perspective, Lab. Invest. 56:4-22, 1987.

49. P.C. Nowell, The clonal evolution of tumor cell populations. Science 194:23-28, 1976.

50. P.E. Schwarze, E.O. Petterson, M.C. Shoaib, and P.O. Seglen, Emergence of a population of small diploid hepatocytes during hepatocarcinogenesis, Carcinogenesis 5:1267-1275, 1984.

51. O. Sudilovsky and T.K. Hei, Aneuploid nuclear DNA content in some enzyme-altered foci during chemical hepatocarcinognesis, Fed. Proc. 42:7, 1983.

52. O. Sudilovsky and T.K. Hei, Prestaining of membrane markers to identify specific areas for Feulgen cytospectrophotometric determinations in a single section, Anal. Quant. Cytol. Histol. 9:323-327, 1987.

53. P. Bannasch, Preneoplastic lesions as end points in carcinogenicity testing. I. Hepatic preneoplasia, Carcinogenesis 7:689-695, 1986.

54. Y.S. Fu and T.L. Hall, DNA ploidy measurements in tissue sections, Anal. Quant. Cytol. Histol. 7:90-96, 1985.

55. J.D. Crissman and Y.S. Fu, Intraepithelial neoplasia (CIS) of the larynx. A clinicopathological study of six cases with DNA analysis, Arch. Otorinolaryngol. Head Neck Surg. 111:522-528, 1985.

56. G. Saeter, P.E. Schwarze, J.M. Nesland, and P.O. Seglen. Diploid nature of hepatocellular tumors developing from transplanted preneoplastic liver cells, Brit. J. Cancer 59:198-205, 1989.

57. P.E. Schwarze, G. Saeter, D. Armstrong, R.G. Cameron, E. Laconi, D.S.R. Sarma, V. Preat, P.O. Seglen, Diploid growth pattern of hepatocellular tumours induced by various carcinogenic treatments, Carcinogenesis, 12:325-327, 1991.

58. J.W. Grisham, M.-S. Tsao, L.W. Lee, and G.J. Smith, Clonal analysis of neoplastic transformation in cultured diploid rat liver epithelial cells. In: O. Sudilovsky, L.A. Liotta, and H.C. Pitot (eds.), The Boundaries Between Promotiona nd Progression during Carcinogenesis (this volume: article and discussion), New York, Plenum Press, 1991.

59. H. Danielson, H.B. Steen, T. Lindmo, and A. Reith, Ploidy distribution in experimental liver carcinogenesis in mice, Carcinogenesis 9:59-63, 1988.

60. S. Haesen, T. Derijke, A. Deleneer, P. Castelain, H. Alexandre, V. Preat, and M. Kirsch-Volders, The influence of phenobarbital and butylated hydroxytoluene on the ploidy rate in rat hepatocarcinogenesis, Carcinogenesis 9:1755-1761, 1988.

61. A.G. Knudson and L.C. Strong, Mutation and cancer: a model for Wilm's tumor of the kidney, J. Natl. Cancer Inst. 48:313-316, 1972.

CLONAL ANALYSIS OF NEOPLASTIC TRANSFORMATION

IN CULTURED DIPLOID RAT LIVER EPITHELIAL CELLS

J. W. Grisham[1], M.-S. Tsao[2],
L. W. Lee[1], and G. J. Smith[1]

[1]Department of Pathology, CB 7525
University of North Carolina at Chapel Hill
303 Brinkhous-Bullitt Building
Chapel Hill, NC 27599

and

[2]Department of Pathology
Montreal General Hospital and McGill University
Montreal, PQ H3G 1A4 Canada

INTRODUCTION

The organizers of this symposium posed specific queries to the participants. In this session we were asked to address this question: "At which
point in time does genetic instability occur in the natural history of
cancer?" We were also asked to consider "mechanistic similarities between
progression and promotion" and "to suggest research needed to elucidate these
issues." Although complete answers to these questions are not yet available,
we will examine these general issues in the context of studies being performed in our laboratory to clonally analyze the process of transformation in
vitro in cultured diploid rat liver epithelial cells[1-10]. Thus far our
studies have concentrated on identifying the essential tumorigenic phenotype/
genotype in cells of this model system by analyzing the clonal cosegregation
of phenotypic/genotypic properties with tumorigenicity. Clonal analysis also
provides a powerful strategy for identifying a lineal linkage between various
nontumorigenic and the tumorigenic variants. Although we have not yet accomplished complete or detailed lineage tracing, we are working toward this
goal after establishing some of the critical features of the tumorigenic
phenotype/genotype. Our early results allow the outline of a possible lineage to be inferred.

Abbreviations: The following abbreviations are used: alp and TGF-alpha,
Transforming growth factor-alpha; Diap, DT-diaphorase; EGF, epidermal growth
factor; GGT, gamma glutamyl transpeptidase; G6PD, glucose 6-phosphate dehydrogenase: GLC, capacity to form colonies in medium containing low levels
of calcium; LDH, lactate dehydrogenase; MNNG, N-methyl-N'-nitro-N-nitroso-
guanidine; mRNA, polyadenlyated messenger ribonucleic acid; myc, gene coding
for genomic sequences that are homologous with some of those of the avian
MC29 (myelocytomatosis) virus; PK, pyruvate kinase; SAG, capacity to form
colonies in soft agar.

CLONAL ANALYSIS

  Before presenting our results, we want to introduce the strategy and some
of the major technical requirements of clonal analysis of complex cellular
processes, to present an example of the power of this technique, and to
elucidate critical features of the process of carcinogenesis. The essential
technical requirements for clonal analysis are to isolate, characterize and
propagate individual index cells from the heterogenous population of cells
involved in the process under study; to establish clonal progeny populations
from each index cell; and to determine the extent to which cells of each of
the derived clonal populations co-express key properties that typify some step
of the process under study (Fig. 1). The ability to determine the consistent
cosegregation of two or more cellular properties among clonal populations
derived form a parental population of cells that expresses these properties
heterogeneously is a powerful technique. Using such a seemingly simple
strategy, one can determine by exclusion features of the phenotype/genotype
that are required for expression of a particular cellular property;
phenotypic/genotypic features that do not cosegregate uniformly with the
property among several clones may be excluded as a requirement.
Phenotypic/genotypic properties which occur in cells that do not express the
index property, but which cosegregate consistently with the index property may
reflect changes that are necessary but insufficient to bring about the index
property. Cells expressing such phenotypes may represent precursor stages
along a lineage leading to the cells that express the index property. By
replicating such studies at different points during the evolution of a complex
cellular process, one may be able to identify lineages that connect the cell
in which changes occur that start a process with the cell that expresses the
features of the complete process.

  Clonal analysis has yielded useful insights into many complex, multicellular developmental processes, including the description of lineages in

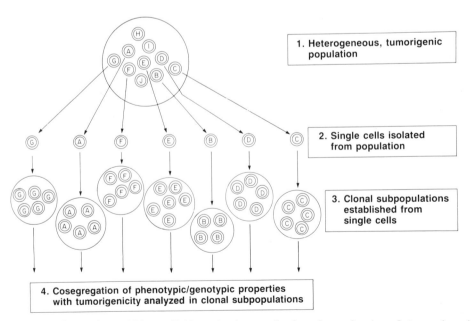

Fig. 1. Schematic outline of the strategy of clonal analysis of tumorigenicity
        and expression of paratumorigenic phenotypic properties.

the developing embryo[11] and in maturing bone marrow[12]. The characterization of the lineages of the various differentiated derivatives of bone
marrow, as well as the identification and purification of many growth and
differentiation factors for bone marrow cells have been the byproducts of
clonal analysis[12]. The initial achievement that made these scientific
feats possible was the development of the spleen colony assay for bone marrow
stem cells[13], coupled with the critical demonstrations that a spleen colony
was the clonal progeny of a single bone marrow stem cell[14] and that cells
in the colony were pluripotent[15].

CLONAL ANALYSIS OF CARCINOGENESIS

     Full understanding of the essential genetic components that underlies
the tumorigenic phenotype or of the complex cellular process that leads to
the formation of tumorigenic cells (carcinogenesis) will not be possible, in
our opinion, until this process can be analyzed fully on the basis of
individual cells and their descendants. Unfortunately, clonal analysis of
the carcinogenic process is not yet possible in in vivo settings (other than
leukemias and lymphomas), including experimental model systems; it is not yet
possible to isolate, characterize, and propagate clonally individual cells
that are progenitor cells for subsequent altered populations, including the
final tumorigenic population. In the well-studied rat liver experimental
carcinogenesis system in vivo[16-19], many thousands of phenotypically altered cells isolated from foci or nodules, or from tumors in livers of carcinogen-treated rats must be transplanted to livers or spleens of recipient
rats to yield a colony[20,21]. This is not sufficient: one must be able to
trace the fates of individual cells. Inability to clonally analyze the process of carcinogenesis in vivo prevents precise identification of the critical cellular phenotypes/genotypes that entrain the process and enable it to
progress. In the absence of clonal analysis, neither the precise lineage
that leads to a tumorigenic cell nor the essential mechanisms that are necessary to enable a diploid cell to progress through the lineage (nor, indeed,
the minimal essential genotype of the tumorigenic cell) can be identified.

     Attempts to trace transforming lineages in complex cellular populations
in tissues in vivo have used marker phenotypes, such as the histochemical
expression of gamma glutamyl transpeptidase or other enzymes. Such markers
may lack sufficient specificity in their association with the tumorigenic
phenotype to allow reliable tracing of the transformation lineage. Quantitation of the ratios of the number of nontumorigenic hepatocytes in phenotypically altered foci and nodules to the number of cells of carcinomas that subsequently develop in the rat liver carcinogenesis model system shows that the
vast majority of the cells that express altered enzyme levels never progress
to become tumorigenic. Hundreds of altered foci or nodules are produced for
each tumor that develops and each focus or nodule contains hundreds or thousands of cells[22-23]. The recent statement that the ratio between foci or
nodules and carcinomas can be reduced to five or six in optimal rat liver
models[24] does not obviate the problem for lineage analysis caused by the
multiplicity of putative precursor cells in relation to tumors. Whether any
of the cells in foci and nodules that express altered enzyme activities
become tumorigenic is a point of controversy that has been addressed only by
the rare observation of hepatocellular carcinomas arising within nodules of
phenotypically altered cells[16,24]. Such an association does not prove that
a particular lineage exists; a lineage can be established by clonal analysis.
Although it is not possible to prove a lineal linkage between the cells of
phenotypically altered foci and tumor cells in the rat liver model, the yield
of such foci clearly is related quantitatively to the dose of carcinogen[23].
Rather than revealing mechanistic associations that are critical for tumor
development, aberrant expression of enzymes in phenotypically altered cells
may reflect merely the "noise" induced more-or-less randomly in the genome by

carcinogen exposure, or it may evince epigenetically driven phenotypic adaptive responses made by cells to the altered intrahepatic microenvironment caused by the regimen of carcinogen and/or promoter. Clonal analysis is necessary also to identify phenotypes that are unrelated to carcinogenesis.

Currently, tissue culture models of carcinogenesis, which show developmental stages that resemble in vivo models[3,25], offer major advantages over in vivo models as systems for clonal analysis. Unlike in vivo transformation systems currently available, in vitro systems allow cells to be clonally manipulated with relative ease[2,5,25]. Index cells can be isolated from a transforming population in vitro at different stages of the process, and their progeny propagated and characterized by combining in vitro and in vivo analyses. Because of the current inability to analyze clonally the process of neoplastic transformation in vivo, we have begun the clonal analysis of carcinogenesis in a model in vitro system that employs diploid rat liver epithelial cells[3,5,6]. We are cognizant that tissue culture models do not provide exact reproductions of processes of carcinogenesis in animals. Insights gained from in vitro transformation systems ultimately must be validated by studies in animals. We hope that our in vitro studies will lead to similar studies in vivo. The in vitro rat liver epithelial cell transformation system that we study has features that resemble the much-studied in vivo experimental carcinogenesis model in rat liver[3], raising the possibility that useful insights from each system may be exchanged. It appears possible ultimately to clonally analyze the rat liver model carcinogenesis system by combining clonal growth and characterization in vitro of cells isolated from a liver in which the cells are transforming in vivo, with back-transplantation to in vivo sites for evaluation of growth and differentiation properties and for assessment of tumorigenicity of cells from clonal cultures. Combined in vivo/in vitro/in vivo studies that may lead to such an accomplishment already are being attempted[26,27].

COMPARISON OF CELL LINEAGES OF NORMAL TISSUES AND TUMORS

Renewing cell populations that are components of normal tissues[28,29] and the population of cells that undergo transformation and form a tumor[30] are quite different in structure, although the terminology applied to both normal tissues and tumors is similar. Normal, renewing populations originate from stem cells that have the capacity to produce many descendants and, at the same time, to maintain their own number. The pathway leading from undifferentiated stem cell to functional, terminally differentiated cells is termed a lineage. A lineage pathway may contain several compartments, including the stem cell compartment, an amplifying (proliferating) transit compartment, and a functionally differentiated compartment[28,29]. In normal tissue lineages, cells lose the capacity to grow clonally and repopulate as they progress through the lineage pathway and acquire differentiated functions. Stem cells are not identifiable morphologically and must be detected indirectly, by their ability to establish colonies (clonogenicity) in vivo after depletion of a lineage (such as the spleen colony assay for bone marrow stem cells in irradiated animals), to repopulate a lineage after its depletion in vivo or to establish colonies in vitro.

Cells may be isolated from a tumor that are able to re-establish the tumor (after transplantation in vivo) or to form colonies in soft agar. These cells are often termed tumor stem cells[30]. The normal cell in which the process of carcinogenesis begins is viewed appropriately as the stem cell of a transformation lineage, and not the ultimate tumorigenic cell. The tumorigenic cell (some of which can re-establish the tumor) is the terminal cell in a transformation lineage, analogous to the fully differentiated functional cell of a normal lineage. The lineage of tumor development, thus, goes from a normal progenitor (stem cell) to a tumorigenic cell. This view of a transformation lineage is similar to the concept of the tumor stem line[31], in which the precursor (stem) cell and the population of tumor cells that arise from the stem cell bear similar chromosomal abnormalities.

THE WB-F344 LIVER EPITHELIAL CELL LINE

The WB-F344 liver epithelial cell line (WB cells), which was isolated by primary cloning of cells recovered from the liver of an adult Fischer-344 rat, expresses the phenotype of immature hepatocytes or "oval" cells[1]. The morphological and phenotypic characteristics of WB cells have been described[1]. Other unpublished properties of wild-type WB cells include the capacity to synthesize and secrete constitutively both insulin-like growth factor II and transforming growth factor-beta. The phenotypic characteristics of WB cells indicate their relationship to fetal hepatocytes and biliary ductular cells. Unpublished studies on the transplantation of wild-type WB cells into the intrascapular fat pads of syngeneic animals have shown that transplanted cells occasionally form aggregates of hepatocyte-like cells or biliary ducts in vivo. The action on WB cells of various biological response modifiers, including growth factors, hormones, and tumor promoters, also has been examined[32-38].

Early passage WB cells contain a diploid complement of male chromosomes lacking any visible structural abnormalities. Wild-type WB cells have been cultured continuously for up to 35 to 45 passages (115 to 130 doublings) without a consistent alteration in major phenotypic properties or consistent changes in karyotype. Occasional WB sublines acquire an additional, apparently normal copy of chromosome 1 after long-term, continuous passaging in culture. Wild-type WB cells, including variants with trisomy of chromosome 1, have not produced tumors when transplanted under standard conditions (1×10^6 cells in 0.2 ml medium) into either syngeneic newborn Fischer 344 rats or nude mice prior to 115 to 130 doublings after establishment in culture[2,3]. After 115 to 130 doublings in continuous culture, WB cells transform spontaneously with a high frequency[39]; maintenance of WB cells at confluence causes them to transform much earlier, both spontaneously and following exposure to MNNG[39].

EXPERIMENTAL DESIGN

WB cells at the 6th passage (about 20 doublings) after establishment were used for this study. Five groups of cells, each group in triplicate, were maintained in culture. Three of the groups were exposed to N-methyl-N'-nitro-N-nitrosoguanidine (MNNG) at concentrations of 2.5 µg/ml, 5.0 µg/ml, and 10.0 µg/ml in serum-free medium for 24 hr at each exposure, following which time the medium containing MNNG was removed, cultures were washed and complete, serum-containing medium was added. Two groups of control cultures, maintained in parallel with MNNG-treated cultures, included medium-only controls and medium-vehicle (acetone) controls. Studies reviewed here utilized only the 5 µg/ml treatment group and medium-only controls. Cells were assessed for acute toxicity after each exposure to MNNG and maintained until cultures reached confluence. When cultures reached confluence, they were trypsinized and cells (both MNNG-treated and controls) were replated at a split-ratio of 3 to 1. At each passage, an aliquot of cells from each group was frozen and other aliquots were tested for selected phenotypic properties (colony forming efficiency, doubling time, DNA content/cell, ability to grow in medium containing a low level of calcium, ability to grow in soft agar, activities of selected enzymes and forms of isoenzymes, and tumorigenicity)[3].

Control cultures reached confluence by 10 to 12 days after plating at the beginning of the study, and by 8 days at the end; cultures exposed to MNNG grew at rates comparable to those of controls for all passages after the third. Following the first two exposures to MNNG, there was a 1- to 3-day delay in reaching confluence due to toxicity, but there was no diminution in rates of growth of treated cultures after the third exposure to MNNG. Tumorigenicity was first noted in MNNG-exposed cultures after the 11th exposure to 5 µg/ml of MNNG (11th passage). When 1×10^6 cells from the population exposed to MNNG 11 times were injected subcutaneously into one-day-old Fischer-44 rats, 34% (21 of 61) developed progressively growing tumors within one year[2,3]. Control cells were never tumorigenic.

PHENOTYPIC CHANGES IN TRANSFORMING POPULATIONS

Prior to the acquisition of tumorigenicity, the MNNG-treated populations developed extensive phenotypic heterogeneity, as compared to the control populations[3]. Control populations remained phenotypically stable, save for decreased doubling times and increased colony forming efficiencies during continuous passaging while in exponential growth. Major phenotypic changes affecting the populations of MNNG-treated cells included decreased activity of alkaline phosphatase and increased activities of glucose 6-phosphate dehydrogenase, lactate dehydrogenase, DT-diaphorase and pyruvate kinase, with peak activities occurring near the conclusion of the exposure sequence. Increases in activities of enzymes ranged from about 2-fold (glucose 6-phosphate dehydrogenase, lactate dehydrogenase, and DT-diaphorase) to nearly 4-fold (pyruvate kinase). Minor changes occurred in the distribution of isoenzymes for some of these enzymes. The activity of GGT showed variable increases in cell populations exposed to MNNG, ranging as high as 20 ± 5 mU/mg protein (GGT activity in control WB cells = 3.5 ± 1.0 mU/mg protein). Associated with increased GGT activity was the development of histochemical positivity in a small fraction of the cells in the MNNG-exposed cultures. Control WB populations were uniformly negative for cells that expressed GGT to a level of activity that could be detected histochemically, but after the 11th cycle of exposure to MNNG about 10% of the population was histochemically positive.

Other phenotypic changes reflected alterations in cellular growth control[3], including an improved ability to grow in medium containing a low level of calcium (0.02 mM as compared to basal medium which contained 2.0 mM) and to form colonies in soft agar. Although control populations were unable to grow in medium containing 0.02 mM calcium, the population exposed 11 times to MNNG grew about 80% as efficiently in this medium as it did in medium containing 2.0 mM calcium. Wild-type WB cells and control populations were never able to form colonies in soft agar. Populations exposed 7 times to MNNG were able to form a few colonies in soft agar (0.01 to 0.03% colony forming efficiency when 1.5×10^4 cells were plated). Supplementation of soft agar medium with epidermal growth factor markedly increased the efficiency of colony formation in soft agar by populations exposed to MNNG more than 6 times (0.3 to 0.7% colony forming efficiency when 1.5×10^4 cells were plated), but had no effect on control cells[4].

Populations of WB cells multiply treated with MNNG showed a progressively increasing fraction of aneuploid cells, reaching a maximal level of 82% of the population after the 11th exposure[3]. Paradiploid aneuploidy was noted initially, followed by a gradually increasing fraction of subtetraploid cells. Tumors derived from the population exposed 11 times to MNNG were poorly differentiated carcinomas that contained occasional adenocarcinomatous foci. Strikingly, over 80% of the cells from tumors were histochemically positive for GGT, as compared to about 10% of the cells in the population exposed 11-times to MNNG from which the tumors arose[3].

These observations showed that the transforming population of WB cells progressively became heterogeneous for several phenotypic properties. Among the phenotypic properties that were expressed by the MNNG-exposed transforming populations (but not control populations) were the histochemical expression of GGT, aneuploidy, anchorage-independent growth, and ability to grow in medium containing 0.02 mM calcium. These phenotypic properties have been considered to be markers for transforming cells, including transforming hepatocytes in the rat liver carcinogenesis model system in vivo[16-19] and in cultured liver epithelial cells[40-44]. Because the population of cultured liver epithelial cells exposed 11 times to MNNG demonstrated similarities to the in vivo liver carcinogenesis model[3], this population seemed to provide a useful subject for clonal analysis of phenotypes/genotypes that segregate with tumorigenicity.

CLONING OF CELL LINES FROM THE TUMORIGENIC POPULATION

We established several clonal sublines from single cells isolated from the phenotypically heterogeneous, tumorigenic population that resulted from 11 exposures to MNNG. Clones were selected by their ability to form colonies in medium containing 0.02 mM calcium[2] and by whether or not they expressed histochemically detectable GGT[5]. Thirteen clones able to grow in low calcium medium, 11 clones histochemically positive for GGT, and 7 clones histochemically negative for GGT were established. Data on the expression of various phenotypic properties, including tumorigenicity of these clonal lines, and of lines established from tumors produced by tumorigenic cells, have been published[2,5,6]. These data form the basis of this analysis. Data from the "B" series tumorigenicity studies are used here, since the "A" series appears to be flawed, as noted[6]. Additional tumorigenicity studies are in process and will be reported elsewhere. The new tumorigenicity studies do not appear to affect qualitative tumorigenicity (fraction of test animal groups in which some tumors occur), which is analyzed here, although the quantitative tumorigenicity (fraction of animals in each test group that have tumors) may change. In addition, this analysis makes use of additional data on the qualitative expression of mRNAs for myc protein and transforming growth factor-alpha[10].

CLONAL EXPRESSION OF PHENOTYPIC PROPERTIES

Clonal lines that were selected for the expression of a particular index property have maintained that index phenotype during multiple subcultures; in other words, significant drift of the index phenotype has not occurred in these cultured cells. In cell lines selected for a particular index phenotype, most of the other phenotypic properties measured were expressed heterogeneously among the various lines[2,5,6]. For example, clones selected in low calcium medium variably expressed histochemical positivity for GGT (10 of 13 clones histochemically positive for GGT) and ability to form colonies in soft agar (6 of 10 clones anchorage-independent). Similarly, clonal lines that were selected because they were histochemically positive or negative for GGT, expressed many other phenotypic properties heterogeneously (for example, 3 of 9 GGT-positive clones and 3 of 6 GGT-negative clones grew poorly in low calcium medium). As with the index phenotype, the other phenotypic properties of clonal lines have shown little drift during extensive subcultures.

Expression of several cellular properties clearly was associated with the ploidy of cells (Table 1). Although cells of all clonal lines contained aneuploid levels of DNA, GGT-positive cells had near-tetraploid contents of DNA while GGT-negative cells had near-diploid DNA levels[5]. Cell volumes

Table 1. Ploidy-dependent Expression of Several Phenotypic Properties Including Tumorigenicity, by Clones Isolated from the Phenotypically Heterogeneous, Tumorigenic Population Resulting from 11 Exposures to MNNG

Property	WB Cells	GN Clones (7)	GP Clones (11)	References
DNA (rel.)	100	104 ± 5	146 ± 12	5
VOLUME	632 ± 134	537 ± 0.8	990 ± 235	5
ENZYMES				
GGT (mU/mgP)	3.5 ± 1.0	1.0 ± 0.9	674 ± 809	
LDH (U/mgP)	2.6 ± 0.50	2.2 ± 0.4	5.0 ± 1.6	
PK (mU/μgP)	3.2 ± 0.50	4.1 ± 1.4	10.8 ± 1.4	5
Diap (mU/μgP)	0.59 ± 0.70	0.58 ± 0.24	5.5 ± 2.4	
G6PD (mU/mgP)	0.16 ± 0.08	0.13 ± 0.04	0.46 ± 0.03	
EGF BINDING (rel.)	100 (18.4 pg/10^6 cells)	94.2 ± 16.1	52.3 ± 33.1	4
TGF-alpha mRNA (rel.)	0	2.9 ± 2.1	48.1 ± 57.6	10
myc mRNA (rel.)	1	5.6 ± 2.8	18.2 ± 26.4	10
SOFT AGAR GROWTH				
-EGF	0	0.02 ± 0.04%	1.8 ± 4.7%	4
+EGF	0	3.9 ± 5.4%	5.8 ± 6.2%	
TUMORIGENICITY				
(% test groups with tumors)	0	83%	80%	6
(% animals with tumors)	0	36.8 ± 45.3	64.3 ± 44.4	6

and the activities of many enzymes in addition to GGT (lactate dehydrogenase, glucose 6-phosphate dehydrogenase, DT-diaphorase and pyruvate kinase) were related to ploidy level, being higher in cells with more DNA. Near-tetraploid cells secreted more EGF-like material (presumptively TGF-alpha) into the medium[9] and expressed higher levels of mRNA for transforming growth factor-alpha than did near-diploid cells[10]. Near-tetraploid cells produced tumors in a larger fraction of the recipients than did near-diploid cells, suggesting that tumor yield was affected by gene dosage. In the analysis reported here, we are concerned only with the qualitative aspects of tumorigenicity.

The histology of tumors also varied among and within clones. Tumors generally were epithelial, and expressed various differentiation phenotypes known for liver cell lineages, including hepatocellular carcinomas, hepatoblastomas, adenocarcinomas, and epidermoid carcinomas, as well as anaplastic tumors[7].

The heterogeneous expression of tumorigenicity and of various cellular phenotypic attributes among multiple clonal lines, independent of index (selected) phenotype, provides a setting in which the phenotypic properties

Table 2. Qualitative Expression of Selected Phenotypes by
Various Clonal Lines of Liver Epithelial Cells

Cell Clone	Phenotype					
	TUM	GGT	SAG	GLC	myc	alp
WB1	-	-	-	-	ND	ND
WB2	-	-	-	-	-	-
GP1	+	+	-	-	+	ND
GP2	+	+	+	+	-	+
GP3	-	+	+	+	+	+
GP4	+	+	-	+	-	ND
GP6	+	+	+	+	+	+
GP7	+	+	+	ND	+	ND
GP8	+	+	+	+	+	+
GP9	+	+	-	+	+	+
GP10	+	+	+	+	+	+
GP11	+	+	+	+	ND	ND
GN1	+	-	-	+	-	ND
GN2	-	-	-	-	+	+
GN3	+	-	-	-	+	ND
GN4	+	-	-	-	+	ND
GN5	+	-	-	+	+	+
GN6	+	-	-	+	+	+
Cl3	+	+	+	+	ND	ND
Cl4	+	+	+	+	ND	ND
Cl5	+	+	+	+	ND	ND
Cl6	-	-	-	+	ND	ND
Cl8	+	+	+	+	ND	ND
GP6T*	+	+	+	+	+	+
GP8T	+	+	+	+	+	+
GP10T	+	+	+	+	+	+
GN6T*	+	+	+	+	+	+

*Cell line designations ending in the letter "T" represent
cells derived from the tumors produced by the appropriate
clonal line.

required for tumorigenicity may be identified by analyzing their clonal cosegregation with tumorigenicity (Fig. 1).

CLONAL COSEGREGATION OF PHENOTYPES, INCLUDING TUMORIGENICITY

Simple inspection of the phenotypic properties expressed by various
clonal lines indicated the lack of tight correlation between tumorigenicity
and each of the measured phenotypes, or between various pairs of individual
phenotypes[2,5,6]. It is possible to quantify the relationships between different phenotypes, and between different phenotypes and tumorigenicity, by
analyzing the data on the segregation of these cellular properties among independent clonal lines using the conditional probability model of Bayes[45].
For this analysis phenotypic properties, including tumorigenicity, are considered in qualitative, binary terms, i.e., as present (+) or absent (-). In

subsequent reports, we plan to analyze the involvement of quantitative variations in expression of selected phenotypes and tumorigenicity. For the analysis reported here, we have examined the following phenotypes in cells of
clonal lines -histochemical expression of GGT (positive stain indicated by
red color after standard reaction[46]), ability to grow in soft agar (positive growth in soft agar indicated by the formation of colonies by more than
0.01% of 150,000 cells plated under standard conditions), ability to form
colonies in medium containing 0.02 mM Ca^{++} (positive growth in low calcium
medium is set at a relative colony forming ability of greater than 10% of the
ability of the same cells to form colonies in medium containing 2.0 mM Ca^{++}),
the relative abundance of myc mRNA (positive level is set at greater than 4-
fold the level of expression by wild-type WB cells as judged by densitometric
tracing of Northern blots), the relative abundance of TGF-alpha mRNA (positive level is set at greater than 2.5-fold the level of expression of TGF-
alpha mRNA by cells of clone GN1; wild-type WB cells did not express TGF-
alpha mRNA) and tumorigenicity (a positive tumorigenic response is defined as
one or more tumors occurring in a group of 6 to 22 test animals). The qualitative results of expression of these selected phenotypes among 27 lines are
presented in Table 2.

Cosegregation of Marker Phenotypes

We examined the clonal cosegregation of selected phenotypic properties
that have been used or might serve as markers for tumorigenicity and lineage
tracing, including expression of GGT, anchorage independence, ability to grow
in medium containing low calcium, and expression of myc and TGF-alpha mRNAs
and/or products (Tables 3 to 7). GGT was a fair predictor of expression of
TGF-alpha mRNA, soft agar growth capacity, and ability to grow in medium containing low calcium, correctly assigning 77% to 85% of the test cells to the
appropriate category (Table 3). However, GGT was a very poor predictor of
myc expression, leading to correct assignments to appropriate myc expression
categories of only 56% to 65% of test cells.

Soft agar growth capacity demonstrated good to fair correlation with GGT
expression and TGF-alpha expression, and poor correlation with low calcium
growth capacity and myc expression (Table 4). Ability to grow in low calcium
medium showed perfect correlation with TGF-alpha expression, fair correlation
with GGT expression, and poor correlation with soft agar growth capacity and
myc expression (Table 5). These impressions are corroborated by analysis of
the cosegregation of myc and TGF-alpha expression with other phenotypes
(Table 6 and 7).

Cosegregation of Phenotypes with Tumorigenicity

Only the expression of TGF-alpha was a reasonable predictor of the ability of clonal lines to produce tumors in at least one of the test animals of
each group (90% to 93% correct assignments) (Table 8). Soft agar growth
capacity, histochemical expression of GGT, low calcium growth capacity, and
myc expression alone individually correctly assigned only 60% to 77% of
clones to the appropriate tumorigenic category. For each of these phenotypes, positive expression was a much better indicator of tumorigenicity than
was negative expression, an indicator of lack of tumorigenicity, i.e., there
were many false positives. It is of note that elevated expression of myc
mRNA was a good predictor of tumorigenic clones, but lack of elevated myc
expression did not predict nontumorigenic clones accurately. Our data on the
quantitative relationships among myc and TGF-alpha message expression and
tumorigenicity suggest that myc and TGF-alpha interact[10]. TGF-alpha expression appears to be most highly correlated with tumorigenicity quantitatively when expression is coordinated with elevated myc expression. However,
elevated myc expression in the absence of expression of TGF-alpha is not
related to tumorigenicity.

Table 3. Clonal Cosegregation of Histochemical Positivity For Gamma-Glutamyl Transpeptidase with Other Phenotypes*

	Phenotype			
	alp	SAG	GLC	myc
Frequency of GGT positivity among clones that express index phenotype	.75 (.83)	.92 (.94)	.75 (.80)	.60 (.71)
Frequency of GGT negativity among clones that do not express index phenotype	1.00 (1.00)	.73 (.73)	.83 (.83)	.50 (.50)
Predictive value of positive GGT	1.00 (1.00)	.79 (.83)	.92 (.94)	.67 (.67)
Predictive value of negative GGT	.67 (.67)	.89 (.89)	.56 (.56)	.43 (.43)
Fraction of clones correctly assigned to positive or negative index category by GGT status	.80 (.86)	.83 (.85)	.77 (.81)	.56 (.65)

*The first number in each entry includes data from all lines listed in Table 1 except those derived from tumors. The second number in each entry, enclosed in parentheses, incorporates data from all lines listed in Table 1, including lines derived from tumors.

Table 4. Clonal Cosegregation of Soft Agar Growth Capacity with Other Phenotypes*

	Phenotype			
	GGT	alp	GLC	myc
Frequency of SAG positivity among clones that express index phenotype	.79 (.83)	.63 (.75)	.63 (.70)	.50 (.64)
Frequency of SAG negativity among clones that do not express index phenotype	.89 (.89)	1.00 (1.00)	.83 (.83)	.67 (.67)
Predictive value of positive SAG	.92 (.94)	1.00 (1.00)	.91 (.93)	.71 (.82)
Predictive value of negative SAG	.73 (.73)	.40 (.40)	.45 (.45)	.44 (.44)
Fraction of clones correctly assigned to positive or negative index category by SAG status	.83 (.85)	.70 (.79)	.68 (.73)	.56 (.65)

*The first number in each entry includes data from all lines listed in Table 1 except those derived from tumors. The second number in each entry, enclosed in parentheses, incorporates data from all lines listed in Table 1, including lines derived from tumors.

Table 5. Clonal Cosegregation of Ability to Grow in Medium Containing Low Calcium with Other Phenotypes*

| | Phenotype | | | |
	alp	GGT	SAG	myc
Frequency of GLC positivity among clones that express index phenotype	1.00 (1.00)	.92 (.94)	.91 (.93)	.56 (.69)
Frequency of GLC negativity among clones that do not express index phenotype	1.00 (1.00)	.56 (.56)	.45 (.45)	.20 (.20)
Predictive value of positive GLC	1.00 (1.00)	.75 (.80)	.63 (.70)	.50 (.64)
Predictive value of negative GLC	1.00 (1.00)	.83 (.83)	.83 (.83)	.20 (.20)
Fraction of clones correctly assigned to positive or negative index category by GLC status	1.00 (1.00)	.77 (.81)	.68 (.73)	.40 (.53)

*The first number in each entry includes data from all lines listed in Table 1 except those derived from tumors. The second number in each entry, enclosed in parentheses, incorporates data from all lines listed in Table 1, including lines derived from tumors.

Table 6. Clonal Cosegregation of Elevated Expression of myc mRNA with Other Phenotypes*

| | Phenotype | | | |
	alp	GGT	SAG	GLC
Frequency of elevated myc mRNA expression among clones that express index phenotype	.63 (.71)	.67 (.79)	.71 (.83)	.50 (.64)
Frequency of absence of elevated myc mRNA expression among clones that do not express index phenotype	.50 (.50)	.43 (.43)	.44 (.44)	.20 (.20)
Predictive value of positive myc expression	.83 (.90)	.60 (.71)	.50 (.64)	.56 (.69)
Predictive value of negative myc expression	.40 (.40)	.50 (.50)	.67 (.67)	.17 (.17)
Fraction of clones correctly assigned to positive or negative index category by myc mRNA expression	.60 (.71)	.56 (.65)	.56 (.65)	.40 (.53)

*The first number in each entry includes data from all lines listed in Table 1 except those derived from tumors. The second number in each entry, enclosed in parentheses, incorporates data from all lines listed in Table 1, including lines derived from tumors.

Table 7. Clonal Cosegregation of Elevated Expression of TGF-alpha mRNA with
Other Phenotypes*

	Phenotype			
	GGT	GLC	SAG	myc
Frequency of alp mRNA expression among clones that express index phenotype	1.00 (1.00)	1.00 (1.00)	1.00 (1.00)	.83 (.90)
Frequency of absence of elevated alp mRNA expression among clones that do not express index phenotype	.50 (.50)	1.00 (1.00)	.40 (.40)	.25 (.25)
Predictive value of positive alp expression	1.00 (1.00)	1.00 (1.00)	.63 (.75)	.63 (.75)
Predictive value of negative alp expression	1.00 (1.00)	1.00 (1.00)	1.00 (1.00)	.50 (.50)
Fraction of clones correctly assigned to positive or negative index category by alp mRNA expression	.80 (.86)	1.00 (1.00)	.70 (.79)	.60 (.71)

*The first number in each entry includes data from all lines listed in Table
1 except those derived from tumors. The second number in each entry, enclosed in parentheses, incorporates data from all lines listed in Table 1,
including lines derived from tumors.

Table 8. Clonal Cosegregation of Selected Phenotypes with Tumorigenicity*

	Phenotype				
	alp	GGT	GLC	SAG	myc
Frequency of marker in tumorigenic clones	1.00 (1.00)	.81 (.85)	.69 (.76)	.59 (.67)	.68 (.79)
Frequency of absence of marker in non-tumorigenic clones	1.00 (1.00)	.50 (.50)	.67 (.67)	.67 (.67)	.67 (.67)
Predictive value of positive marker	.88 (.92)	.81 (.85)	.90 (.98)	.83 (.88)	.86 (.89)
Predictive value of negative marker	.67 (.67)	.50 (.50)	.33 (.33)	.36 (.36)	.44 (.44)
Fraction of clones correctly assigned by marker	.90 (.93)	.73 (.77)	.69 (.75)	.61 (.67)	.70 (.74)

*The first number in each entry includes data from all lines listed in Table
1 except those derived from tumors. The second number in each entry, enclosed in parentheses, incorporates data from all lines listed in Table 1,
including lines derived from tumors.

OTHER CELLULAR PROPERTIES AND THEIR ASSOCIATION WITH TUMORIGENICITY

Aneuploidy and Nonrandom Karyotypic Changes

All cells cloned from the phenotypically heterogeneous, tumorigenic population that was exposed 11 times to MNNG were aneuploid, and contained consistent structural alterations (including translocations) involving chromosomes 1, 4, 7, and 10. The chromosomal structural changes will be detailed
elsewhere. These consistent aberrations involved all cells cloned from the
11-times-treated population and appear to represent a stem line. The emergence of the stem line must have antedated the acquisition of tumorigenicity
since both tumorigenic and nontumorigenic clones show identical structural
changes in chromosomes. If this karyotype is causally involved in acquisition of tumorigenicity, then it must represent a permissive step, with additional cellular changes being required for tumorigenicity to be expressed.
An alternative explanation for common structural changes in chromosomes of
tumorigenic cells is convergence to a common pattern of chromosomal abnormalities that is associated with tumorigenicity[47]. However, convergence would
not be expected to produce identical karyotypes in both tumorigenic and nontumorigenic clones. Therefore, the presence of a common karyotype is most
consistent with the origin of all clonal lines from a single altered precursor (stem) cell early in the transformation process. Tracing of this stem
line backward through populations that were exposed to MNNG and passaged
fewer times should enable the elucidation of this interesting situation.

Polyploidy

Both paradiploid and hypotetraploid clones have been isolated from the
heterogeneous tumorigenic population[5]. Paradiploid and hypotetraploid cells
contained virtually identical chromosomal structural abnormalities, and differed only in the hypotetraploid cells being partially polyploid. A similar
fraction of paradiploid and hypotetraploid clones were qualitatively tumorigenic (i.e., tumors produced in at least one of the animals of test groups),
but the hypotetraploid clones produced a higher yield of tumors (i.e., tumors
produced in a larger fraction of the test animals of each group). One paradiploid clone produced tumors containing hypotetraploid cells, and this clone
subsequently has shifted to higher ploidy in culture. These observations
allow the speculation that the paradiploid cells represent an earlier stage
in the tumorigenic stem line than do the partially polyploid, hypotetraploid
cells. Polyploidy appears to increase tumorigenicity of cells, possibly
through alterations in gene dosage or gene "balance".

Transforming DNA

We have evaluated the ability of DNA from 7 of the lines cloned from the
tumorigenic population exposed 11 times to MNNG to transform NIH 3T3 cells by
transfection. The clones examined represent both tumorigenic and nontumorigenic variants, and include both paradiploid and hypotetraploid cells. DNA
from all 7 clones tested was able to transform NIH 3T3 cells, unrelated to
the ability of the clones to produce tumors in isogenic Fischer 344 rats. In
contrast, DNA from wild-type WB cells did not transform NIH 3T3 cells. Although we have detected rat DNA in the transformed NIH 3T3 cells, we have
not been able yet to identify the gene that has been transfected, but we have
excluded ras and raf genes as candidates. These results suggest that cloned
lines contain an unknown "activated" gene that is able to transform NIH 3T3
cells, but that this "activated" gene is not sufficient to make the liver
epithelial cells tumorigenic in isogenic test animals. As is possible for
the consistent chromosomal structural changes, this putatively "activated"
gene may be necessary but not sufficient for tumorigenicity in isogenic
Fischer 344 rats, which may require an additional change.

SUMMARY AND CONCLUSION

Expression of TGF-alpha mRNA, which correlates well with the ability of cells to condition medium with an EGF-like activity[9], clonally segregates best with tumorigenicity among the several single phenotypes considered in this study. The results of unreported studies in which we have analyzed the quantitative relationships between the expression of selected phenotypes and tumorigenicity, suggest that the elevated expression of myc and TGF-alpha mRNAs interact in their associations with tumor yield. These results suggest that elevated myc expression sensitizes hepatic epithelial cells to the possible tumorigenic action of TGF-alpha[10]. This observation may explain why the correlation between the qualitative expression of TGF-alpha and tumorigenicity, described here, is not perfect.

Conventionally applied markers of transformation in hepatocytes in vivo[16-19] and in cultured liver epithelial cells in vitro[40-44] that we studied -histochemical expression of GGT, ability to grow in medium containing low levels of calcium, and ability to grow in soft agar- clonally segregated with tumorigenicity poorly in liver epithelial cells transformed in vitro. We conclude that these phenotypes are not adequate markers for determining the lineage of hepatic epithelial neoplasms (including, probably hepatocellular cancers arising in vivo). This study appears to be the first to attempt to analyze clonally the association of these markers with tumorigenicity, and to quantify the sensitivity, specificity, and predictive value of the associations. Our study suggests that the relatively weak associations of these phenotypes with tumorigenicity may be related only to their stronger associations with expression of TGF-alpha, or to some other property that is strongly associated with tumorigenicity. Expression of TGF-alpha is more strongly associated with expression of GLC, for example, than is the GLC phenotype with tumorigenicity. At least for GLC, autocrine stimulation by TGF-alpha is likely, since EGF increases growth of WB cells in low calcium medium[33]. This observation may explain the perfect correlation between expression of TGF-alpha and GLC. EGF also stimulates lactate dehydrogenase, pyruvate kinase, and glucose 6-phosphate dehydrogenase in WB cells[33]. However, quantitative correlation between GGT activity and TGF-alpha is less strong.

Thus, our data from these studies suggest that the tumorigenic phenotype of cultured hepatic epithelial cells is intimately dependent on the expression of the TGF-alpha gene, possibly producing autocrine stimulation of growth via the cells' EGF receptors. This is the most simple view of the potential relationship between TGF-alpha expression and tumorigenicity in liver epithelial cells. However, it is evident also from our unpublished studies that additional factors modify this relationship, including EGF receptor number, secretion and activation of TGF-beta, and, possibly, TGF-beta receptors, as well as the elevated expression of some proto-oncogenes (especially myc and ras) and, possibly, tumor suppressor genes. Furthermore, our unpublished data suggest that there may be additional transformation pathways for rat liver epithelial cells that do not involve TGF-alpha and EGF receptors.

Cells of all clonal lines contain identical chromosomal structural changes, and the DNA's of all lines tested, whether tumorigenic or not, are able to transform NIH 3T3 cells by transfection. The transforming ability of the DNA suggests that all clones contain an activated gene that can transform NIH 3T3 cells. The fact that this trait is also expressed by DNA from liver epithelial clones that are not tumorigenic in Fischer 344 rats suggests that this activated gene is not sufficient for tumorigenicity of the liver epithelial cells in isogenic hosts. We have not identified the gene from hepatic epithelial cells that is responsible for transformation of NIH 3T3 cells. Although these hepatic epithelial cell clones variably overexpress several proto-oncogenes, including Harvey and Kirsten <u>ras</u> genes[10], neither of these genes was transfected into the genome of focus-forming NIH 3T3 cells by DNA from the liver epithelial cells. Similar observations have been made by others about the transfection of NIH 3T3 cells with DNA from rat liver tumors that developed in vivo following carcinogen exposure[48].

The consistent nonrandom chromosomal structural changes may bear a similar relationship to tumorigenicity as does the transforming DNA, since these structural alterations are found in cells of both tumorigenic and nontumorigenic clones. In any event, the consistent karyotypic abnormalities indicate that all of the clonal lines are derived from a common clonal precursor in which the altered karyotype first emerged, i.e., that they share a common stem line. Presence of this stem line, together with the other properties that correlate with the stem line or clonally segregate with tumorigenicity, allow the outlines of a transformation lineage leading from a diploid hepatic epithelial cell to a tumor cell to be suggested (Fig. 2). Diploid cells of the starting population of liver epithelial cells undergo chromosomal structural and numerical changes soon after exposure to MNNG[3], which may occur cataclysmically or sequentially over time. Ultimately, a cell develops that contains a stable configuration of chromosomal abnormalities allowing the affected cell to supplant the normal population, possibly by giving the progeny of the altered cell a growth advantage under the conditions of MNNG-exposure and culture. Since both paradiploid and hypotetraploid cells are present in the heterogeneous population, since both bear common chromosomal structural aberrations, and since the hypotetraploid variants are more numerous (and more tumorigenic) than are the hypodiploid variants, we speculate that the stem line began as a paradiploid variant, subsequently polyploidizing and possibly gaining a further growth advantage. Neither paradiploidy or hypotetraploidy alone confer tumorigenicity, although the chromosomal structural aberrations may be responsible for activating the unknown gene that is able to transform NIH 3T3 cells by transfection. Additional changes are required, most notably the expression of the normally quiescent TGF-alpha gene and augmented expression of the <u>myc</u> gene. Coordinate elevation of expression of these genes clonally segregates with tumorigenicity, but their expression levels do not segregate consistently with the common stem line. It is likely

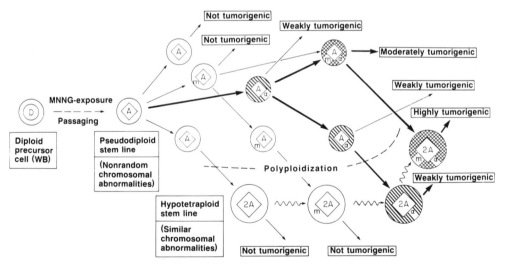

Fig. 2. A tentative, MNNG-induced transformation lineage leading from diploid WB rat liver epithelial cells to tumorigenic cells. The letters m and α signify cells that express elevated levels of <u>myc</u> and/or TGF-alpha mRNAs.

that other genetic alterations are also involved in predicating tumorigenicity. Stimulation of heightened expression of TGF-alpha and myc genes must have occurred after the development of the common stem line in some of the stem line cells. Occurrence of clones that bear the stem line karyotype, but lack elevated expression of either the TGF-alpha gene or the myc gene, or that express only one or the other of these genes in an elevated fashion, indicates that separate events are responsible for their activations and that the chromosomal changes occurred earlier. Availability of clones that express high levels of neither myc mRNA nor TGF-alpha mRNA or that express only one of these genes offers a substrate for studying the mechanisms of their activation and of their action in predicating tumorigenicity. Examination of the populations that preceded the 11-times-cultured and MNNG-exposed tumorigenic population by techniques similar to those used here should allow us to substantiate this tentative lineage (Fig. 2) and to fill in some of the missing details.

In concluding, we want to consider how our studies may bear on the questions that were asked by the conference organizers. It is clear from our observations (and, obviously, from many other studies as well) that carcinogenesis involves multiple genetic changes, and that these separate genetic alterations may provide the mechanistic basis for multistep carcinogenic processes. Phenotypically aberrant clones emerge repeatedly and yield altered populations that have competitive growth advantages. Under the conditions of our studies, genetic instability occurs throughout the carcinogenic process, in part possibly because we have used an experimental protocol in which cell populations were exposed repeatedly to a carcinogen. Nevertheless, in recent studies in which we have used a protocol that exposes cells only once to MNNG[39], we have found a similar pattern of evolution of the carcinogenic process in the transforming population.

It is impossible to identify separate and distinct stages of initiation, promotion, and progression among the complex pattern of changes that occur in the transforming populations of liver epithelial cells that we have studied. In our system the stages of initiation, promotion, and progression do not appear to occur in an ordered linear sequence except, possibly, at the level of an individual cell. At the level of the heterogeneous population of transforming cells, all three stages occur simultaneously among various clonal subpopulations. The entire sequence of genetic changes that are involved in the development of tumorigenicity in a population of liver epithelial cells in vitro may be compressed into a brief period of time[39]. Requirement for multiple genetic alterations and expansion of several clonal populations does not mean necessarily that a long, slowly evolving process always precedes the acquisition of tumorigenicity. In the in vitro setting, at least, the multiple genetic alterations may take place virtually simultaneously, and the tumorigenic population can emerge with astonishing rapidity when all the stages are concatenated.

Regarding the types of genes that are involved, we suspect that a variety of specific genes may have a role in tumorigenicity, including protooncogenes, tumor suppressor genes, genes for growth factors/receptors, as well as other genes as yet unidentified. A variety of mechanisms, ranging from single-base gene mutations to more complex chromosomal aberrations, may be involved in altering gene expression. We doubt that the starting of the process of carcinogenesis ("initiation") can be correlated with any particular type of genetic alteration, and we favor as a working hypothesis a recent model in which there is no required order to the genetic changes, including chromosomal aberrations, that enable the tumorigenic phenotype to be acquired[49].

As we stated at the beginning of this paper, we believe that the process of carcinogenesis must be clonally analyzed before one can discern the essential components of the tumorigenic phenotype/genotype or understand the mechanism(s) by which the tumorigenic phenotype is acquired during the process of carcinogenesis. Therefore, we believe that clonal analysis of transformation and tumorigenicity in various tissues and cells in vivo and in vitro should be pursued vigorously as a high priority research objective.

ACKNOWLEDGEMENTS

We have been helped by many outstanding technical assistants, but especially by William Bell, who established the WB-F344 line, and by Judith Smith, who carried out the carcinogen-exposure studies. M. Theo Cantwell prepared the manuscript. The work reported here was supported by NIH Grant CA 29323.

REFERENCES

1. M.-S. Tsao, J. D. Smith, K. G. Nelson, and J. W. Grisham, A diploid epithelial cell line from normal adult rat liver with phenotypic properties of "oval" cells, Exp. Cell Res. 154:38-52 (1984).
2. J. W. Grisham, J. D. Smith, and M.-S. Tsao, Colony-forming ability in calcium-poor medium in vitro and tumorigenicity in vivo not coupled in clones of transformed rat hepatic epithelial cells, Cancer Res. 44:2821-2834 (1984).
3. M.-S. Tsao, J. W. Grisham, K. G. Nelson, and J. D. Smith, Phenotypic and karyotypic changes induced in cultured rat hepatic epithelial cells that express the "oval" cell phenotype by exposure to N-methyl-N'-nitro-N-nitrosoguanidine, Am. J. Pathol. 118:306-315 (1985).
4. M.-S. Tsao, H. S. Earp, and J. W. Grisham, Gradation of carcinogen-induced capacity for anchorage-independent growth in cultured rat liver epithelial cells, Cancer Res. 45:4428-4432 (1985).
5. M.-S. Tsao, J. W. Grisham, B. B. Chou, and J. D. Smith, Clonal isolation of populations of gamma-glutamyl transpeptidase-positive and -negative cells from rat liver epithelial cells chemically transformed in vitro, Cancer Res. 45:5134-5138 (1985).
6. M.-S. Tsao, J. W. Grisham, and K. G. Nelson, Clonal analysis of tumorigenicity and paratumorigenic phenotypes in epithelial cells chemically transformed in vitro, Cancer Res. 45:5139-5144 (1985).
7. M.-S. Tsao and J. W. Grisham, Hepatocarcinomas, cholangiocarcinomas and hepatoblastomas produced by chemically transformed cultured rat liver epithelial cells, Am. J. Pathol. 127:168-181 (1987).
8. M.-S. Tsao and J. W. Grisham, Phenotypic modulation during tumorigenesis by clones of transformed rat liver epithelial cells, Cancer Res. 47:1282-1286 (1987).
9. C. Liu, M.-S. Tsao, and J. W. Grisham, Transforming growth factors produced by normal and neoplastically transformed rat liver epithelial cells in culture, Cancer Res. 48:850-855 (1988).
10. L. W. Lee, J. Harris, D. C. Lee, H. S. Earp, and J. W. Grisham, Relationship between expression of H-ras, K-ras, myc, alpha-transforming growth factor, and epidermal growth factor (EGF)-receptor mRNAs and tumorigenicity among carcinogen-transformed rat liver epithelial cells, Proc. Am. Assoc. Cancer Res. 30:415 (1989) (abstract).
11. A. S. Wilkins, "Genetic Analysis of Animal Development," John Wiley, New York (1986).
12. P. Quisenberry and L. Levitt, Hematopoietic stem cells, N. Engl. J. Med. 301:755-760, 819-823 (1979).
13. J. E. Till and E. A. McCulloch, A direct measurement of the radiation sensitivity of normal mouse bone marrow cells, Radiat. Res. 14:213-222 (1961).

14. A. J. Becker, E. A. McCulloch, and J. E. Till, Cytological demonstration of the clonal nature of spleen colonies derived from mouse marrow cells, Nature 197:452-454 (1963).

15. A. M. Wu, J. E. Till, L. Siminovitch, and E. A. McCulloch, A cytological study of the capacity for differentiation of normal hematopoietic colony forming cells, J. Cell. Physiol. 69:177-184 (1967).

16. E. Farber, The sequential analysis of liver cancer induction, Biochim. Biophys. Acta 605:149-166 (1980).

17. E. Farber, Cellular biochemistry of the stepwise development of cancer with chemicals, Cancer Res. 44:5463-5474 (1984).

18. P. Emmelot and E. Scherer, The first relevant cell stage in rat liver carcinogenesis: A quantitative approach, Biochim. Biophys. Acta 605:247-304 (1980).

19. H. C. Pitot and A. Sirica, The stages of initiation and promotion in hepatocarcinogenesis, Biochim. Biophys. Acta 605:191-215 (1980).

20. B. A. Laishes, L. Fink, and B. I. Carr, A liver colony assay for a new hepatocyte phenotype as a step towards purifying new cellular pheno-types that arise during carcinogenesis, Ann. NY Acad. Sci. 349:373-382 (1980).

21. B. A. Laishes and P. B. Rolfe, Quantitative assessment of liver colony formation and hepatocellular carcinoma incidence in rats receiving injections of isogenic liver cells isolated during hepatocarcino-genesis, Cancer Res. 40:4133-4143 (1980).

22. C. Peraino, E. F. Staffeldt, B. A. Carnes, V. A. Lunderman, J. A. Blomquist, and S. D. Vesselinovitch, Characterization of histochemi-cally detectable altered hepatocyte foci and their relationship to hepatic tumorigenesis in rats treated once with diethylnitrosamine or benzo(a)pyrene within one day after birth, Cancer Res. 44:3340-3347 (1984).

23. W. K. Kaufmann, S. A. MacKenzie, and D. G. Kaufman, Quantitative rela-tionship between hepatocytic neoplasms and islands of cellular al-teration during hepatocarcinogenesis in male F344 rats, Am. J. Pathol. 119:171-174 (1985).

24. E. Farber and D. S. R. Sarma, Hepatocarcinogenesis: A dynamic cellular perspective, Lab. Invest. 56:4-22 (1987).

25. J. C. Barrett, T. W. Hesterberg, and D. G. Thomassen, Use of cell trans-formation systems for carcinogenicity testing and mechanistic studies of carcinogenesis, Pharmacol. Rev. 36:53S-70S (1984).

26. T. Kitigawa, R. Watanabe, T. Kayano, and H. Sugano, In vitro carcino-genesis of hepatocytes obtained from acetylaminofluorene-treated rat liver and promotion of their growth by phenobarbital, Japan. J. Cancer Res. (Gann) 71:747-754 (1980).

27. W. K. Kaufmann, M.-S. Tsao, and D. L. Novicki, In vitro colonization ability appears soon after initiation of hepatocarcinogenesis in the rat, Carcinogenesis 7:669-671 (1986).

28. B. I. Lord, C. S. Potten, and R. J. Cole, "Stem Cells and Tissue Homeo-stasis," Cambridge U. Press, Cambridge, U.K. (1978).

29. N. Wright and M. Allison, "The Biology of Epithelial Populations," Clarendon Press, Oxford, U.K. (1984).

30. G. G. Steel, "Growth Kinetics in Tumors," Clarendon Press, Oxford, U.K. (1977).

31. T. S. Hauschka, The chromosome in ontogeny and oncogeny, Cancer Res. 21:957-974 (1961).

32. M.-S. Tsao, K. G. Nelson, and J. W. Grisham, Biochemical effects of 12-0-tetradecanoylphorbol-13-acetate, retinoic acid, phenobarbital, and 5-azacytidine on a normal rat liver epithelial cell line, J. Cell. Physiol. 121:1-6 (1984).

33. M.-S. Tsao, J. D. Smith, and J. W. Grisham, The modulation of growth of normal rat liver epithelial cells in calcium-poor medium by epidermal growth factor, phenobarbital, phorbol ester, and retinoic acid, In Vitro Cell. Dev. Biol. 21:249-253 (1985).

34. M.-S. Tsao, H. S. Earp, and J. W. Grisham, The effects of epidermal growth factor and the state of confluence on enzymatic activities of cultured rat liver epithelial cells, J. Cell. Physiol. 126:167-173 (1986).

35. H. S. Earp, K. A. Austin, J. Blaisdell, R. Rubin, K. G. Nelson, L. W. Lee, and J. W. Grisham, Epidermal growth factor (EGF) stimulates EGF receptor, J. Biol. Chem. 261:4777-4780 (1986).

36. P. Lin, C. Liu, M.-S. Tsao, and J. W. Grisham, Inhibition of proliferation of cultured rat liver epithelial cells at specific cell cycle stages by transforming growth factor-beta, Biochim. Biophys. Res. Commun. 143:26-30 (1987).

37. M.-S. Tsao, G. H. S. Sanders, and J. W. Grisham, Regulation of growth of cultured hepatic epithelial cells by transferrin, Exp. Cell Res. 171:52-62 (1987).

38. H. S. Earp, J. S. Hepler, L. A. Petch, A. Miller, A. R. Berry, J. Harris, V. W. Raymond, B. A. McCune, L. W. Lee, J. W. Grisham, and T. K. Harden, Epidermal growth factor (EGF) and hormones stimulate phosphoinositide hydrolysis and increase EGF receptor protein synthesis and mRNA levels in rat liver epithelial cells, J. Biol. Chem. 263:13868-13874 (1988).

39. L. W. Lee, G. J. Smith, M.-S. Tsao, and J. W. Grisham, Emergence of aneuploidy is correlated with tumorigenicity in carcinogen-transformed rat liver epithelial cell populations, but not in spontaneously transformed cell populations, Proc. Am. Assoc. Cancer Res. 30:153 (1989).

40. R. Montesano, C. Drevon, T. Kuroki, L. Saint Vincent, S. Handleman, K. K. Sanford, D. DeFeo, and I. B. Weinstein, Test for malignant transformation of rat liver cells in culture: Cytology, growth in soft agar, and production of plasminogen activator, J. Nat. Cancer Inst. 59:1651-1658 (1977).

41. E. Huberman, R. Montesano, C. Drevon, T. Kuroki, L. Saint Vincent, T. D. Pugh, and S. Goldfarb, Gamma-glutamyl transpeptidase and transformation of cultured liver cells, Cancer Res. 39:269-272 (1979).

42. R. H. C. San, M. F. Laspia, A. I. Soieffa, C. J. Maslansky, J. M. Rice, and G. M. Williams, A survey of growth in soft agar and cell surface properties as markers for transplantation in adult rat liver epithelial-like cells in cultures, Cancer Res. 39:1026-1034 (1979).

43. R. H. C. San, T. Shimada, C. J. Maslansky, D. M. Kreiser, M. F. Lasfea, J. M. Rice, and G. M. Williams, Growth characteristics and enzyme activities in a survey of transformation markers in adult rat liver epithelial-like cell cultures, Cancer Res. 39:4441-4448 (1979).

44. P. T. Iype, S. Turner, and M. S. Siddiqui, Markers for transformation in rat liver epithelial cells in culture, Ann. NY Acad. Sci. 349:312-322 (1980).

45. W. Feller, "An Introduction to Probability Theory and Its Applications," Volume One, 3rd Ed., Rev., John Wiley, New York (1968).

46. A. M. Rutenberg, H. Kim, J. W. Fishbein, J. S. Hanker, H. L. Wasserkrug, and A. M. Seligman, Histochemical and ultrastructural demonstration of gamma-glutamyl transpeptidase activity, J. Histochem. Cytochem. 7:189-201 (1959).

47. S. Heim and F. Mittelman, "Cancer Cytogenetics," Liss, New York (1987).

48. M. Goyette, M. Dolan, W. Kaufmann, D. Kaufman, P. B. Shank, and N. Fausto, Transforming activity of DNA from rat liver tumors induced by the carcinogen methyl(azoxymethyl)nitrosamine, Mol. Carcinogenesis 1:26-32 (1988).

49. P. Cerutti, Response modification creates promotability in multistage carcinogenesis, Carcinogenesis 9:519-526 (1988).

DISCUSSION

George Michalopoulos: Do you have any information on the size of the oncogene from rat liver cells that transfects NIH 3T3 cells?

Joe Grisham: We think it's fairly large. Since we couldn't identify it by hybridization with conventional probes, we did studies in which we tried to inactivate it by restriction enzyme digestion. After digestion with restriction enzymes, we determined whether the restricted DNA was still able to transfect NIH 3T3 cells. Every restriction enzyme we used, except one very uncommon one, inactivated the DNA. That result might suggest that the putative oncogene is fairly large, but I'm not sure these results are good evidence for size.

Michalopoulos: This is very interesting because the work from Marshall Anderson's lab also shows that in some mouse hepatomas there are some very large transfecting oncogenes that have been detected, but whose nature is unknown. I wonder whether we are dealing with a whole novel family of large transfecting oncogenes.

Grisham: I don't know. As you know, the story of the rat liver oncogene, if there is one, is obscure. Although there seems to be something in rat liver tumor cell DNA that can transform NIH 3T3 cells by transfection, no one yet has identified what is being transfected.

Oscar Sudilovsky: Joe, how did you determine the hepatocarcinoma's ploidy?

Grisham: We did it both by flow cytometry and by karyotyping them, Oscar. Flow cytometry is a lot easier, but it is not as precise for detecting small changes in DNA content. Obviously, you get different information from the two techniques.

Sudilovsky: I know that this is the $64,000 question: How do you answer the conclusion by Seglen that all of the cells of hepatocellular carcinomas are diploid?

Grisham: All of our tumorigenic cells are aneuploid, either near-diploid or near-tetraploid. We have to tune our flow cytometer so that the CVs are around 3-4% before we can distinguish the near-diploid aneuploid cells from true diploids. The near-diploid aneuploid cells have lost a very small amount of total DNA. I have asked Dr. Seglen about the sensitivity of the flow cytometric techniques he uses, and he said that his CVs are around 5-6%. Of course, a cell can be aneuploid and have a nearly normal amount of DNA. Karyotyping is more sensitive to loss of small amounts of genome, but it is more tedious than flow cytometry.

Sudilovsky: Did Seglen isolate the cells on the basis of their positivity for the gamma-glutamyltranspeptidase marker?

Grisham: I think he was using whole cell or bare nuclear preparations, as I recall. If so, he may have had a mixture containing stromal cells. He presented his studies at Copper Mountain a couple of weeks ago, but I've forgotten the details.

In our studies on MNNG-induced transformation, cells from every one of the tumorigenic clones from heterogeneous variable populations and cells from all of the tumors which they form are aneuploid, either near-diploid or near-tetraploid. Changes in chromosomal structure and number are nonrandom.

We've recently studied spontaneous transformation in these liver epithelial cells, which occurs when the cells are put in selective environments. For example, if you leave rat liver cells at confluence for a considerable period of time, spontaneous transformants will emerge because they continue to grow, whereas growth of the more normal cells is suppressed. The spontaneously transformed cells don't show exactly the same pattern of karyotypic change as to MNNG-induced transformants, but there are certain nonrandom changes in chromosomes that are virtually identical in terms of structure. We are continuing to study these chromosomal changes.

EPIGENETIC FEATURES OF SPONTANEOUS TRANSFORMATION

IN THE NIH 3T3 LINE OF MOUSE CELLS

H. Rubin and Kang Xu*

Department of Molecular Biology and Virus Laboratory
University of California
Berkeley, CA 94720

INTRODUCTION

There is a longstanding debate in cancer research about the primary cause of malignant cell growth: is cancer the result of genetic events or the outcome of epigenetic processes? The weight of opinion seems to shift with research trends of biology in general. It is, of course, central to the resolution of such a problem that the concepts at issue be defined. Genetic events are of two basic kinds, mutation and chromosome recombination. Mutations result from a change in the sequence of nucleotides in DNA. They are generally assumed to occur at random with a frequency of less than 10^{-6} per cell division with little or no evidence of specificity[1]. Chromosome recombination normally occurs in an orderly way in sexual reproduction. It also occurs in disorderly fashion in somatic cells of aging individuals[2,3], in tumors[4,5] and in cell culture[6,7]. Except for certain leukemias[8], abnormalities in cell chromosome structure or number in common adult cancers show little evidence of a specific causal relation to the origin of the tumor. However, genetic change is conceptually simple and has been vigorously analyzed in this area of molecular biology. Concurrently, there has been a strong shift toward acceptance of genetic change in somatic cells as the cause of most cancers, and at least part of this shift stems from the combination of conceptual simplicity plus the availability of a highly developed molecular technology for genetic analysis.

Epigenetic processes underlie the changes in cellular phenotype associated with differentiation in multicellular organisms. The characteristics of differentiated cells are stably inherited as long as the cells retain their three dimensional tissue associations in the body, but are gradually lost when cells are dissociated for tissue culture. The term epigenesis arose in antiquity with Aristotle as the antithesis of preformation, and therefore signified the appearance of entirely new structures in development (Dictionary of Biology), without regard to mechanisms. It was interpreted in this century by Waddington as the interaction between the genome and its environment which results in the development of the organism[9]. The definition was

*Current address: West China University of Medical Sciences, Chengdu,
 Sichuan, China
Abbreviations: MCDB 402, molecular, cellular and developmental biology
medium 402; CS, calf serum; FBS, fetal bovine serum.

Boundaries between Promotion and Progression during Carcinogenesis
Edited by O. Sudilovsky *et al.*, Plenum Press, New York, 1991

broadened by protozoologists to include single cells as a result of the analysis of heritable changes induced in mating type, serotype and cortical
structures of ciliate protozoa[10,11]. Some characteristics of epigenetic
events leading to the formation of new structures are the regularity of their
induction by specific environmental conditions and the existence of a wide
range of stabilities when the inducing conditions are removed[10]. They also
require an appropriate level of competence on the part of the responding
cells. Contrary to widespread belief, the distinction between genetic and
epigenetic events does not rest with their location in the nucleus or cytoplasm, since genetic material occurs in the cytoplasm[12], and nuclei undergo
differentiation[13]. In general, genetic events can be traced to stable
changes in the base sequence of DNA, whereas epigenetic events involve envi-
ronmentally sensitive changes in rates of complex cellular processes. Epigenesis may even include changes in DNA structure as seen in the development
of antibody producing cells, but it does not require such changes.

Theories of the genetic origin of cancer in somatic cells have been re-
inforced in recent years by the discovery in some tumors of altered cellular
genes or oncogenes which induce neoplastic behavior when transfected into
target cells, preeminently NIH 3T3 cells[14]. However, tumors can be considered new structures and therefore the product of epigenetic changes, without
specifying changes in DNA or other macromolecules. Evidence supporting the
epigenetic origin of tumors includes the findings that a) up to 100% of
target cells are altered into a cancer-susceptible state by treatment with
carcinogens[15,16,17], and b) the neoplastic state is reversible, depending
on the local environment in chemically-induced cancer of newts[18] and in
teratocarcinomas of mice[19]. Both epigenetic and genetic events are thought
to underlie transformation of carcinogen-treated tracheal epithelium in
vitro[20]. Since the discovery of the molecular mechanism for replication of
DNA, cancer research has focussed ever more strongly on the genetic origin of
the malignant change, which holds the promise of isolating and characterizing
the putatively causal DNA and protein molecules. By contrast, epigenetic
change presents an intimidating prospect to the molecularly minded, since it
involves dynamic processes with no guarantee of a crucial role for informa-
tional molecules. To complicate matters further, there has been no reliable
quantitative epigenetic system in which the neoplastic change could be induced with regularity by specific environmental conditions.

We now describe a system which presents the major features of epigenetic
change leading to neoplastic growth, although we cannot rule out some change
in nucleic acids. Interestingly enough, the system is the NIH 3T3 mouse cell
line which is the conventional target for transfection by cellular oncogenes.
We feel this system provides a basis for generating and testing concepts of
epigenetic change.

MATERIALS AND METHODS

Cells and Culture Methods

NIH 3T3 cells were originally derived from NIH Swiss mouse embryo cultures by selecting clones that had a high plating efficiency and a low saturation density[21]. They were kindly supplied by Dr. S. A. Aaronson and
cryo-preserved after passage in Dulbecco's Modified Eagle's Medium plus 10%
FBS. Our first observation of transformed foci was with cells which had been
passaged in MCDB 402[22] plus 10% FBS without cryopreservation. The cryo-
preserved cells were used in subsequent experiments. They were passaged
every 7 days at a concentration of 10^4 cells per 60 mm dish in MCDB 402
plus 10% FBS[23]. Foci were recognized in confluent cultures as discrete
colonies of elongated, refractile, multiplying cells set against a background of flat, isometric non-multiplying cells. The foci were highlighted

by Giemsa staining or by labeling the cells with ³H-thymidine (1 μC per ml) for one hour and processing for autoradiography[24]. Giemsa-stained foci were counted in a stereomicroscope at a 10x or 20x magnification. ³H-thymidine labeled cells were counted in a standard upright microscope at 125x or 500x magnification. CS and FBS were supplied by HyClone Laboratories, Logan, Utah. MCDB 402 medium was prepared in the laboratory.

RESULTS

The Occurrence and Nature of Transformed Foci in NIH 3T3 Cultures

The NIH 3T3 cells upon their receipt were cultured in FBS at a variety of seeding densities and under several transfer schedules. No remarkable changes in appearance occurred in cultures that were transferred once or twice a week but in those which were left undisturbed, except for medium changes, foci of altered cells were noticed at 17 days. The foci consisted of slender, elongated cells which grew in multiple layers and infiltrated the surrounding flat monolayer (Fig. 1A). After staining, foci could be readily scored as dark colonies on a light background (Fig. 1B). When labeled with

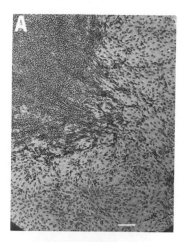

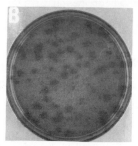

Fig. 1A. A Giemsa-stained focus of transformed NIH 3T3 cells. 7 week
old multiple-passage cells
seeded at 10⁵ per 60 mm dish
in 10% FBS and incubated with
medium changes for 14 days.
Stained with Giemsa. Bar =
200 μm.
1B. An unmagnified view of the same
dish showing the countable
colonies.

³H-thymidine, the percentage of labeled cells in the transformed foci (32%) was much higher than the percentage of labeled cells in the surrounding areas (2.5%) (Fig. 2).

Dependence of Focus Formation on Number of Passages in Culture and On Serum Type and Concentration

We wished to determine the optimal serum concentration for production of transformed foci, and also whether the competence of cells for transformation changed with repeated transfer. Cells which had been in weekly passage for 2 weeks and 20 weeks after thawing the cryopreserved stock were cultured in varying concentrations of CS and FBS and the number of transformed foci in each serum variation was determined at 13 and 30 days. In cells raised in CS, there were about 10 times as many foci in the 20 week cultures as in the 2 week cultures (Table 1). The CS concentration producing the greatest number of foci in both sets of cultures was 2% at 13 days. There was a marked suppression of focus formation in 10% CS at 13 days in the 20 week cultures, but less suppression in the 2 week cultures, where even the maximum number of foci was low. However, there was a sharp increase in foci by 20 days in the same 10% CS cultures. Focus formation in 5% and 10% FBS, para-doxically enough, was higher in the 2 week than in the 20 week cultures. No foci were seen at any time in cells cultured in 20% FBS. When the same experiment was done with cultures which had been passaged for 4 weeks, they

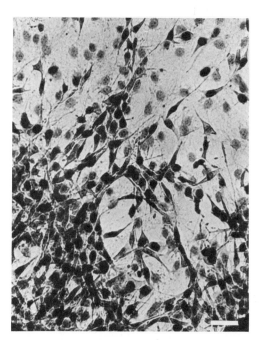

Fig. 2. Autoradiograph of a transformed focus and surrounding cells labeled with a one hour pulse of ³H-thymidine. 6 week old multiple-passage cells seeded at 10⁵ per 60 mm dish in 10% FBS, incubated with medium changes for 18 days, exposed for 1 hour to 1 μC per ml ³H-thymidine and processed for autoradiography. Bar = 50 μm.

Table 1. Dependence of Focus Formation on Number of Cell Passages
in Culture and on Serum Type and Concentration

Serum Type	Concentration %	Foci in 2 week cultures at 13 days	Foci in 2 week cultures at 20 days	Foci in 20 week cultures at 13 days	Foci in 20 week cultures at 20 days
CS	0.5	14	29	171	146
	1	19	31	278	TNTC
	2	64	71	468	TNTC
	5	49	66	308	TNTC
	10	31	TNTC	21	280
FBS	5	76	186	43	109
	10	68	160	9	195
	20	0	0	Detached	

TNTC = too numerous to count.

responded to serum variations in about the same way as the cultures passaged for 20 weeks. Thus, the sensitivity of the NIH 3T3 cells to transformation increased with successive passages of the original stock of cells for about 4 weeks and remained relatively constant thereafter. The fact that foci only appeared under the growth inhibitory infuence of confluency, and that they were markedly delayed in high concentrations of growth stimulatory serum, suggested that the foci were induced by the inhibition of cell growth and/or metabolism.

Effect of Seeding Density on the Occurrence of Transformed Foci

To determine the role of confluency on transformation, the 2 week and 20 week cultures were seeded at varying concentrations from 10^4 to 10^6 per dish and maintained in 2% CS. The time at which confluency was reached was noted and foci were counted 7 and 13 days after seeding. (7 days were required for the lowest density seeding of 10^4 cells per dish to become confluent.) No foci were seen at 7 days in the 2 week cultures seeded with 1

Fig. 3. Effect of seeding density of cells on the occurrence of transformed foci. Cells from 2 week and 20 week old, multiple-passage cultures were seeded at concentrations of 1, 2, 5, 10, 50 and 100 x 10^4 per 60 mm dish in 10% FBS for one day, then shifted to 2% CS the next day. They were incubated for 7 and 13 days with medium changes, fixed, stained, and the foci were counted. A): 2 week cultures counted at: O, 7 days; ●, 13 days. B): 20 week cultures counted at: □, 7 days; ■, 13 days. *, faint colonies.

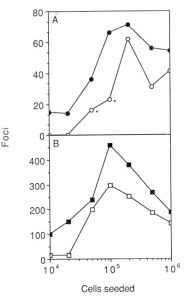

x 10^4 and 2 x 10^4 cells, and only about 15 foci were seen in the 20 week
cultures under the same conditions (Fig. 3). There was a sharp increase of
7-day foci when the seeding was increased to 5 x 10^4 cells per culture, and
further increases of foci in cultures seeded with 1 to 2 x 10^5 cells. With
further increases in seeding density up to 10^6 cells per culture, the
number of 7-day foci tended to decrease. Between 7 and 13 days, there was a
sharp increase in the numbers of foci among cultures seeded with 1 and 2 x
10^4 cells, and a much smaller relative increase over the same interval in
the cultures seeded at higher densities. The pattern seen at 7 days of
decreasing number of foci with seeding densities higher than 1 to 2 x 10^5
was maintained in the 13-day counts. Daily observation revealed that the
formation of most foci in the low serum concentrations we used here began
about two days after the cultures had become confluent. Focus formation
reached a maximum with seedings of 1 to 2 x 10^5 cells and tended to
decrease at higher seeding densities, indicating that there was not a fixed
proportion of transformed cells already existing in the population at the
time of seeding, and that transformation was actually induced after the cultures became confluent.

Autoradiographic Labeling of Foci and Background with ^{3}H-thymidine

The apparent reduction of focus formation with high serum concentrations
seen in Table 1 could have been an illusion created by the high population
density of these cultures, which interfered with discrimination between foci
and background. If this were merely a question of morphological discrimination, it could be sidestepped by using ^{3}H-thymidine autoradiography to determine whether there were foci of cells which continued to multiply at a
relatively high rate in heavily confluent cultures. Cells were grown in
2% and 10% CS, or in 10% FBS. At 5 day intervals they were labeled with
^{3}H-thymidine and a count was made of the number of sharply circumscribed
areas (foci) with distinctly higher proportions of labeled cells than the
surrounding areas. In addition, the percentage of labeled cells in the foci
and in the surrounding areas was determined. In the case of cultures in 10%
CS, where no discrete areas of high labeling index were detected, the
percentage of labeled cells was determined in randomly chosen fields. The
greatest numbers of labeled foci were produced in this experiment by cultures
in 10% FBS (Fig. 4). No foci were detected in cultures with 10% CS, and an
intermediate number was detected in cultures with 2% CS. There were 30% to
43% labeled cells in the foci of cultures in 2% CS and 10% FBS, compared with
2% to 7% labeled cells in the monolayer areas surrounding the foci of these
cultures (Fig. 5). Among the cultures in 10% CS, the highest proportion of
labeled cells in any of the randomly chosen fields was 14%. This was lower

Fig. 4. Counting foci with high per-
centages of ^{3}H-thymidine
labeled nuclei. 10^5
single passage cells were
incubated in 2% and 10% CS
or in 10% FBS. They were
labeled for one hour with
^{3}H-thymidine at 5 day
intervals and prepared for
autoradiography. Counts
were made of focal areas
which had a much higher
percentage of labeled
nuclei than did the sur-
rounding areas (see Fig.
5). □, 2% CS; △, 10% CS;
O, 10% FBS.

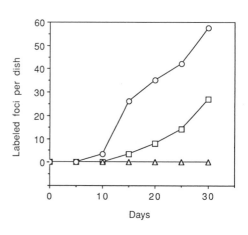

than the percentage of labeled cells in all but 2 of 72 foci counted in 2%
and 10% FBS. We conclude therefore, that there were few, if any, groups of
transformed cells in the 10% CS cultures of this experiment.

Stability of the Transformed State on Cell Transfer

Genetic mutations are perpetuated indefinitely in cell populations from
the originally altered cell, while epigenetic alterations exhibit a wide
range of stability. However, to be considered more than just a fleeting
physiological response to a particular effector, an epigenetic system should
persist for some time during growth which occurs after removal of the original inducing conditions. To determine the stability of the transformed
state, small numbers of cells from heavily transformed dishes were mixed with
much larger numbers of non-transformed cells. At intervals, the number of
foci was determined without staining and compared to the number in parallel
cultures initiated with only non-transformed cells. Foci could be clearly
discerned in the mixed cultures at 6 days, and were easily counted at 8 days
(Fig. 6). The number of foci increased till day 11, with no further increase
at day 13. By contrast, reliable identification of foci could not be made in
the control cultures until day 11. At day 13, there were only half as many
foci in the controls as in the mixed cultures. Subtraction of the number of
control foci at 13 days from the experimental foci at the same time showed
that at least 10% of the cells from the original transformed culture
maintained the capacity to transmit the transformed state to their progeny
through the 7 to 10 cell divisions it took to generate the excess of
recognizable foci. The actual percentage is probably higher since not all
the cells on the donor dish were necessarily transformed, and the cloning
efficiency of the transformed cells on a plastic surface with no indicator
cells to form a background is usually about 30%.

However, we have found that the early stages of transformation are reversible in some of our cell populations maintained in low serum concentration. In those cases, 50 to 100 very small but clear-cut foci were seen
macroscopically and confirmed microscopically one week after the seeding of
cells in 2% calf serum. This number was reduced by two- to five-fold when

Fig. 5. Percentage of labeled
nuclei in transformed
foci and in the sur-
rounding areas. This
represents a count of
the number of nuclei
labeled autoradio-
graphically with ^{3}H-
thymidine in Fig. 4.
Cells in foci were iden-
tified by the elongated
shape and dark staining
of their nuclei. Cells
in interfocal areas
were identified by
their flat, pale stain-
ing nuclei. The verti-
cal bars indicate the
standard errors of the
mean. Foci: ■, 2% CS;
●, 10% FBS. Interfocal
areas: □, 2% CS; △,
10% CS; O, 10% FBS.

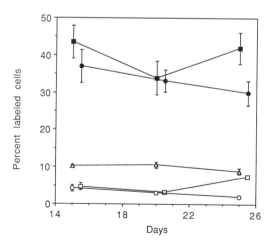

scored one week later. It seems as though there is an optimal depression of growth and/or metabolism for transformation which occurs as the culture becomes confluent. With continued incubation in the confluent state, the metabolic depression deepens, and many of the small groups of cells which had

become morphologically transformed disappear. At present, we cannot say with certainty whether they have reverted to a non-transformed morphology or have detached from the dish. Since we have not observed detachment, we favor the reversion hypothesis. This does not happen invariably, as seen in Fig. 3, where the number of foci increases between one and two weeks at all points. The occasional reduction in foci seems to depend on a transient state of the particular cell population in use, but when it occurs it is unequivocal. It is obviously necessary to better define the conditions that favor the reduction in foci in order to trace its path.

DISCUSSION

We now consider whether the morphological transformation observed in the NIH 3T3 cells corresponds to Nanney's criteria for epigenetic events[10], plus those associated with tissue differentiation such as cellular competence. These are to be distinguished from genetic events or mutations, which classically were considered to be sporadic and to occur at a low frequency, i.e., $\leq 10^{-6}$ per cell per division[1]. The frequency of mutations can be increased by mutagenic agents, but these agents are generally non-specific in their genetic target, and the mutations are stably propagated in progeny cells[1,10,11]. Epigenetic changes, as typified in differentiating systems, occur with regularity when certain conditions apply, e.g., inductive tissue interactions between prospective epithelial structures and mesenchyme[25], or

Fig. 6. Assay for number of transformed cells by seeding on a background of non-transformed cells. 5 week old, multiple passage cells were transformed by seeding 10^5 cells in 10% FBS for 20 days. They were then transferred at 10^5 cells per dish and incubated an additional 8 days for further transformation. Then 5 x 10^4 non-transformed NIH 3T3 cells were seeded in 10% FBS on empty dishes, followed the next day (= 0 days) by addition of 1000 of the transformed cells to one-half of the dishes and 1000 non-transformed cells to the remaining control dishes. $\triangle$, 5 x 10^4 non-transformed cells plus 1000 non-transformed cells the next day; $\square$, 5 x 10^4 non-transformed cells plus 1000 cells from a heavily transformed culture the next day.

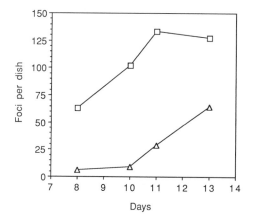

the effect of cell confluency on the in vitro induction of adipocyte formation[26]. They have a wide range of stability characteristics, e.g., the
differentiated state of epithelial tissues is gradually lost when the tissue
is dispersed into individual cells for culture[27]. Only a limited number of
epigenetic changes are available to a particular cell type, whereas a great
diversity of mutational changes are possible. Cells must be in a special
state of competence for epigenetic change to occur.

We show here that neoplastic transformation is regularly induced after a
cell population becomes confluent and the time of transformation is dependent
on the type and concentration of bovine serum in the medium. These are the
same conditions which produce adipocytes from Swiss 3T3 cells[26] and C3H
10T1/2 cells (unpublished); and keratinocytes from human epidermal cells[28].
Thus, NIH 3T3 cells exhibit a specific competence for transformation. Early
stage foci formed in low serum concentrations disappear with further incubation, suggesting they can revert to the non-transformed state when the conditions for induction become suboptimal. This combination of characteristics
provides strong evidence that the "spontaneous" transformation of NIH 3T3
cells is primarily an epigenetic phenomenon.

There are several parallels between the spontaneous transformation of
NIH 3T3 cells and the induction of transformation in C3H 10T1/2 cells by
chemical carcinogens and radiation. The treated C3H 10T1/2 cells do not
undergo transformation until several weeks after the cultures reach confluency[16,29], and the transformation is suppressed by high concentrations
of serum[30]. Morphological transformation of these cells can be stably
reversed if newly transformed cells are subcultured in the presence of ascorbic acid[31]. An additional epigenetic-like feature of the chemically and
physically induced transformation of these cells is that the potential for
transformation is induced in most or all of the cells in the treated population[15,16,17], instead of the very small fraction that is characteristic of
a genetic event[1].

There are also parallels between the NIH 3T3 transformation and chemical
induction of liver cancer in rats[32,33]. The first sign of tumor formation
in the liver of treated rats is the appearance of nodules of abnormally organized hepatocytes with a variety of characteristic enzymatic changes. Over
95% of these nodules eventually redifferentiate and become indistinguishable
from the surrounding tissue. The nodules fail to produce cancer upon transplantation. Nodule formation is thus considered a physiological adaptation
to the carcinogenic treatment, and would be considered an epigenetic process
by our criteria. Fewer than 5% of the nodules persist and serve as the origin of cells which lead to malignant growth. The persistent nodules could
result from progressive accumulation of epigenetic events, or from genetic
changes in the affected cells. It has been pointed out that the promotion of
liver carcinogenesis is as much an effect on the whole organ or tissue as on
the initiated cells[33]. This effect of cellular environment is, of course,
characteristic of epigenetic phenomena. Farber notes similarities between
carcinogenesis in the liver and in a variety of other organs, including skin,
urinary bladder and colon[33]. The epigenetic model of carcinogenesis,
therefore, appears to have general significance.

It is of more than passing interest that NIH 3T3 cells are the target of
choice for assaying cellular oncogenes by transfection[14]. Their readiness
to undergo oncogene-induced transformation may be related to their propensity
for spontaneous transformation. This propensity is expressed not only in
focus formation in cell culture, but in the production of tumors in nude mice
when a sufficiently large number of cells is inoculated[34,35,36]. The effect of transfected cellular oncogenes might simply be an acceleration or
amplification of a canalized path of response, i.e., the promotion rather
than the initiation of tumor development. Since progress to the transformed

state in these cells is readily manipulated and quantified, it may prove to be a good model to study the biology of progression. The emphasis in such a study would be on the interactions between the cells and their environment which bring about the new structures and functions, rather than the intracellular mechanisms presumed to underlie the change. Ultimately, it will be necessary to image epigenetic change within an appropriate theoretical structure. The most appropriate theoretical structure available for dealing with the problem of epigenesis is Elsasser's theory of organisms[37]. This theory is based on the premise that epigenesis is far too complex to be fully reduced to any particular molecular mechanism. It must be described in terms of living units -the cell and higher orders- and represented in abstract structures. Only then can it be considered a fully scientific study[38].

ACKNOWLEDGEMENT

We thank Berbie Chu and David Ord for excellent technical assistance, and Jason Yang for help with the autoradiographic techniques. This research was supported by grants from the National Institutes of Health CA 15744, and the Council for Tobacco Research 1984.

REFERENCES

1. T. T. Puck, Roundtable: Definition of criteria to define a genetic event, in: "Banbury Report 2. Mammalian Mutagenesis: The Maturation of Test Systems," Cold Spring Harbor Laboratory, Cold Spring Harbor (1979).
2. P. A. Jacobs, M. Brunton, W. M. Court Brown, R. Doll, and R. Goldstein, Change of human chromosome count distributions with age: Evidence for a sex difference, Nature 197:1080-1081 (1963).
3. D. T. Hughes, Cytogenetical polymorphism and evolution in mammalian somatic cell populations in vivo and in vitro, Nature 217:518-523 (1968).
4. J. R. Shapiro, W.-K. A. Yung, and W. R. Shapiro, Isolatin, karyotype and clonal growth of heterogeneous subpopulations of human malignant gliomas, Cancer Res. 41:2349-2359 (1981).
5. S. R. Wolman, T. F. Phillips, and F. F. Becker, Fluorescent banding patterns of rat chromosomes in normal cells and primary hepatocellular carcinomas, Science 175:1267-1269 (1972).
6. M. Terzi, Chromosomal variation and the establishment of somatic cell lines in vitro, Nature 253:361-362 (1975).
7. L. S. Cram, M. F. Bartholdi, A. F. Ray, G. L. Travis, and P. M. Kraemer, Spontaneous neoplastic evolution of Chinese hamster cells in culture: Multistep progression of karyotype, Cancer Res. 43:4828-4837 (1983).
8. J. D. Rowley, Biological implications of consistent chromosomal rearrangements in leukemia and lymphoma, Cancer Res. 44:3159-3168 (1984).
9. C. H. Waddington, "The Strategy of the Genes: A Discussion of Some Aspects of Theoretical Biology", Allen and Unwin, London (1957).
10. D. L. Nanney, Epigenetic control systems, Proc. Natl. Acad. Sci. USA 44:712-717 (1958).
11. D. L. Nanney, Molecules and morphologies: The perpetuation of pattern in the ciliated protozoa, J. Protozool. 24:27-35 (1977).
12. D. A. Clayton, Transcription of the mammalian mitochondrial genome, Ann. Rev. Biochem. 53:573-594 (1984).
13. T. J. King and R. Briggs, Serial transplantation of embryonic nuclei, Cold Spring Harbor Symposia Quant. Biol. 21:271-290 (1956).
14. H. E. Varmus, The molecular genetics of cellular oncogenes, Ann. Rev. Genet. 18:553-612 (1984).

15. S. Mondal and C. Heidelberger, In vitro malignant transformation by methylcholanthrene of the progeny of single cells derived from C3H mouse prostate, Proc. Natl. Acad. Sci. USA 65:219-229 (1970).

16. A. R. Kennedy, M. Fox, G. Murphy, and J. B. Little, Relationship between X-ray exposure and malignant transformation in C3H 10T1/2 cells, Proc. Natl. Acad. Sci. USA 77:7262-7266 (1980).

17. A. R. Kennedy, J. Cairns, and J. B. Little, Timing of the steps in transformation of C3H 10T1/2 cells by x-irradiation, Nature 307:85-86 (1984).

18. F. Seilern-Aspang and K. Kratochwil, Relation between regeneration and tumor growth, in: "Regeneration in Animals and Related Problems," V. Kiortsis and H. Trampusch, eds., North Holland, Amsterdam (1965).

19. B. Mintz and K. Illmensee, Normal genetically mosaic mice produced from malignant teratocarcinoma cells, Proc. Natl. Acad. Sci. USA 72:3585-3589 (1975).

20. M. Oshimura, D. J. Fitzgerald, H. Kitamura, P. Nettesheim, and J. C. Barrett, Cytogenetic changes in rat tracheal epithelial cells during early stages of carcinogen-induced neoplastic progression, Cancer Res. 48:702-708 (1988).

21. J. L. Jainchill, S. A. Aaronson, and G. J. Todaro, Murine sarcoma and leukemia viruses: assay using clonal lines of contact-inhibited mouse cells, J. Virol. 4:549-553 (1969).

22. G. D. Shipley and R. G. Ham, Attachment and growth of Swiss and Balb/c 3T3 cells in a completely serum-free medium, In Vitro 16:218 (1980).

23. H. Rubin, B. M. Chu, and P. Arnstein, Dynamics of tumor growth and cellular adaptation after inoculation into nude mice of varying number of transformed 3T3 cells and of readaptation to culture of the tumor cells, Cancer Res. 46:2027-2034 (1986).

24. T. Gurney, Local stimulation of growth in primary cultures of chick embryo fibroblasts, Proc. Natl. Acad. Sci. USA 62:906-911 (1969).

25. R. Fleischmajer and R. E. Billingham, eds. "Epithelial-Mesenchymal Interactions," The Williams and Wilkins Company, Baltimore (1968).

26. R. E. Scott, B. J. Hoerl, J. J. Wille, Jr., D. L. Florine, B. R. Krawisz, and K. Yun, Coupling of proadipocyte growth arrest and differentiation II. A cell cycle model for the physiological control of cell proliferation, J. Cell. Biol. 94:400-405 (1982).

27. D. I. DePomerai, F.-H. Kotecha, C. Fullick, A. Young, and M. A. H. Gali, Expression of differentiation markers by chick embryo neuroretinal cells in vivo and in culture, J. Embryol. Exp. Morph. 77:201-220 (1983).

28. P. R. Cline and R. H. Rice, Modulation of involucrin and envelope competence in human keratinocytes by hydrocortisone, retinyl acetate and growth arrest, Cancer Res. 43:3203-3207 (1983).

29. D. A. Haber, D. A. Fox, W. S. Dynan, and W. G. Thilly, Cell density dependence of focus formation in the C3H 10T1/2 transformation assay, Cancer Res. 37:1644-1648 (1982).

30. J. S. Bertram, Effects of serum concentration on the expression of carcinogen-induced transformation in the C3H/10T1/2 CL8 cell line, Cancer Res. 37:514-523 (1977).

31. W. F. Benedict, W. L. Wheatley, and P. A. Jones, Inhibition of chemically induced morphological transformation and reversion of the transformed phenotype by ascorbic acid in C3H 10T1/2 cells, Cancer Res. 40:2796-2801 (1980).

32. E. Farber, Pre-cancerous steps in carcinogenesis: Their physiological adaptive nature, Biochim. Biophys. Acta 738:171-180 (1984).

33. E. Farber and D. S. R. Sarma, Biology of disease. Hepatocarcinogenesis: A dynamic cellular perspective, Lab. Invest. 56:4-22 (1987).

34. D. G. Blair, C. S. Cooper, M. K. Oskarsson, L. A. Eader, and G. F. Vande Woude, New method for detecting cellular transforming genes, Science 218:1122-1124 (1982).

35. M. A. Tainsky, F. L. Shamansky, D. Blair, and G. Vande Woude, Human recipient cell for oncogene transfection studies, Mol. Cell. Biol. 7:1280-1284 (1987).

36. R. G. Greig, T. P. Koestler, D. L. Trainer, S. P. Corwin, L. Miles, T. Kline, R. Sweet, S. Yokoyama, and G. Poste, Tumorigenic and metastatic properties of "normal" and ras-transformed NIH 3T3 cells, Proc. Natl. Acad. Sci. USA 82:3698-3701 (1985).

37. W. M. Elsasser, Reflections on a theory of organisms, Éditions Orbis Publishing, Frelighsburg (1987).

38. A. Eddington, "The Philosophy of Science," reprinted by the University of Michigan Press, Ann Arbor (1939).

DISCUSSION

Unidentified Speaker: Is the spontaneous morphologic transformation that you observe in these cells transferrable by transfection? I would suggest that that is one operational definition of genetic versus epigenetic change.

Rubin: We have not done the experiment. To do it properly, the results would have to be evaluated by comparison with the inevitable spontaneous transformants of the control. There is confusion about whether or not they produce tumors in nude mice. The standard oncogene papers say they do not. Four papers say they do, although sometimes with a longer latent period than tumors produced by oncogene-transfected NIH 3T3 cells. Perhaps Dr. Charles Boone would like to comment, since he is an expert on tumor production by ostensibly non-transformed cells.

Charles Boone: What passage level did you get these from, when Stu Aronson gave them to you.

Rubin: I do not know the precise passage level, but I do know they were the earliest passage material available to anyone who uses them, for transfection with oncogenes or any other purpose.

Boone: That's about five to eight passages, then.

Rubin: I can only tell you how they were derived. The NIH 3T3 line of cells were derived by culturing embryonic cells of partially inbred Swiss mice at cloning densities and picking flat clones with low saturation densities. This operation was repeated five times, so the cell line was certainly clonal in origin.

Boone: I'd like to propose a new view of Balb 3T3 cells and cells like them. Because they are genetically heteroploid, very heteroploid, and because conditions of culture can change the karyotypic patterns in a reproducible way, these cells are cancerous. I think the Balb/3T3 and C3H/10%1/2 lines are long since neoplastic except for the property of anchorage dependence. Some years ago (1-3) we showed that these lines were immediately tumorigenic at doses as low as 10^4 cells if they were implanted subcutaneously attached to a solid substrate, such as small glass beads or plastic plates. The tumors were clonal, each one possessing a different karyotype and tumor transplantation antigen. Thus when searching for the specific mechanism of transformation of NIH 3T3 by the various oncogenes, one should look in the direction of abrogation of anchorage dependence, i.e., alteration of cytoskeletal formation and function, actin, tubules, etc.

Moderator (Russel G. Greig): Before we close I want to say that, as suspected, I don't think we have succeeded in answering completely the specific questions posed by the title of this colloquium. I believe that we all agree that genetic instability does become tumor progression, but that the time at which instability occurs (and this is particularly important for the patient) is left to be established in the next few years.

REFERENCES

1. Science, 188:68-70 (1975).
2. Cancer Res. 36:1626-1633, (1976).
3. J. Supramolecular Structure 5:131-137 (1976).

THE HUMAN MELANOCYTE SYSTEM AS A MODEL

FOR STUDIES ON TUMOR PROGRESSION*

Istvan Valyi-Nagy*, Ulrich Rodeck*, Roland Kath*,
Maria Laura Mancianti*, Wallace H. Clark+,
and Meenhard Herlyn*ᵃ

*The Wistar Institute for Anatomy and Biology
Philadelphia, PA 19104

+The Pigmented Lesion Clinic
University of Pennsylvania
Philadelphia, PA 19104

INTRODUCTION

Melanocytes are distinctive cells in the basal layer of the epidermis, the choroid of the eye, certain mucous membranes, and the leptomeninges. Melanocytes arise during embryonal development from pluripotent cells migrating out of the neural crest. Functional maturation, i.e., the process by which cells express specific properties characteristic of the cell type, may progress in melanocytes through several, as yet undefined, stages (Fig. 1). Precursor cells for melanocytes (premelanocytes or melanoblasts) have been identified in human skin[1], but these cells have been only preliminarily characterized. The phenotypic and functional characteristics of melanocytes are: a) melanin synthesis through the action of the tyrosinase enzyme; b) dendritic morphology; c) pigment donation to surrounding keratinocytes and d) no detectable proliferation in situ. Despite the undetectable proliferation, a stable 5-6:1 ratio between basal keratinocytes and melanocytes is maintained throughout the life of an individual suggesting a constant renewal of melanocytes.

---

*These studies were supported in part by grants from the NIH, CA-25874, CA-44877, and from the Herzog Foundation.
ᵃTo whom requests for reprints and all correspondence should be addressed, at The Wistar Institute, 36th Street at Spruce, Philadelphia, PA 19104.

Abbreviations used are: alpha-MSH, alpha-melanocyte stimulating hormone; bFGF, basic fibroblast growth factor; BPE, bovine pituitary extract; CFE, colony forming efficiency; EGF, epidermal growth factor; FCS, fetal calf serum; FSH, follicle stimulating hormone; IBMX, isobutyl methyl xanthine; IGF, insulin-like growth factor; MAb, monoclonal antibody; MGF, melanocyte growth factor; MSGA, melanocyte-stimulating growth activity; NGF, nerve growth factor; PDGF, platelet derived growth factor; RGP, radial growth phase; TPA, 12-0-tetradecanoyl phorbol-13-acetate; VGP, vertical growth phase.

Boundaries between Promotion and Progression during Carcinogenesis
Edited by O. Sudilovsky *et al.*, Plenum Press, New York, 1991

315

Tumor Progression In Vivo

Tumor progression of melanocytes may follow direct and indirect pathways. Direct tumor progression is characterized by the appearance of malignant cells as the first manifest lesion without evidence of precursor steps. Malignant transformation of melanocytes may lead to the development of either radial growth phase melanomas[2], or nodular melanomas[3]. More often, in vivo transformation is viewed as an indirect process consisting of five consecutive steps of tumor progression (Fig. 1).

Extensive clinical and histopathological studies have led to the delineation of these steps[3,4].

Common acquired melanocytic nevi (Step 1) have no architectural or cytologic atypia by histological examination, and usually appear in the first 20 years of life. The induction of melanocytic nevi in the early years seems to be related to UV light exposure. The higher the number of moles (nevi), the greater the relative risk for developing melanoma. Over the course of several decades, the moles usually exhibit a programmed clinical life history with corresponding histological features. They may cease growth and pigment synthesis and persist as a stable lesion or gradually disappear by differentiating along a Schwannian pathway.

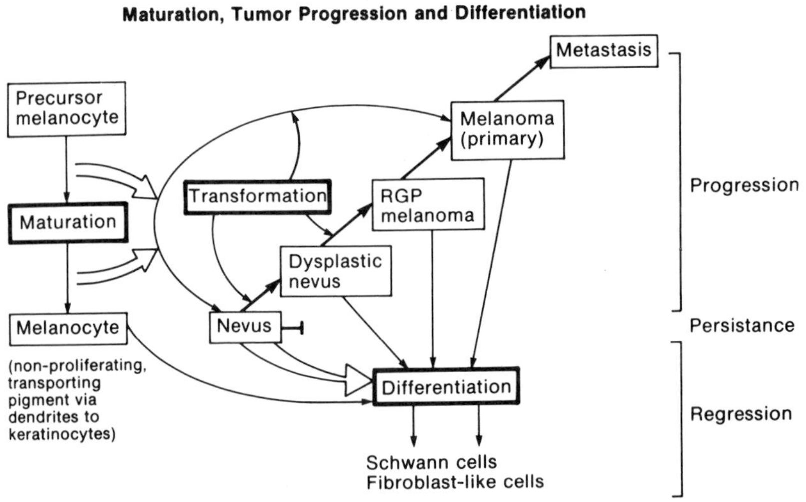

Maturation, Tumor Progression and Differentiation

Fig. 1. Precursor cells of melanocytes mature functionally into cells that can transport pigment through dendritic processes to keratinocytes. Generally, mature melanocytes in situ show no detectable proliferation. In vitro melanocytes proliferate and may differentiate into a fibroblast-like cell type. Progression of normal cell to melanoma cells may occur at any maturation stage. Tumor progression may be sequential, nevus to dysplasia to melanoma, or occur without intermediate steps. Lesions composed of cells undergoing differentiation will regress until they disappear. Nevus cells may persist in situ or spontaneously differentiate along pathways which have been histopathologically defined as Schwannian differentiation and experimentally as fibroblast-like differentiation. Differentiation of melanoma cells is rare but may occur along the same pathways.

Dysplastic nevi (Step 2) reveal architectural and cytological atypia and are histogenetic precursors of melanoma, although they do not represent an obligatory precancerous state and may persist with stability or differentiate and disappear. Steps 1 and 2 are regarded as facultative precursors of melanoma. Approximately 94% of familial melanomas and 42% of non-familial (sporadic) melanomas arise from precursor nevi[4].

Radial growth phase (RGP) melanomas (Step 3) show a strong tendency of local invasiveness, architectural and cytologic atypia, but have no competence for metastasis[5]. RGP may be confined to the epidermis ("in situ RGP") or the radial growth may include a few small nests of tumor cells in the upper papillary dermis ("invasive RGP")[2].

Vertical growth phase (VGP) melanomas (Step 4) arise as foci within a nevus or RGP as a new population of tumor cells that show growth perpendicular to the skin surface and acquire competence for metastasis. There is a positive correlation between the depth of dermal invasion and the likelihood of developing metastatic disease[5].

Metastatic melanoma (Step 5) represents the last step in tumor progression. The primary lesion tends to give rise to metastases in the regional lymph nodes before the further dissemination of the disease.

In conclusion, clinical and histopathological evidence suggests that the majority (92%) of melanomas develop along the described sequential progression pathway and in only 8% of melanomas does the tumor arise de novo without precursor steps. Several stages of tumor progression may be present simultaneously within a single lesion.

Differentiation of Melanocytic Cells In Situ and In Vitro

Cells of early, and rarely late, stages may undergo differentiation. In situ, Schwannian differentiation may occur in lesions of all stages of tumor progression including metastatic melanoma[4]. In culture, differentiation of melanocytic cells may be characterized by: a) loss of melanocyte- or melanoma-associated antigens specific for a given progression stage; b) loss of distinctive morphology; c) acquisition of markers of non-melanocytic cells, as in fibroblasts or neuronal cells, and d) loss of other markers specific for melanocytic cells at a given progression stage (Fig. 1).

Characterization of Cultured Melanocytic Cells Isolated from Normal Skin and Pigmented Lesions at Different Stages of Tumor Progression

Normal Melanocytes: Melanocytes isolated from newborn foreskins grow for up to 60 doublings with doubling times of 2 to 6 days (Table 1). They have a bipolar to tripolar morphology (Fig. 2), and the extent of their pigmentation depends on the race of the donor. Cultures reach only low cell densities ($6-8 \times 10^4$ cells/cm^2). Addition of dibutyryl cAMP to the medium leads to the formation of dendrites which may extend to several times the length of the cell body with extensive branching of dendritic processes[6]. After the induction of dendrite outgrowth, the cells may slow in their proliferation but are not terminally differentiated. Even highly dendritic melanocytes can replicate in vitro. The optimal growth medium for normal melanocytes contains fetal calf serum (FCS), bovine pituitary extract (BPE), basic fibroblast growth factor (bFGF), the protein kinase C activator 12-0-tetradecanoyl phorbol-13-acetate (TPA), and insulin. We recently developed a chemically defined medium in which our standard W489 medium is supplemented with IGF-1 or insulin, TPA, and bFGF[6]. Cell growth can be further stimulated by enhancers of intracellular cAMP level such as the pituitary hormones alpha-MSH and follicular stimulating hormone (FSH), tetanus, pertussis, and cholera toxins, prostaglandin $F_{2\alpha}$, forskolin, and isobutyl

Table 1. Characteristics of Normal Epidermal Melanocytes In Vitro

Morphology:	Bi- to tripolar; Pigmented; Little morphologic heterogeneity, induction of dendrites by dibutyryl cAMP is reversible.
Growth rate:	Newborn: up to 60 doublings (2-6 days doubling time); Adult: up to 10 doublings (7-14 days doubling time).

Growth requirements
in serum-free medium
(essentials in
W489 medium):
	IGF-1 (or insulin);
	  bFGF;
	  Phorbol ester (TPA, BDBu);
	  α-MSH.

Effect of Phorbol ester:	Growth stimulation; Maintenance of phenotype, e.g. high tyrosinase activity, pigmentation, antigen expression.
Production of growth factors:	None.
Anchorage independent growth in soft agar:	None.
Tumorigenicity:	None.

Antigens:	Expression of melanoma-associated antigens (NGF
	    receptor, GD_3, Proteoglycan) in culture but not
	    in situ;
	No expression of HLA-DR and GD_2 in culture and in
	    situ.

Stimulation of autologous T-lymphocytes:	No data available.
Chromosomal abnormalities:	None.

methyl xanthine (IBMX). Withdrawal of phorbol ester from the culture medium
for 1 to 3 months results in terminal differentiation (a fibroblast-like
differentiation) which is characterized by: a) loss of growth potential; b)
loss of pigmentation and tyrosinase activity; c) flattened morphology, and
d) loss of expression of nerve growth factor (NGF) receptor and p97 (melanotransferrin)[7]. Melanocytes in culture but not in situ express melanoma-
associated antigens including NGF receptor, p97 (melanotransferrin), chondroitin sulfate proteoglycan, and gangliosides GD_3 and 9-O-acetyl GD_3[8].
HLA-DR and GD_2 are found on cells in situ but are not found on cultured
melanocytes[9,10]. Despite the expression of melanoma-associated antigens on
cultured melanocytes these cells are not transformed. Therefore, expression
of melanoma-associated antigens seems to be growth- and not transformation-
related.

Nevus Cells: Nevus cells from common acquired and congenital nevi grow for 20 to 50 doublings with doubling times between 20 h and 14 days (Table 2). In contrast to melanocytes, nevus cells grow anchorage independently in semisolid media (with approximately 0.9% colony forming efficiency (CFE) in soft agar), but they show no chromosomal abnormalities and are not tumorigenic in nude mice[11,12]. Nevus cells isolated from the basal layer of the epidermis and from the dermal/epidermal junction have similar morphology but are more heterogenous compared to normal melanocytes (Fig. 3)[12]. Epidermal nevus cells exhibit lower growth rates than nevus cells isolated from the dermis. A chemically defined medium has been developed containing TPA, insulin, bFGF, epidermal growth factor (EGF), and alpha-MSH, which allows long term (> 6 months) growth of nevus cells (Mancianti et al., manuscript in preparation). Nevus cells are less dependent on bFGF and TPA, although their growth can be stimulated by these compounds. The decreased dependence on bFGF in nevus cells is apparently due to production of this factor which may act in an autocrine fashion (Menssen et al., manuscript in preparation).

When nevus culture medium is depleted of TPA, cells differentiate within 4 to 8 wk along the fibroblast-like pathway exhibiting changes in antigen expression similar to that observed in melanocytes[12].

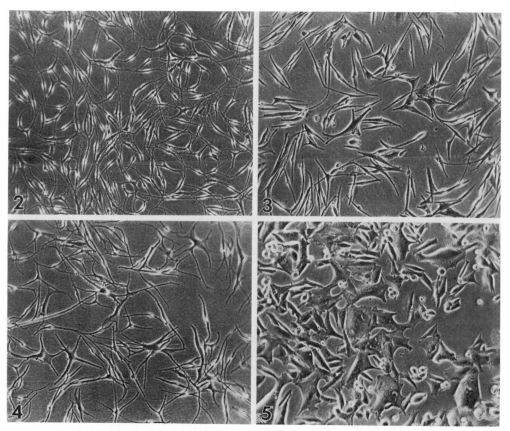

Fig. 2-5. Morphology of melanocytes isolated from different stages of tumor progression.
Fig. 2. Normal melanocytes.
Fig. 3. Common acquired nevus cells.
Fig. 4. Radial growth phase primary melanoma cells.
Fig. 5. Vertical growth phase primary melanoma cells.

Table 2. Characteristics of Common Acquired and Congenital
          Nevus Cells in Vitro

Morphology:	Bipolar; Pigmented; Morphologic heterogeneity; Induction of dendrites by dibutyryl cAMP is less dramatic than in melanocytes and not consistent.
Growth rate:	20-50 doublings (20 hours to 14 days doubling time depending on age of donor).
Growth requirements in serum-free medium (essentials in W489 medium):	IGF-1 (or insulin); bFGF, but most cultures are independent Phorbol ester (TPA, PDBu); α-MSH.

Effect of
Phorbol ester:
	Growth stimulation;
	Maintenance of phenotype, e.g. high tyrosinase
	    activity, pigmentation, antigen expression,
	    however prolonged stability in absence of TPA.

Production of growth factors:	bFGF.
Anchorage independent growth in soft agar:	0.001-3.0%, average of 0.9% CFE.
Tumorigenicity:	None.

Antigens:	Expression of melanoma-associated antigens (NGF
	    receptor, HLA-DR, GD_3, Proteoglycan) in culture
	    and weakly in situ;
	Antigenic heterogeneity;
	No expression of GD_2.

Stimulation of autologous T-lymphocytes:	None.
Chromosomal abnormalities:	None.

Nevus cells in culture, like normal melanocytes, express melanoma-associated antigens including HLA-DR (but not GD_2). In situ only a small percentage of nevus cells shows weak expression of melanoma-associated antigens (Elder et al., manuscript in preparation). No differences could be detected in this study between congenital and common acquired nevi.

Dysplastic Nevus Cells and Radial Growth Phase (RGP) Primary Melanoma Cells. Cells from both lesions have not been fully characterized despite intense efforts to establish long-term cultures. Approximately 80% of specimens show initial growth and cells from both lesions are initially bipolar and of melanocytic morphology (Table 3). However, after 1 or 2 passages,

Table 3. Characteristics of Dysplastic Nevus and Radial Growth Phase
          (RGP) Primary Melanoma Cells In Vitro

Morphology:	Bipolar; Morphologic heterogeneity with individual cells having a "transformed" phenotype; Pigmented, often hyperpigmented; Spontaneous differentiation frequently into non-pigmented, flat cells with no tyrosinase activity; Spontaneous transformation in culture is possible.
Growth rate:	Poor growth, prolonged survival but no infinite growth.
Growth requirements in serum-free medium (essentials in W489 medium):	No data available.
Effect of Phorbol ester:	Response heterogenous, often no effect.
Production of growth factors:	No data available.
Anchorage independent growth in soft agar:	1-8% CFE.
Tumorigenicity:	None.
Antigens:	Expression of melanoma-associated antigens (NGF receptor, HLA-DR, GD_3, Proteoglycan) and melanocyte-associated antigen p98 kDA; Antigenic heterogeneity; No expression of GD_2.
Stimulation of autologous T-lymphocytes:	Stimulation by cells of both lesions.
Chromosomal abnormalities:	Random (and possibly non-random of chromosome 10).

they often develop a flat, cuboidal morphology. Simultaneously, these cells lose expression of pigment cell-associated antigens such as NGF receptor, and levels of tyrosinase activity are undetectable. Differentiated nevus cells may maintain expression of chondroitin sulphate proteoglycan and melanotransferrin[7]. These similarities of differentiated cells to fibroblasts make it difficult to distinguish between these cell types, especially since differentiated nevus cells appear to express the "fibroblast marker" leucine aminopeptidase. The morphology of dysplastic nevus cells and RGP primary melanoma cells is similar to common acquired and congenital nevus cells but individual cells may be highly polymorphic and individually indistinguishable from vertical growth phase primary or metastatic melanoma cells (Fig. 4). Limited studies indicate that RGP primary melanoma cells are not tumorigenic in nude mice although they grow better in soft agar than do the non-tumorigenic nevus cells (6% versus 0.9%). Dysplastic nevus and RGP primary melanoma cultures

Table 4. Characteristics of Vertical Growth Phase (VGP) Primary
          Melanoma Cells In Vitro

---

3 groups:	I. Early (no evidence of recurrence, >72 months) II. Intermediate (delayed recurrence, 9-87 months) III. Advanced (simultaneous metastasis)

Morphology:	Spindle or cuboidal, similar to metastases;
	Pigmented in 20-30% of cases;
	Morphologic heterogeneity between cultures but not
	  within cultures;
	Rarely spontaneous differentiation (15-30% mostly
	  from early lesions).

Growth rate:	Success rate for establishing permanent cell lines
	  are: 30% (early), 30-40% (intermediate), and 70%
	  (late).

Growth requirements in serum-free medium (essentials in W489 medium):	IGF-1 (or insulin) only for cells from early and intermediate lesions.
Effect of Phorbol ester:	Growth inhibition.
Production of growth factors:	bFGF, PDGF A, PDGF B, TGF-α, TGF-ß, IL-1.
Anchorage independent growth in soft agar:	5-15% CFE.
Tumorigenicity:	100% in permanent cell lines.

Antigens:	Expression of GD$_2$, quantitatively less expression
	  of other melanoma-associated antigens compared to
	  metastatic cells. No expression of melanocyte-
	  associated antigens p145 kDa and p98 kDa.

Stimulation of autologous T-lymphocytes:	Cells of early and mixed lesions but not of late lesions.
Chromosomal abnormalities:	Non-random (chromosomes 1, 6, and 7).

Presence of cells with metastatic capacity in early and intermediate lesions.

---

may have a diploid or a pseudodiploid karyotype with limited aberrations involving chromosome 6[13,14] and possibly 10. The response to phorbol ester
is heterogenous; most cultures are not stimulated.

RGP primary melanoma cells may spontaneously transform to a more malignant phenotype. Such cells have an unlimited lifespan and are tumorigenic in
nude mice. An example is cell line WM 35[15]. Dysplastic nevus cells and

RGP primary melanoma cells stimulate autologous T lymphocytes if they express HLA-DR on their surface[16]. HLA-DR may be induced by γ-interferon on melanocytic cells at all stages of tumor progression[9]. The expression of melanoma-associated antigens is similar on dysplastic nevus cells and RGP primary melanoma cells[11].

VGP Primary Melanoma Cells: VGP primary melanoma cells may be divided into three groups according to the clinical history of patients (Table 4). Cells of group I are derived from patients who, after removal of the primary lesion, lived for more than 6 years without evidence of metastases. Group II cells are derived from patients who, at the time of surgical removal of the primary lesion, had no evidence of metastases but who later developed metastatic melanoma. Group III cells are from patients whose primary tumor was removed simultaneously with metastases in the subcutis or in the regional lymph nodes. Cells from all three groups exhibit more similarities to metastatic melanoma cells than to RGP primary melanoma cells (Fig. 5). Morphology, growth rate, expression of melanoma-associated antigens, chromosomal abnormalities and growth inhibition by TPA or VGP cells of all groups are similar to metastatic cells. Early and intermediate VGP primary melanoma cells grow slower, reach lower densities and express quantitatively fewer melanoma-associated antigens than metastatic cells[11], whereas advanced VGP primary melanoma cells are indistinguishable from metastatic cells, suggesting that cells representing the metastatic phenotype have overgrown cells representing earlier stages of tumor progression in the primary lesions at the time of surgical removal[17].

VGP primary melanoma cells express GD_2 ganglioside[10], but not the melanocyte- and nevus associated antigens p145 kDa[8] and p98 kDa[11]. Cells from early and intermediate lesions can be adapted to grow in serum-free medium, but they still require at least one growth factor, IGF-1 or insulin, for continuous growth[18]. These growth factors are interchangeable since insulin at high concentrations (5 μg/ml) acts predominantly via the IGF-1 receptor. The need for exogenous growth factors clearly distinguishes early and intermediate VGP primary melanoma cells from metastatic melanoma cells. Conversely, advanced VGP primary melanoma cells are able to grow continuously in protein-free W489 medium[17]. Primary melanoma cells from all three groups produce growth factors, including PDGF[19], bFGF (Menssen et al., manuscript in preparation), TGF-alpha, TGF-beta (unpublished), and IL-1[20]. It is possible that these growth factors contribute to the increased growth autonomy of VGP primary melanoma cells.

Metastatic Melanoma Cells: Metastatic melanoma cells may show morphologic heterogeneity between cultures but rarely within cultures. Metastatic melanoma cells, in general, grow more rapidly than primary cells, reach higher densities, more often detach spontaneously from substrate, and form tumors rapidly in nude mice[11] (Table 5). Non-random chromosomal abnormalities have been detected in chromosomes 1, 6 and 7. Our cytogenetic studies point to the clonal evolution of metastases[13,14]. Random abnormalities are also often found in metastatic cells (in contrast to primary melanoma cells).

Differences between the exogenous growth factor requirements of primary and metastatic cells are very clear. Seventeen out of 18 metastatic melanoma cell lines could be adapted to continuous (> 8 weeks) growth in W489 medium wihout any exogenous growth factors or other proteins[17]. Metastatic melanoma cells produce a range of growth factors including bFGF[21,22] (and Menssen et al., manuscript in preparation), PDGF[19], TGF-alpha, TGF-beta (unpublished data), IL-1[20], MSGA[23], and MGF[24]. It is expected, but not experimentally proven, that these growth factors contribute to the growth autonomy of metastatic cells. Since we have not found, by Southern blot analyses, any genetic amplifications, rearrangements, deletions or other structural abnormalities for growth factors IGF-1, IGF-2, PDGF A, PDGF B, or EGF, nor for the

Table 5. Characteristics of Metastatic Melanoma Cells In Vitro

---

Morphology:

Spindle or cuboidal;
Morphologic heterogeneity between cultures but not
   within cultures;
High cell density/cm^2.

Growth rate:

Success rate for establishing permanent cell lines
   is 50-70% (15 h to 5 days doubling time).

Growth requirements
 in serum-free medium
 (essentials in
 W489 medium):

None.

Effect of
 Phorbol ester:

Growth inhibition.

Production of growth
 factors:

bFGF, PDGF A, PDGF B, TGF-α, TGF-β, MSGA, IL-1.

Anchorage independent
 growth in soft agar:

5-70% CFE.

Tumorigenicity:

100% in permanent cell lines (more rapid than VGP
   cells).

Antigens:

Quantitatively highest expression of
   melanoma-associated antigens;
No expression of melanocyte-associated antigens
   e.g. p145 kDa and p98 kDa.

Stimulation of
 autologous
 T-lymphocytes:

None.

Chromosomal
 abnormalities:

Non-random (chromosomes 1, 6, and 7) and random.

Instability of invasive phenotype but rapid development of cell variants with
invasive properties in vivo.

---

growth factor receptors for EGF, and IGF-1[25], it is possible that growth
autonomy in melanoma is controlled by gene regulatory elements.

CONCLUSIONS

     The human melanocyte system is ideally suited to study cells isolated
from different stages of tumor progression. The ability to maintain cells
from each stage in culture has allowed biologic, immunologic, genetic and
molecular studies. Normal melanocytes and common acquired and congenital
nevus cells have many similarities including morphology, the expression of
melanocyte- and melanoma-associated antigens, the requirements for exogenous
growth factors, and a limited life span in culture. However, growth of nevus
cells is less dependent on bFGF and TPA, and such cells are able to grow

anchorage independently in soft agar. VGP primary melanoma cells, on the other hand, possess all properties of malignant cells including chromosomal abnormalities, tumorigenicity in athymic nude mice, growth in soft agar, independence from protein kinase C-activating phorbol esters, high extracellular calcium concentration, and enhancers of intracellular levels of cAMP. Cells of dysplastic nevi and RGP primary melanoma may represent the preinvasive steps in melanoma development. Unfortunately, their characterization has been limited by poor growth in vitro. The reasons for this poor proliferation and apparent spontaneous differentiation into a non-melanocytic phenotype are not clear. VGP primary melanoma cells share a number of common properties with metastatic cells. In fact, advanced VGP primary melanoma cells are indistinguishable from metastatic cells. The dependence of early and intermediate VGP primary melanoma cells on exogenous growth factors and the growth factor independence of advanced VGP cells may help to delineate characteristics associated with increased risk for recurrence. Further studies on the growth autonomy of metastatic cells can then lead to the development of novel strategies for cancer therapy.

REFERENCES

1. D. C. Bennet, K. Bridges, and I. A. McKay, Clonal separation of mature melanocytes from premelanocytes in a diploid human cell strain: Spontaneous and induced pigmentation of premelanocytes, J. Cell Sci. 77:167-183 (1985).
2. D. E. Elder and W. H. Clark, Jr., Developmental biology of malignant melanoma, in: "Pigment Cell", Vol. 8, R. M. MacKie, ed., Karger, Basel (1987).
3. W. H. Clark, Jr., D. E. Elder, and M. Van Horn, The biologic forms of malignant melanoma, Hum. Pathol. 17:443-450 (1986).
4. W. H. Clark, Jr., D. E. Elder, D. Guerry, IV, M. N. Epstein, M. H. Greene, and M. Van Horn, A study of tumor progression: the precursor lesions of superficial spreading and nodular melanoma, Hum. Pathol. 15:1147-1165 (1984).
5. D. E. Elder, D. Guerry, IV, M. N. Epstein, L. Zehngebot, E. Lusk, M. Van Horn, and W. H. Clark, Jr. Invasive malignant melanomas lacking competence for metastasis, Am. J. Dermatopathol. 6:55-62 (1984).
6. M. Herlyn, M. L. Mancianti, J. Jambrosic, J. B. Bolen, and H. Koprowski, Regulatory factors that determine growth and phenotype of normal human melanocytes, Exp. Cell. Res. 179:322-331 (1988).
7. M. Herlyn, W. H. Clark, U. Rodeck, M. L. Mancianti, J. Jambrosic, and H. Koprowski, Biology of tumor progression in human melanocytes, Lab. Invest. 56:461-474 (1987).
8. M. Herlyn, U. Rodeck, M. L. Mancianti, F. M. Cardillo, A. Lang, A. H. Ross, J. Jambrosic, and H. Koprowski, Expression of melanoma-associated antigens in rapidly dividing human melanocytes in culture, Cancer Res. 47:3057-3061 (1987).
9. M. Herlyn, D. Guerry, and H. Koprowski, Recombinant γ-interferon induces changes in expression and shedding of antigens associated with normal human melanocytes, nevus cells, and primary and metastatic melanoma cells, J. Immunol. 134:4226-4230 (1985).
10. J. Thurin, M. Thurin, M. Herlyn, D. E. Elder, Z. Steplewski, W. H. Clark, Jr., and H. Koprowski. GD2 ganglioside biosynthesis is a distinct biochemical event in human melanoma tumor progression. FEBS Lett., 208:17-22 (1986).
11. M. Herlyn, J. Thurin, G. Balaban, L. J. Bennicelli, D. Herlyn, D. E. Elder, E. Bondi, D. Guerry, P. C. Nowell, W. H. Clark, and H. Koprowski, Characteristics of cultured human melanocytes isolated from different stages of tumor progression, Cancer Res. 45:5670-5676 (1985).

12. M. L. Mancianti, M. Herlyn, D. Weil, J. Jambrosic, U. Rodeck, D. Becker, L. Diamond, W. H. Clark, and H. Koprowski, Growth and phenotypic characteristics of human nevus cells in culture. J. Invest. Dermatol. 90:134-141 (1988).

13. G. Balaban, M. Herlyn, D. Guerry, R. Bartolo, H. Koprowski, W. H. Clark, and P. C. Nowell, Cytogenetics of human malignant melanoma and pre-malignant lesions, Cancer Genet. Cytogenet. 11:429-439 (1984).

14. G. B. Balaban, M. Herlyn, W. H. Clark, Jr., and P. C. Nowell, Karyotypic evolution in human malignant melanoma, Cancer Genet. Cytogenet. 19:113-122 (1986).

15. M. Herlyn, W. H. Clark, Jr., M. J. Mastrangelo, D. Guerry, IV, D. E. Elder, D. LaRossa, R. Hamilton, E. Bondi, R. Tuthill, Z. Steplewski, and H. Koprowski, Specific immunoreactivity of hybridoma-secreted monoclonal anti-melanoma antibodies to cultured cells and freshly derived human cells, Cancer Res. 40:3602-3609 (1980).

16. D. Guerry, IV, M. A. Alexander, D. E. Elder, and M. Herlyn, Interferon-γ regulates the T cell response to precursor nevi and biologically early melanoma, J. Immunol., 139:305-312 (1987).

17. R. Kath, U. Rodeck, J. Jambrosic, and M. Herlyn, Growth factor independence in vitro of primary melanoma cells from advanced but not early or intermediate lesions (Submitted for publication).

18. U. Rodeck, M. Herlyn, H. D. Menssen, R. W. Furlanetto, and H. Koprowski, Metastatic but not primary melanoma cells grow in vitro independently from exogenous growth factors, Int. J. Cancer 40:687-690 (1987).

19. B. Westermark, A. Johnsson, Y. Paulsson, C. Betsholtz, C.H. Heldin, M. Herlyn, U. Rodeck, and H. Koprowski, Human melanoma cells lines of primary and metastatic origin express the genes encoding the chains of platelet-derived growth factor (PDGF) and produce a PDGF-like growth factor, Proc. Natl. Acad. Sci. USA 83:7197-7200 (1986).

20. J. L. Bennicelli, J. Elias, J. Kern, and D. Guerry, IV, Production of interleukin 1 activity by cultured human melanoma cells, Cancer Res. (in press).

21. R. Halaban, B. S. Kwon, S. Ghosh, P. S. Delli Bovi, and A. Baird, bFGF as an autocrine growth factor for human melanomas, Oncogene Res. 3:177-186 (1988).

22. D. Moscatelli, M. Presta, J. Joseph-Silverstein, and D. B. Rifkin, Both normal and tumor cells produce basic fibroblast growth factor, J. Cell. Physiol. 129:273-276 (1986).

23. A. Richmond, D. H. Lawson, D. W. Nixon, and R. K. Chawla, Characterization of autostimulatory and transforming growth factors from human melanoma cells, Cancer Res. 45:6390-6394 (1985).

24. M. Eisinger, O. Marko, S.-I. Ogata, and L. J. Old, Growth regulation of human melanocytes: Mitogenic factors in extracts of melanoma, astrocytoma, and fibroblast cell lines, Science 229:984-986 (1985).

25. A. J. Linnenbach, K. Huebner, E. Premkumar Reddy, M. Herlyn, A. H. Parmiter, P. C. Nowell, and H. Koprowski, Structural alteration in the MYB protooncogene and deletion within the gene encoding α-type protein kinase C in human melanoma cell lines, Proc. Natl. Acad. Sci. USA 85:74-78 (1988).

DISCUSSION

George Milo: I found this to be a very interesting presentation and I wanted to ask you in your Northerns that you were looking at as far as the differences in expression of some of those cellular genes and oncogenes, have you given consideration to the fact that possibly going from the in vivo to the in vitro situation and using that technology that the conditions in changing the environment may result in the loss of expression of those particular genes? Have you seen any evidence for this?

Meenhard Herlyn: Of course, every culture system has its limitations. However, the best studied in this system is the EGF receptor and the EGF receptor on metastatic melanoma cells like in other tumor systems we have heard
yesterday is highly amplified in the expression. We have some of the in situ
hybridization studies are in the process. We have shied a little bit away
from them but I think the technology is getting better and we are going after
it, yes.

Milo: You turn on exactly the point. In situ hybridization in the systems
that we're looking at, human squamous cell carcinoma, we find the upregulation of some of those genes in the human tumors; but when we separate the
cells out and put them in culture, you get an immediate downregulation of
some of these genes. There appears to be a requirement for a cooperative
upregulation of a couple of them in order to have metastasis expressed, and I
was just wondering if you would have seen it in the melanoma system. It
appeared to be on your slide that it was possible by virtue of the fact that
you did find some cultures that did not have the upregulation or overexpression of some of those genes, and so it is at least suggestive that the
environment surrounding these cells, namely in vivo or in vitro, can affect
the expression of those genes. So that is what prompted the question.

Herlyn: We actually find the most differences in the nonmalignant cells,
meaning a normal melanocyte, as I explained, does not proliferate in general
with some exceptions. Now in culture we push them to proliferate as fast as
they can. These cells show differences. These differences are most obvious
in antigens and we find antigens on melanocytes in culture on the cell surface which are not there in situ. Then, on the other hand, in melanoma cells
these differences are much less obvious. On the other hand, the conditions
in which we grow them now is the protein-free medium. They have to make
everything on their own. All they are getting are amino acids and sugars and
vitamins and some salts. This system is, we find, quite related to the cells
as they are in situ.

Jose Russo: My question is, you said that the cells are synthesizing chondroitin sulfate. Is this kind of proteoglycan synthesized in the nevus status
or in the melanoma status? Can you specify a little more about this?

Herlyn: Yes. In general, the extracellular matrix proteins are very little
produced by normal cells. Normal melanocytes do make some fibronectin.
Normal melanocytes make very little of a 250,000 dalton extracellular matrix
protein, and it starts with the nevi. Nevus cells in situ as well as in culture produce more. It is a gradual increase in production. The melanoma
cells would then start producing collagens as well as a little bit of laminin. We see quantitative difference. The absolute difference is only
between normal cells and tumor cells.

Russo: But you're saying fibronectin, laminin, or proteoglycans like heparan
sulfate or chrondoitin sulfate are also produced? Because you mentioned chondroitin sulfate and I would like for you to give specifics. Do they produce
these kinds of proteoglycans?

Herlyn: That's right. It's also expressed on the surface, and they secrete
it.

Thomas Pretlow: If you take those cells that you've selected for in your
chemo attraction system and mix them with cells prior to separation, do they
still have the capacity to grow in protein-free medium?

Herlyn: That's an interesting question. We have not done it.

Harry Rubin: When you select in protein-free medium, the cells that you are selecting from are not multiplied. Is that right?

Herlyn: We selected from cell lines. They were proliferating.

Rubin: In the absence of serum?

Herlyn: In the absence of serum. I should show you some growth curves.
Once they get used to making everything on their own, they are as good as in
serum.

Rubin: That's what you select for, but you select from a population that
cannot grow well in serum; is that right?

Herlyn: I have to make a distinction. Maybe you haven't gotten the information initially. Metastatic cells can be switched immediately over a period
of 2-3 weeks, whereas in primary melanomas they still need at least IGF-1 or
insulin.

Rubin: But do you then select by leaving those components out? Do you
select then some cells that can grow without...?

Herlyn: Right, from the primaries.

Rubin: The question I'm really asking is, do you think that what you're
selecting in that way already preexists in that population as a rare, let's
say mutated cell, or does that conversion get induced by those conditions, by
the absence of serum.

Herlyn: From the evidence we have, we think it preexists at a very low
percentage.

EARLY AND LATE EVENTS IN THE DEVELOPMENT

OF HUMAN BREAST CANCER*

Helene S. Smith[1], Robert Stern[2]
Edison Liu[3], and Chris Benz[4]

[1]Peralta Cancer Research Institute
3023 Summit Street
Oakland, CA 94609

[2]Department of Pathology
University of California, School of Medicine
San Francisco, CA 94143

[3]University of North Carolina, School of Medicine
Lineburger Cancer Research Center
Chapel Hill, NC 27512

[4]Cancer Research Institute
University of California, School of Medicine
San Francisco, CA 94143

I. STROMAL CONTRIBUTIONS TO TUMOR CARCINOGENESIS:
    POSITIVE EARLY EVENTS IN MALIGNANT PROGRESSION

There is a large body of literature using various model systems to address early events in neoplastic transformation. These studies (which encom-
pass various suggested etiologic agents such as viruses, carcinogens, hormones and growth factors, oncogenes, radiation, etc.) all focus on the target
cell itself. However, carcinomas arise in organized tissues where there is a
close association with mesenchymal cells and their secreted products. Hence,
it is reasonable to consider the possibility that abnormal stromal tissue may
actively participate in some events of the malignant process. A number of
recent studies suggest that this view may be particularly relevant for the
induction of breast cancer. These studies provide evidence at the cellular
and biochemical level that the fibroblasts obtained from breast cancer patients differ from those of normal women.

Two laboratories have described abnormalities in the in vitro properties
of skin fibroblasts from several patients with breast cancer[1-4]. The implication from these results is that breast cancer may be a systemic disease
involving all stroma in a woman's body, not just the neoplastic epithelial
cells in her breast. Azzarone et al.[1] found that cultured skin fibroblasts
from breast cancer patients, unlike fibroblasts obtained from normal

*This work was supported by DHHS grant P01 CA-44768 and a grant from the
 Susan G. Komen Foundation.

Boundaries between Promotion and Progression during Carcinogenesis
Edited by O. Sudilovsky *et al.*, Plenum Press, New York, 1991

donors, formed colonies in soft agar and on monolayers of normal human epithelial cells. In addition, these fibroblasts showed increased saturation densities and invaded embryonic heart tissue. Skin fibroblasts from one patient with a benign breast lesion also displayed the same abnormal growth properties characteristic of fibroblasts from breast cancer patients; three years later a breast carcinoma was detected in this patient. This report was the first to demonstrate that expression of an abnormal fibroblast phenotype could precede the appearance of clinically detectable invasive breast cancer and suggested that stromal changes might be early events in the malignant process. The major criticism of this study from Azzarone's group is that the sample size was small (7 skin specimens from each patient group: carcinoma vs. benign disease). Clearly, additional studies are necessary to confirm and extend these important observations.

Schor and his colleagues have also studied skin fibroblasts from breast cancer patients[2-4]. They devised an assay based on the property of density-dependent cell migration. Skin fibroblasts from normal adults migrate into collagen gels more readily when plated at low density than those plated at high density. Fetal fibroblasts display the opposite behavior, migrating more readily at high plating density rather than at low density. In studies involving skin fibroblasts derived from breast cancer patients, Schor et al. find that 50-70% display fetal-like migratory behavior. In contrast, only 8-10% of normal adult samples display fetal behavior. These investigators have also found that when fetal fibroblasts are passaged in culture they undergo a transition toward more adult migratory behavior, usually occurring after 50 population doublings. Of interest, the fibroblasts from breast cancer patients never acquire this adult migratory behavior. Schor and his colleagues have hypothesized that fetal fibroblasts undergo a normal transition during development from fetal to adult migratory pattern; the failure to undergo this transition in some individuals appears to put them at an elevated risk of developing breast cancer.

Other studies have compared the stromal cells cultured from normal and malignant breast tissue. For the interpretation of these studies, it is not necessary to hypothesize that breast cancer has a systemic mesenchymal manifestation. Alternatively, stromal abnormalities might be localized only to the region adjacent to the epithelial breast cancer. Adams et al.[5] found that fibroblasts derived from normal breast tissue secrete factors which inhibit the growth of cultured breast cancer cells, whereas fibroblasts from malignant breast tumors secrete factors which stimulate the growth of these same cancer cells.

We have begun studies to examine the stromal contribution to breast cancer at a biochemical level. Our attention has focused on hyaluronic acid (HA) because this secreted proteoglycan stimulates cell detachment and motility. The water of hydration of HA opens up tissue spaces enabling cells to migrate easily. HA apparently interacts by a receptor attached to the cytoskeleton to promote this migration[6-10]. Normal cell migration during embryogenesis is associated with an extracellular environment rich in HA[11]; in neoplasia, increased HA correlates with invasiveness and tumor aggressiveness[12-15], and high levels of serum HA have been found in patients with disseminated malignancy[16].

The mechanism by which HA is deposited around tumors is not known. However, one likely inducing agent is tumor growth factor-beta (TGF-ß), because this growth factor is produced by many normal and malignant epithelial cells, including breast cancers[17-19]. Additionally, TGF-ß can stimulate the synthesis of molecules associated with the extracellular matrix including collagens and proteoglycans[20-24], as well as adhesion protein receptor[25]. We have shown that the peritumor fibroblasts from 6 of 14 breast cancer patients were stimulated to accumulate an increased level of HA, by all concentrations

Table 1. Effect of TGF-ß on Hyaluronic Acid (HA)
Accumulation by Cultured Fibroblasts

Specimen	Samples Stimulated in HA Accumulation/ Total Samples Analyzed	
	1 ng/ml TGF-ß	10 ng/ml TGF-ß
Fibroblasts cultured from nonmalignant breast tissues	0/8	1/8
Fibroblasts cultured from breast cancers	9/14	8/14

of TGF-ß tested. Five more fibroblast samples were stimulated to accumulate
HA at some, but not all TGF-ß concentrations. In contrast, TGF-ß inhibited
HA accumulation in normal mammary tissue fibroblasts obtained from 7 of 8
normal mammary tissue specimens[26] (Table 1).

This differential response in HA production between tumor and normal
mammary fibroblasts was specific for TGF-ß. Incubation of fibroblasts with
either tumor growth factor alpha (TGFα) or epidermal growth factor (EGF) did
not modify levels of HA, and there were no additional effects when either of
these factors were added together with TGF-ß. Other growth factors were also
examined in this assay, including platelet-derived growth factor (PDGF),
nerve growth factor (NGF), insulin-like growth factor II (IGF-II), and both
acidic and basic forms of fibroblast growth factor (a + bFGF). None of these
other growth factors affected fibroblast accumulation of HA in our assay.
These results indicate that the fibroblasts contained in many breast cancers
can respond to TGF-ß by increasing their production of HA. Since extracellular environments rich in HA are known to stimulate cell migration, we propose
that the stromal component of some breast cancers participate actively in the
malignant process by creating an environment conducive for malignant epithelial cell invasion. We also found the differences in response to TGF-ß
between normal and tumor-derived fibroblasts could be maintained for more
than 14 passages in culture (42 population doublings), suggesting that this
stimulatory response to TGF-ß is a permanent characteristic of the peritumor
fibroblasts. This mesenchymal abnormality may result from epithelial tumor
cells directing expansion of a sub-population of fibroblast stem cells with
fetal-like properties; there is recent evidence to suggest that putative
fibroblast stem cells from fetal tissue have different responses to growth
factors[27]. Another possibility is that the mammary epithelial tumor cells
induce stem cells to differentiate into mesenchymal cells with myofibroblast-
like properties[28-31].

In summary, these are the first observations at the biochemical level to
suggest that stromal cells play an active role in the malignant process of
human breast cancer. They extend the earlier morphologic and biologic observations that first implicated the stroma in mammary carcinogenesis[32].
Fibroblasts may participate in one or several of the various steps that
result in an epithelial malignancy: initiation, promotion, progression, or,
perhaps, the maintenance of the transformed state.

II. GENETIC CONTRIBUTIONS TO HUMAN BREAST CARCINOGENESIS:
     LATER EVENTS IN MALIGNANT PROGRESSION

Aberrations of proto-oncogenes have been implicated in both the gene-

Table 2. Incidence of Activating ras Mutations in Human Breast Cancer

Specimen	Incidence (% Positive)	Gene	Codon Number	Type
		Codon Mutation		
Primary Breast Cancer	1/40 (3%)	Ki	13	gly-->asp
Soft Tissue Metastases	0/7 (0%)			
Metastatic Effusion	1/9 (11%)	Ki	12	gly-->val
Established Cell Lines	2/5 (40%)	Ki	12	gly-->val
		Ki	12	gly-->asp

sis and progression of human breast cancers. Amplification of c-erb-B2/ HER-2 (the human homologue of neu) is seen in 16 to 30% of primary breast tumors[33,34], and the loss of heterozygosity at the c-Ha-ras locus is present in approximately 30% of breast cancers[35-36]. Both of these genetic lesions are associated with aggressive tumor behavior and unfavorable patient outcome. While c-myc amplification or rearrangement has also been observed in 32% of primary tumors, it has less prognostic significance[37]. Other genetic aberrations which are frequent in human epithelial cancers and could potentially play a role in mammary tumorigenesis include point mutations in the ras family of proto-oncogenes[38,39]. Although there exists some association between these oncogenes and mammary carcinomas, the exact role of specific oncogene aberrations in the progression of human breast cancers remains unclear. Thus, we undertook the characterization of human breast cancers for two genetic abnormalities frequently associated with malignancies in other organ systems: 1) ras activation by point mutation, and 2) allelic loss at the c-Ha-ras chromosome locus, 11p15.

The overall incidence of ras activation in human cancer has been estimated to be 10-15%[40]. However, this figure is much higher for specific solid tumors such as adenocarcinomas of the lung (50%) and gastrointestinal tract (40%)[41-43] or acute myeloid leukemia (25%)[44,45], where Kirsten (Ki-) or Harvey (Ha-) or N-ras activation has been shown to occur during malignant initiation. With respect to breast cancer, very few primary tumors have been analyzed and only occasional breast cancer cell lines have been found to contain activated ras oncogenes[28,46,47]. To establish the incidence for ras activation in all stages of breast cancer we analyzed tumor DNA derived from 40 invasive primary breast tumors, 7 lymph node and skin metastases, 9 metastic effusions, and 5 established breast cancer cell lines[48]. To look for ras mutations we used the polymerase chain reaction (PCR) technique to amplify DNA fragments containing Ki-, Ha-, and N-ras codons 12, 13, and 61 which were then probed on slot-blots with labeled synthetic oligomers to detect all possible non-conservative single base mutations. We found activating mutations in only 1 of 40 primary tumors and in 1 of 9 metastatic effusions (Ki-ras codons 12 and 13, data summarized in Table 2). These results indicate that activating ras mutations are rarely involved in either the initiation or metastatic progression of human breast cancer. This analysis represents the first comprehensive search for activating ras mutations in a broad spectrum of human breast carcinomas. The negative findings are significant in at least two regards: i) they provide evidence for etiologic differences between spontaneous human breast tumors and carcinogen induced animal models of breast cancer which bear a high incidence of ras mutations[40]; ii) they illustrate the molecular-biological differences between breast adenocarcinoma and morphologically similar adenocarcinomas of the lung and gastrointestinal tract, which have a high incidence of ras mutations[41-43]. With a less than 7% incidence of ras mutations found in

Table 3. Sequential Effusion Metastases from a Breast Cancer Patient

Specimen	Clinical History	Establishment of Cell Lines No. Successful Lines/No. Attempts[2]	Karyotypic Markers	Allelic Loss H-ras locus	Mutated Ras
Specimen 1	no prior chemotherapy	0/5	yes	no	no
Specimen 2	14 days later, after initiation of CMF[1] therapy	0/4	yes	no	no
Specimen 3	190 days later, after initiation of doxorubicin therapy	2/2	yes	yes	yes

[1]C = cytoxan, M = methotrexate, F = 5-fluorouracil
[2]each attempt involved plating 2×10^7 cells

breast tumor from 60 different individuals, it is unlikely that ras activation by gene mutation has any significant role in the initiation or metastic progression of human breast cancer.

We further examined the function of ras gene activation and also of allelic loss at the c-Ha-ras locus in human mammary tumor progression by determining the temporal occurrence of these changes during the clinical course of a patient with progressive breast cancer[50]. Three breast cancer effusions that occurred sequentially in a single patient were examined. Though common cytogenetic abnormalities were found in all effusion samples, only the last effusion exhibited a loss of heterozygosity at the c-Ha-ras locus and a mutated Ki-ras gene (Table 3). These observations, together with previous work showing that allelic loss at the c-Ha-ras locus in primary cancers correlated with poor prognosis[35,36] suggest that this genetic alteration is associated with a more aggressive tumor phenotype and not necessarily with the initiation of primary breast cancer, or the establishment of metastatic disease.

Only the last effusion metastasis exhibited improved in vitro growth of the primary cells and was consistently able to develop into a permanent cell line (Table 3). Since activating ras mutations were more commonly seen in breast cancer cell lines than in uncultured primary or metastatic breast cancers (Table 2), it is possible that mutated ras may confer on cells an improved capacity for growth in culture. It is noteworthy that both colon cancers and small cell lung cancers where activating ras mutations are frequently found, also develop into cell lines much more readily than do breast cancers.

SUMMARY AND CONCLUSION

We hypothesize that early events in the development of at least some human breast cancers involve faulty epithelial-mesenchymal interactions and that the stromal cells themselves play an active role in this abnormal process. In contrast, later events accelerating breast tumor progression may occur in association with genetic changes involving only the malignant epithelial cells. These conclusions arise from a review of the literature, our comparative studies of HA metabolism in fibroblasts cultured from either normal or malignant breast tissues, and from molecular-genetic studies performed on sequential specimens from a single patient and on a wide variety of human breast tumor samples.

HA is a proteoglycan component of the ECM which is known to stimulate epithelial cell detachment and motility and is most abundant in fetal and rapidly growing tissues. We find that many breast cancer-derived fibroblasts are stimulated to produce HA in response to TGF-ß under conditions where HA accumulation by normal tissue fibroblasts is almost uniformly inhibited.

In a single patient, we had the opportunity to examine three malignant effusions that occurred sequentially to identify genetic changes associated with the later stages of breast cancer progression. Although, common cyto-genetic abnormalities were found in all the effusion samples, only the last effusion exhibited a loss of heterozygosity at the c-Ha-ras locus. In this case, the allelic loss correlated with improved growth in vitro of the pri-mary cells and with ability to become a permanently established cell line. Thus, the loss of heterozygosity at the c-Ha-ras locus was associated with a more aggressive tumor phenotype and not with the initiation of the primary breast cancer, nor with the establishment of metastatic disease.

Attempting to identify other genetic changes associated with malignant initiation or progression in breast epithelial cells, we examined primary breast cancers, soft tissue and effusion metastases for evidence of Ki-, Ha-, or N-ras activation known to occur frequently in other types of epithe-lial malignancies. Unlike the high incidence of ras mutations found in carcinogen-induced animal models of breast cancer and in human adenocarcino-mas of the lung and gastrointestinal tract, ras point mutations in codon 12, 13 and 61 were rarely seen (less than 5%) in either primary or metastatic human breast carcinomas.

REFERENCES

1. B. Azzarone, M. Mareel, C. Billard, P. Scemama, C. Chaponnier, and A. Macieira-Coellho, Abnormal properties of skin fibroblasts from pa-tients with breast cancer, Int. J. Cancer 33:759-764 (1984).

2. S. L. Schor, A. M. Schor, P. Durning, and G. Rushton, Skin fibroblasts obtained from cancer patients display fetal-like migratory behavior on collagen gels, J. Cell Sci. 73:235-244 (1985).

3. P. Durning, S. L. Schor, and R. A. S. Sellwood, Fibroblasts from pa-tients with breast cancer show abnormal migratory behavior in vitro, Lancet 890-892 (1984).

4. S. L. Schor, A. M. Schor, G. Rushton, and L. Smith, Adult fetal and transformed fibroblasts display different migratory phenotypes on collagen gels: Evidence for an isomorphic transition during fetal development, J. Cell Sci. 73:221-234 (1985).

5. E. F. Adams, C. J. Newton, H. Braunsberg, N. Shaikh, M. Ghilchik, and V. H. T. James, Effects of human breast fibroblasts on growth and 17ß-estradiol dehydrogenase activity of MCF-7 cells in culture, Breast Cancer Res. and Treatment 11:165-172 (1988).

6. B. P. Toole, Chapter 9, in: "Cell Biology of the Extracellular Matrix," E. D. Hay, ed., Plenum Press, New York (1982).

7. E. A. Tourley, J. Torrance, Localization of hyaluronate and hyaluronate-binding protein on motile and non-motile fibroblasts, Exp. Cell Res. 161:17-28 (1984).

8. B. P. Toole, G. Jackson, and J. Gross, Hyaluronate in morphogenesis: inhibition of chondrogenesis in vitro, Proc. Natl. Acad. Sci. USA 69:1384-1386 (1972).

9. M. Brecht, U. Mayer, E. Schlosser, and P. Prehm, Increased hyaluronate synthesis is required for fibroblast detachment and mitosis, Biochem. J. 239:445-450 (1986).

10. N. Mian, Analysis of cell-growth-phase-related variations in hyaluronate synthase activity of isolated plasma-membrane fractions of cultured human skin fibroblasts, Biochem. J. 237:333-342 (1986).

11. B. E. Lacy and C. B. Underhill, The hyaluronate receptor is associated with actin filaments, J. Cell Biol. 105:1394-1404 (1987).

12. J. C. Angello, H. L. Hosick, and L. W. Anderson, Glycosaminoglycan synthesis by a cell line (C1-S1) established from a preneoplastic mouse mammary outgrowth, Cancer Res. 42:4975-4976 (1982).

13. B. P. Toole, C. Biswas, and J. Gross, Hyaluronate and invasiveness of the rabbit V2 carcinoma, Proc. Natl. Acad. Sci. USA 76:6299 (1979).

14. K. Kimata, Y. Honma, M. Okayama, K. Oguri, M. Hozumi, and S. Suzuki, Increased synthesis of hyaluronic acid by mouse mammary carcinoma cell variants with high metastatic potential, Cancer Res. 43:1347-1354 (1983).

15. J. Tekauchi, M. Sobue, E. Sato, M. Shamoto, and K. Miura, Variation in glycosaminoglycan components of breast tumors, Cancer Res. 36:2133-2139 (1976).

16. G. Manley, and C. Warren, Serum hyaluronic acid in patients with disseminated neoplasm, J. Clin. Pathol. 40:626-630 (1987).

17. A. B. Roberts, M. A. Anzano, L. C. Lamb, J. M. Smith, and M. B. Sporn, New class of transforming growth factors potentiated by epidermal growth factor: Isolation from non-neoplastic tissues, Proc. Natl. Acad. Sci. USA 78:5339-5343 (1981).

18. C. Knabbe, M. E. Lippman, L. M. Wakefield, K. C. Flanders, A. Kasid, R. Derynck, and R. B. Dickson, Evidence that transforming growth factor-ß is a hormonally regulated negative growth factor in human breast cancer cells, Cell 48:417-428 (1987).

19. R. B. Dickson, A. Kasid, K. K. Huff, S. E. Bates, C. Knabbe, D. Bronzert, E. P. Gelman, and M. E. Lippman, Activation of growth factor secretion in tumorigenic states of breast cancer induced by 17ß-estradiol or v-Ha-ras oncogene, Proc. Natl. Acad. Sci USA 84:837-841 (1987).

20. R. A. Ignotz and J. Massagué, Transforming growth factor-ß stimulates the expression of fibronectin and collagen and their incorporation into the extracellular matrix, J. Biol. Chem. 261:4337-4345 (1986).

21. J. Massagué, S. Cheifetz, T. Endo, and B. Nadel-Ginard, Type ß transforming growth factor is an inhibitor of myogenic differentiation, Proc. Natl. Acad. Sci. USA 83:8206-8210 (1986).

22. A. B. Roberts, M. B. Sporn, R. K. Assoian, J. M. Smith, N. S. Roche, L. M. Wakefield, U. I. Heine, L. A. Liotta, V. Falanga, J. H. Kehrl, and A. S. Fanci, Transforming growth factor type ß: Rapid induction of fibrosis and angiogenesis in vivo and stimulation of collagen formation in vitro, Proc. Natl. Acad. Sci. USA 83:4167-4171 (1986).

23. A. Bassols, and J. Massagué, Transforming growth factor type ß specifically stimulates synthesis of proteoglycan in human adult arterial smooth muscle cells, Proc. Natl. Acad. Sci. USA 84:5287-5291 (1987).

24. J.-K. Chen, H. Hoshi, and W. L. McKeehan, Transforming growth factor type ß specifically stimulates synthesis of proteoglycan in human adult arterial smooth muscle cells, Proc. Natl. Acad. Sci. USA 84:5287-5291 (1987).

25. R. A. Ignotz and J. Massagué, cell adhesion protein receptors as targets for transforming growth factors-ß action, Cell 51:189-197 (1987).

26. R. Stern, J. T. Huey, J. Hall, and H. S. Smith, Hyaluronic acid production in response to type-ß transforming growth factor distinguishes normal from breast cancer-derived fibroblasts, submitted for publication.

27. W. Wharton, Newborn human skin fibroblasts senesce in vitro without acquiring adult growth factor requirements, Exp. Cell Res. 154:310 (1984).

28. W. Schurch, T. A. Seemayer, and R. Lagace, Stromal myofibroblasts in primary invasive and metastatic carcinomas, Virchows Arch. (Pathol. Anat.) 391:125-139 (1981).

29. S. H. Barsky, W. R. Green, G. R. Grotendorst, and L. Liotta, Desmoplastic breast carcinoma as a source of human myofibroblasts, Am. J. Pathol. 115:329-333 (1983).

30. B. A. Gusterson, M. J. Warbutron, D. Mitchell, M. Ellison, A. M. Neville, and P. S. Rudland, Distribution of myoepithelial cells and basement membrane proteins in the normal breast and in benign and malignant breast diseases, Cancer Res. 42:4763-4770 (1982).

31. A.-P. Sappino, O. Skalli, B. Jackson, W. Schurch, and B. Gabbiani, Smooth muscle differentiation in stromal cell of malignant and non-malignant breast tissues, Int. J. Cancer 41:707-712 (1988).

32. A. van den Hooff, The part played by the stroma in carcinogenesis, Perspec. Biol. 27:498 (1984).

33. D. J. Slamon, G. M. Clark, S. J. Wong, W. J. Levin, A. Ullrich, and W. L. McGuire, Human breast cancer: Correlation of relapse and survival with amplification of the HER-2/neu oncogene, Science 235:177-182 (1987).

34. M. van de Vijver, R. Van de Berssalaar, P. Deville, C. Cornelisse, J. Peterse, and R. Nusse, Amplification of the neu (c-erbB-2) oncogene in human mammary tumors is relatively frequent and is often accompanied by amplification of the linked c-erbA oncogene, Mol. Cell. Biol. 7:2019-2023 (1987).

35. C. Theillet, R. Lidereau, C. Escot, P. Hutzell, M. Brunet, J. Gest, J. Schlom, and R. Callahan, Loss of a c-Ha-ras-1 allele and aggressive human primary breast carcinoma, Cancer Res. 46:4776-4781 (1986).

36. M. J. Cline, H. Battifora, J. Yokota, Proto-oncogene abnormalities in human breast cancer: with anatomic features and clinical course of disease, J. Clin. Oncology 5:999-1006 (1987).

37. C. Escot, C. Theillet, R. Ledereau, F. Spyratos, M.-H. Champeme, J. Gest, and R. Callahan, Genetic alteration of the c-myc proto-oncogene (MYC) in human primary breast carcinoma, Proc. Natl. Acad. Sci. USA 83:4834-4838 (1986).

38. M. H. Kraus, Y. Yuasa, and S. A. Aaronson, A position 12-activated Ha-ras oncogene in all HS578T mammary carcinosarcoma cells but not normal mammary cells of the same patient, Proc. Natl. Acad. Sci. USA 81:5384 (1984).

39. H. Zarbl, S. Sukumar, A. V. Arthur, D. Martin-Zanca, and M. Barbacid, Direct mutagenesis of Ha-ras-1 oncogenes by N-nitroso-N-methylurea during initiation of mammary carcinogenesis in rats, Nature 315:382-385 (1985).

40. M. Barbacid, ras genes, Ann. Rev. Biochem. 56:779-827 (1987).

41. S. Rodenhuis, M. L. van de Wetering, W. J. Moot, S. G. Evers, N. van Zandwizh, J. L. Bos, Mutational activation of the K-ras oncogene: A possible pathogenetic factor in adenocarcinoma of the lung, N. Eng. J. Med. 317:929-935 (1987).

42. J. L. Bos, E. R. Feron, S. R. Hamilton, M. Verlaan-de Veries, J. H. van Boom, A. J. van der Eb, B. Vogelstein, Prevalence of ras gene mutations in human colorectal cancers, Nature 327:293-297 (1987).

43. K. Forrester, C. Almoquera, K. Han, W. E. Gizzle, and M. Perucho, Detection of high incidence of K-ras oncogenes during human colon tumorigenesis, Nature 327:298-303 (1987).

44. J. L. Bos, D. Toksoz, C. J. Marshall, M. Verlaan-de Veries, Amino acid substitutions in codon 13 of the N-ras oncogene in human acute myeloid leukemia, Nature 315:726-730 (1985).

45. J. L. Bos, M. Verlaan-de Veries, A. J. van der Eb, J. W. G. Janssen, R. Delwel, B. Lowenberg and L. P. Colby, Mutations in N-ras predominate in acute myeloid leukemia, Blood 69:1237-1241 (1987).

46. M. T. Prosperi, J. Even, F. Calvo, J. Lebeau, and G. Goubin, Two adjacent mutations at position 12 activate the K-ras-2 oncogene in a human mammary tumor cell line, Oncogene Res. 1:121 (1987).

47. S. C. Kozma, M. E. Bogaard, K. Buser, S. M. Saurer, J. L. Bos, B. Groner, and N. E. Hynes, The human c-Kirsten ras gene is activated by a novel mutation in codon 13 in the breast carcinoma cell line MDA-MB231, Nucl. Acid Res. 15:5963-5971 (1987).

48. C. F. Rochlitz, G. K. Scott, J. M. Dodson, E. Liu, C. Dollbaum, H. S. Smith, and C. C. Benz, Incidence of activated ras oncogene mutations associated with primary and metastatic human breast cancer, Cancer Res. 49:357-360 (1989).

49. E. Liu, C. Dollbaum, G. Scott, C. Rochlitz, C. Benz, and H. S. Smith, Molecular lesions involved in the progression of a human breast cancer, Oncogene 3:323-327 (1988).

50. H. S. Smith, S. R. Wolman, S. H. Dairkee, M. C. Hancock, M. Lippman, A. Leff, and A. J. Hackett, Immortalization in culture: occurrence at a late stage in progression of breast cancer, J. Natl. Cancer Inst. 78:611-615 (1987).

DISCUSSION

Lance A. Liotta: Dr. Smith, that was a fascinating presentation. I'd like to add something to your concept of the role of fibroblasts. We've found recently that IGF-1 and IGF-2 are potent chemoattractants for breast cancer cells, stimulating their movement towards a concentration gradient. We have looked at the human breast cancer tissue for the expression of IGF-1 and IGF-2. Where do they come from? The fibroblasts. It's not made by the tumor cells. The fact that fibroblasts could make IGF-1 and IGF-2 could be one mechanism for causing the directional migration of tumor cells out of the duct, once they've degraded the basal membrane.

Helene S. Smith: Yes, that's really interesting. I certainly don't want to leave you with the impression that hyaluronic acid is the only molecule. I think that perhaps what we've demonstrated is that there is a difference at a biochemical level which is maintained even after you take the stroma away from the stimulating environment of the breast cancer. When you allow these cells to proliferate (and these cells can be cultured 10-15 times) we still see these differences in phenotype. One has to look in situ at the stroma.

Liotta: Another quick question. What about desmoplasia and myofibroblasts which are known to increase tremendously in the desmoplastic areas surrounding breast cancer development.

Smith: There are a number of reports that myofibroblasts are present in breast cancers. What we're culturing may be the undifferentiated derivatives of myofibroblasts. They're not making myosin and they're not behaving like muscle phenotypes, but there's no reason to think that they aren't the derivatives of those. It may be that what are being called myofibroblasts are a different stem population that has been recruited into the tumor. The fact is that even if they've lost their myofibroblastic distinguishing features, they still remember some distinctions that are relevant potentials for study.

George Yoakum: Firstly, I'd like to complement you on some beautiful data. I would like to ask you to think about your data with a different interpretation. That is, your comments that ras may not be involved in human breast cancer. Why couldn't mutations occur early and lead to intolerance and intolerance lead to deletions? Couldn't this simply be what you're following? You know, ras is just as malignant-generating in the cells that you're looking at as they are any place else.

Smith: There is a flaw of the whole approach of characterizing specimens at various stages. When we grind up a primary tumor, we're looking at the end result of a whole bunch of things that have happened and been selected for. There's no way around that, and in breast cancer in particular it's very difficult to talk about what the premalignant lesion was in order to fish it out and see whether it was present. In contrast to colon, where the adenomas are obviously there, in breast cancer pathologists are still quarreling over what is the premalignant lesion. To accept your postulation one would have to assume, that mutations in the ras loci were selected against in breast cancer, but not in other tissues. One can never rule this out but I don't think there's any evidence that creating a ras mutation is a negative growth selection phenomenon. So, there is not yet to my knowledge any reason for evoking that hypothesis. Incidentally, I want to make the point that mutations at the ras locus are different from allelic loss of the Harvey ras locus. These are two different phenomena and it is easy to confuse them. I'm talking about mutations at the ras locus, and that is what we are not finding. Dr. Callahan mentioned that in his work he hasn't found them either and probably there are other groups as well. We are not finding mutated ras at the three loci commonly found in breast cancers. That's different than allelic loss at the Harvey ras locus, which is what I was talking about in the second part of the genetics aspect of my talk. In fact, as Dr. Callahan and his colleagues have shown, the real deletion is a little away from the Harvey ras and closer to betaglobin.

Ben Kim: It seems to me you may be able to discriminate between what may be a primary fibroblast change, alteration in a host versus one induced by the breast carcinoma, if you looked at the other areas of the breast after specimen. Have you done that either in the contralateral breast or the same breast in the same patient with breast carcinoma, to see if the changes are still there?

Smith: No, we haven't done it yet. But yes, we are taking specimens and growing up the stroma from the peripheral tissue of the breast as well. Those are interesting experiments.

Robert Callahan: Helene, have you been able to look at other markers on chromosome 11p like betaglobin or the PTH locus?

Smith: No, we haven't. We just assumed you had.

Callahan: Well, our experience was that there were many tumors in which the Harvey ras locus was not involved in the deletion. The most frequently deleted region actually occurred closer to the centromere.

Thomas Pretlow: You stated that the allelic loss of H-ras seems to be associated with progression. The only data I could see was the slide that showed one patient on one side and the other patient on the other side. Is there more than one cell line that this generalization is based on?

Smith: There are two pieces of evidence that it's based on. One is the fact that, as first reported by Callahan, that people who have this change in their primary breast cancer have a worse prognosis. Then, what I showed is that in one series of specimens from a single patient, that the patient (although she had metastatic disease) didn't have this change but acquired it as the disease worsened. It's very difficult to get sequential specimens from the same patient, and there are technical problems with getting that in a series of cases. But I think the combination of the two suggests to us that it is not an early change, but one associated with allowing some cells to grow even more readily at a metastatic site.

Sandy Markowitz: Mark Whitman's lab, I guess working in model systems, largely MCF7, has suggested that <u>ras</u> mutations can cause progression, not in the linear scheme that we think of but rather from estrogen dependence and estrogen independence. Associated with that progression is an increase in estrogen independent secretion of TGFα and IGF-1. In the samples that you looked at, did you have stratification for estrogen dependent or independent tumors? And if it's not <u>ras</u> that is involved in organs with that progression, do you find the same increase in IGF-1 or TGFα?

Smith: I haven't done any work with IGF-1 or TGFα, and yes, they were both ER positive and negative tumors in the collection that we looked at. And again, we use these model systems to give us clues as to what to do and what to look for in human systems. I think the proof of the pudding is that these things like <u>ras</u> mutations are turning out to be important in some places and in some human systems. I think what we need to do with our models is to use them to help us decide where, in fact, they fit. And sometimes we'll find that the fit was exactly the way we thought, and sometimes nature will be cleverer than we are and we may be surprised that the fit isn't right but it may turn out to be more interesting than we ever thought it was. I certainly didn't mean to say that model systems weren't good. Where would we get the clues to look for? And by refining the model system and by going back and forth between the human and the model system I think we will maybe get it to the point where we can even ask more subtle questions. If some breast cancer carcinogens induce <u>ras</u> mutations, but other kinds of carcinogens didn't induce it, one could perhaps hypothesize that the model was telling us what things were doing it for people. Getting into fantasy, one could put fatty acids in and see if that altered the mouse system the way we see it in humans, and perhaps that could tell us something about what fat is really doing in humans. That's where I see the field going.

Joseph Locker: I was especially pleased to see the <u>ras</u> mutation study because of the complexity of actually carrying it out. What I was wondering is just how comprehensive your survey was of point mutations. I've calculated that to do all the possible mutations would require 81 probes.

Smith: Yes, right. The way we did it was to combine the three different mutations as a degenerate probe and hybridized together, and only if we found anything did we then dissect it with the different probes.

Locker: I'd be a little bit concerned about using the mixed probes because the thermal stability is very sensitive and the different base combinations have quite different thermal stabilities. Is there a possibility that you might not have detected all of the possible mutations with the same sensitivity?

Smith: I'd be surprised, because we always used as a positive control one that had the mutation there. If you noticed, there were two that we didn't do because we didn't have a positive control. So, we analyzed the results only if in the same blots we found hybridization with one where we knew the change had happened. In fact Dr. Benz and his colleagues have published a paper recently talking about how they made mutations of specific sites in order to address that question.

Harry Rubin: It seems to me that your finding about the stroma fibroblasts also raises the question of whether the primary driving force in tumor formation is in the morphogenetic field, since that's what you've changed. All the rest follows, but as long as you've got some degree of "disorganization," or dysfunction in the tissue as a whole, you're going to generate variations. Then what follows ultimately is increased likelihood of getting a tumor. I'm sure you've thought about it, and I wonder if you have some comments about it.

Smith: Obviously. How can I not, being in the same city as Harry Rubin (Berkeley), think about these things? Yes, those are really important ideas and issues. I just don't know how to test that with this system yet. I'd be interested in thoughts on how to. The one thing we have now is a system where we can take the fibroblasts that do express these changes. How could we study interactions in order to examine whether the presence of these abnormal fibroblasts creates a situation where there is genetic instability? I don't know. This is another theme that has come up again and again: How do we actually test genetic instability? What is the appropriate model for testing genetic instability that's relevant for human cancer? I hope we're all going to go home and start thinking about it. Maybe we'll know it next time, during a future "Boundaries" conference.

MALIGNANT PROGRESSION OF HARVEY RAS TRANSFORMED

NORMAL HUMAN BRONCHIAL EPITHELIAL CELLS

G. H. Yoakum[+], L. Malan-Shibley[*], and C. C. Harris[*]

+Radiation Biology Division
Radiation Oncology Department
Albert Einstein College of Medicine
708F Forchheimer Bldg.
1300 Morris Park Ave.
Bronx, NY 10461

*Division of Cancer Etiology
Laboratory of Human Carcinogenesis
Rm 2C05 Bldg 37
National Institutes of Health
Bethesda, MD 20892

INTRODUCTION

Metastasis is a complex process involving the physiology of cellular interactions with environmental factors such as growth factors and biological modifiers, membrane structures, local and humoral immunological effectors[1-9]. These interactions ordinarily shape the regulatory functions that coordinate the individual role of cells in the tissue and organ system. However, when metastasis occurs the following aberrant events must occur to cause this biological transition: 1) a genetic change in a cell within a primary tumor[3] results in at least the following subsequent changes before metastasis of primary tumor cells to distal sites: a) the cell must be capable of growth to produce a clone capable of metastatic expression, b) the cells in this clone produce enzymes that hydrolyze surrounding structures constituting the basement membrane[7-15], c) motility permits the cell to escape its local position and move through anatomical barriers to other locations in the host via the circulatory or lymphatic systems[11,13,16], d) cells with the potential to develop tumors at distant locations must escape immune surveillance[17,18] and e) finally cells from these metastatic clones establish themselves at distal locations and initiate development of a metastasized tumor that damages tissues and organs not associated with the primary tumor site[19-24]. Interestingly, these processes often involve the spread of primary tumors in a given location along fairly restricted guidelines of anatomical location for metastatic tumors[1,5,8]. This observation indicates a specificity of biological control of this process and suggests that the mechanisms regulating metastasis must result from specific changes at the genetic level to permit the development of defined phenotypic expression in these events. Therefore, the specific and consistent pattern of metastatic tumor spread suggests the existence of consistently altered genetic pathways that result in the aberrant physiology required to explain the observation of particular

Boundaries between Promotion and Progression during Carcinogenesis
Edited by O. Sudilovsky *et al.*, Plenum Press, New York, 1991

341

and specific patterns of tumor growth and spread during the metastatic phase of cancer.

The transformation of normal human bronchial epithelial (NHBE) cells by transfection with the vHa-ras oncogene[25] to cell lines capable of continuous growth in cell culture, and the subsequent growth of cell lines from the original transformants that attain the capability to grow as tumors in irradiated athymic nude mice (TBE-1)[25] has resulted in an in vitro biological model for the oncogenic activity of vHa-ras gene in human bronchial epithelial cell carcinogenesis[26]. We have therefore studied the biological, physiological, and genetic characteristics of four NHBE cell lines originally isolated from vHa-ras transfected NHBE cells (TBE-1; TBE-2; TBE-3; TBE-4). In addition, the TBE-1 cell line was used to derive tumorigenic cell lines by: i) selection for anchorage independent growth in soft agar (TBE-1SA); and ii) continued growth in cell culture to approximately 30 passages (TBE-1P30) (estimated 200 population doublings). The initial observation of tumorigenicity, when inoculated subcutaneously in irradiated athymic nude mice, required an extended incubation period of 9 to 13 months[25]. However, subsequent transfer between mice as minced xenotransplanted grafts increased the growth rate significantly, resulting in the observation of TBE-tumors that exceeded 1 cm in diameter within 2 months. This progressive increase in tumorigenicity after transformation with vHa-ras indicates that the carcinogenesis of the Ha-ras pathway in human bronchial epithelial cells may be studied by characterization of TBE-1 and its variant cell lines derived in vitro.

MATERIALS AND METHODS

Cell Lines and Growth Conditions

The TBE-series of cell lines were isolated as non-contact inhibited, transformed foci after transfection of NHBE cells with the vHa-ras oncogene. Although four separate foci were isolated in the initial experiment, an extended period of very slow growth occurred during the transformation of NHBE cells to attain the capacity for continuous growth in cell culture[25]. We have analyzed the karyotype of TBE-1 through 4 and determined TBE-1 through 4 probably originated from a single transforming event because an identical marker chromosome is found in the karyotype from each cell line[26]. Although we have observed differences in the tumorigenic capabilities of cell lines originating from these foci, we have extensively studied the characteristics of TBE-1 and consider this cell line (TBE-1) to be representative of the type of cell lines we would expect to derive from the other sublines which originated following a single transforming event.

The reduced dependence of TBE-1 on growth factors[25] needed for the culture of the NHBE parental cells permits the growth of TBE-1 RPMI 1640 supplemented with 5% fetal bovine serum (FBS), or in HUT medium (RPMI 1640 with 10% FBS) as described[25]. Therefore, TBE-series cell lines were routinely cultured in HUT medium after a 1:3 split and growth to ca. 80% confluence (6-9 days) the cells were harvested by trypsinization; placed on ice, counted, washed, and resuspended in Leibowitz L15 buffer at ca. 2 x 10^7 cells/0.2 ml for tumorigenicity assays. Similar growth conditions were used to prepare cultures for DNA isolation, determination of the expression of isozymes, and preparation of mitotic chromosomes for karyotype analysis.

Tumorigenicity Assays

Subcutaneous inoculation of nude mice. TBE-cells and selected variants of this cell line were assayed for tumorigenicity in athymic Balb/c nude mice from the following two sources: i) Nude Mouse Facility, Hazelton Labs, Rockville, MD; and ii) mice purchased from the NIH extramural animal research contractor to the Albert Einstein College of Medicine Experimental Animal Facilities. Animals were irradiated to 400 Gy gamma radiation 24 hours prior to subcutaneous inoculation of the test cells over the right shoulder (some experiments included left/right sites for each animal) at 2×10^7 cells per site. Test cells were grown and prepared for inoculation as described above, and the mice were observed weekly by palpation at the site of inoculation, and an estimate of tumor diameter was recorded to determine the size and growth response of inoculated TBE-series cells. The cutoff point for TBE-tumors observed in these experiments was operationally defined at 0.7 cm in diameter, since smaller nodules almost always regressed in size without developing the capability to grow with an expected geometric rate of increase in size at some phase of tumor development[26] (Fig. 1). In addition, TBE-tumors observed to achieve 0.7 cm in diameter ultimately developed a capability to grow progressively. Growth of tumors from that point in the experiment to maximum size (1.5-2.0 cm in diameter) occurred within 8 to 24 weeks depending on the relative tumorigenicity of each TBE-derivative. Therefore, tumor observations estimating the size at less than 0.7 cm were considered negative, those which achieved 0.7 cm diameter or greater were positive. When tumors were estimated to be 1.5 to 2.0 cm in diameter animals were sacrificed and tumor tissue removed to: i) disperse tissue and derive TBE-series cell lines from primary tumors; ii) test primary tumors for transfer by xenograft transfer into a second test animal; iii) test for the presence and type of one to six human isozymes as described[25] in the primary tumor

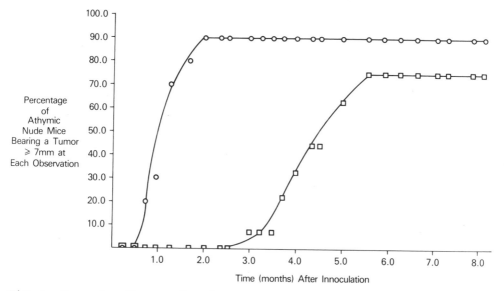

Fig. 1. Comparing the growth and latency period of TBE-1 and TBE-1SA cells when inoculated subcutaneously in irradiated Balb/c nude mice.

The latency period of TBE-1 and TBE-1SA cells was compared by plotting the observations of tumor size on test animals, when monitoring twenty irradiated nude mice which were inoculated with 2×10^7 cells from each cell line. A total of 40 mice were initiated and the surviving animals were observed for a period of 14 months as described in the Materials and Methods. Those animals that developed a tumor greater than or equal to 0.7 cm in diameter and survived developed a progressively growing tumor before the end of this period. However, those with nodules that remained smaller than 0.7 cm diameter very frequently regressed. Therefore, 0.7 cm diameter observations were considered positive for the growth of primary tumors after subcutaneous injection of TBE-cells[26].

tissue; iv) obtain dispersed cell suspensions from the primary tumor to be placed in culture and arrested in mitosis and spread to test for the presence of cells with human chromosomes in primary tumors; v) harvest DNA from primary tumor tissue to test for the presence of human Alu repetitive DNA sequences by Southern analysis; and vi) Southern analysis to test for the structure of the transfected vHa-ras gene. At necropsy all internal organs were grossly examined for metastatic tumor sites, and metastatic tumor tissues were tested for Alu-hybridization, human isozymes, and TBE-histopathology.

Xenograft transplantation between mice. The ability of TBE derived tumors to grow as xenotransplants when transferred between test animals, and the effect of intra-animal transfer on the progressive development of tumor growth properties was tested by transferring tissue from primary TBE-tumors between test animals. When mice were being sacrificed to terminate tumor growth, 1.5-2.0 cm diameter tumors were removed to obtain tissue for pathologic examination and in vitro assays. A randomly selected portion containing approximately 20 to 30% of the tumor mass was placed in L15 at a ratio of 0.2 cm diameter tissue per 0.2 ml L15 buffer, and minced to a consistency capable of passing through a 22 gauge needle. The irradiated nude mouse receiving xenotransplanted TBE-tumor tissue was inoculated with 0.2 ml of this suspension, in a fashion similar to that described above for cell suspensions. The growth of tumors after transplantation was monitored as described above, and tumor tissue was harvested when an estimated diameter of 1.5 to 2.0 cm was reached for pathologic examination, and in vitro assays to: i) derive TBE-series cell lines from tumors transferred as xenotransplanted tissue between test animals; ii) test for the presence of human isozymes in the primary tumor tissue; iii) obtain dispersed cell suspensions from the primary tumor, arrested in mitosis and spread to test for the presence of human cell karyotype in the primary tumors; iv) harvest primary tumor DNA to test for the presence of human Alu repetitive DNA sequences by Southern analysis; and v) Southern analysis to test for the structure of the transfected vHa-ras gene.

DNA isolation and Southern Hybridization analysis. The presence of human DNA in tumor tissue from test animals was determined by freezing at -70°C and pulverizing the frozen tissue from TBE-tumors prior to protease A digestion, phenol extraction, and ethanol precipitation[25,27]. DNA concentration was determined, and 10 μg of tumor tissue DNA was digested with EcoRI and loaded on an 0.8% agarose gel and subsequently transferred to nitrocellulose for Southern hybridization analysis as previously. To determine if human repetitive sequences were present these filters were hybridized with ^{32}P-radiolabeled Blur8 fragment probe (0.3Kb). The 300 bp Blur8 fragment probe contains the human repetitive Alu sequence which hybridizes with human repetitive sequences at high stringency conditions[25-27]. The human Alu fragment probe was labeled with P^{32}-ATP[26,27] to a final specific activity of 2×10^7 to 8×10^7 CPM/μg probe DNA.

Isolation of TBE-tumor cell lines in cell culture. TBE-tumor tissue was harvested in chilled HUT medium, washed with HEPES Buffered Saline (HBS) and minced into a suspension of 1.0 mg/ml collagenase to disassociate the tumor cells from stromal tissue. Single cell suspensions were obtained by incubating the collagenase and minced tumor tissue at 37° for 30 minutes, or until significant degradation of tissue occurred. This dispersed material was decanted through a 100 μm nylon mesh, cell suspensions were washed with HBS, and plated at a density of 1×10^4 cells per cm^2 in HUT medium. When cultures were growing and prior to their achieving 70-80% confluence, viable cells were recovered by trypsinization, washed in HUT medium, and plated after dilution in 92 well plates to isolate cultures of clonal origin. The clones selected in this fashion were characterized for the identifying properties of TBE-1 as discussed above.

Table 1. Tumorigenic Properties of TBE-1 and TBE-Derivative Cell Lines

A. Identification and General Properties of TBE-Cells

Cell Name	1 (TUM)	2 (MET)	3 (XENOG)	4 (COL-IV)	5 LDH G6PD	6 KAR-MARK[a]
TBE-1	2/16	1/38	1/5	nd	human B	human
TBE-1P30	7/10	4/57	1/4	+	human B	human
TBE-2	0/18	0/18	nd	+	human B	human
TBE-3	1/18	nd	nd	+	human B	human
TBE-4	1/19	nd	nd	+	human B	human
TBE-1SA	13/14	1/14	7/9	+	human B	human
TBE-1SAt	4/5	0/5	14/16	+	human B	human

[a]Karyotype was examined to identify human (TBE-series) chromosomes compared to TBE-1.

[nd]The notation nd indicates that the listed determination was not made in that case (not determined).

<u>Column Number</u>: 1(TUM) lists the number of observed tumors first and the total number of test animals surviving to the endpoint; 2 (MET) is the number of metastases observed at necropsy that proved to be human origin tumors, and the number of animals examined from three experiments; 3 (XENOG) is the number of tumors observed on xenograft recipients and the number of animals tested; 4 (COL-IV) + indicates the production of type IV collagenase; 5 LDH G6PD lists the results from isozyme electrophoresis assays that discriminate human and/or mouse LDH (lactate dehydrogenase) and G6PD (Glucose-6-phosphate dehydrogenase).
Note: G6PD type B is consistent with a TBE-cell derivative[25,26].

B. Identification and properties of TBE-tumor tissues[26]

Cell Name	1 (TUM)	2 (MET)	3 (XENOG)	4 PATH	5 HU-ALU (TBE-ID*)	6 LDH G6PD	7 Tu-Cell Line	8[a] Latency Period (weeks)
TBE-1	2/2			+	+	human B	+	48
		1/1		+	+	human B	–	nd
			1/1	+	+	human B	+	8
TBE-1P30	7/7			+	+	human B	+	32
		4/4		+	+	human B	–	nd
			1/1	+	+	human B	+	6
TBE-1SA	4/4			+	+	human B	+	8
		0						
			4/4	+	+	human B	+	6
TBE-1SAt	4/4			+	+	human B	nd	8
		0						
			4/4	+	+	human B	nd	6

[a]The latency period was defined as the period of incubation in the mouse required to grow to a diameter of 0.7 cm.

345

RESULTS

Tumorigenicity of Subcutaneously Injected TBE-1 Cells and Derivatives in Balb/c Nude Mice

The tumorigenicity of TBE-1 is very weakly expressed during the first passage of growth in cell culture. This is indicated by the infrequent occurrence of very long latency period tumors (Table 1A). The first progressively growing TBE-1 tumors followed inoculation of 20 mice after 11 months at a frequency of 2/16. Tumor observations requiring such extended latency period would usually be considered negative since 9 months is the end point for most tumor assays[30]. The frequency of observation of test animals that are positive for TBE-tumors increased when a cell line was selected for anchorage independent growth by culture in soft agar (TBE-1SA[1]) (Table 1). The frequency of tumors observed at the primary site increases to greater than 90% when 2×10^7 TBE-1SA cells are injected sc into athymic nude mice (Table 1). The observation that TBE-1 tumors occur after a long latency period is supported by the following results: i) TBE-1 tumor tissues from long latency tumors are positive for human isozymes (Table 1B); ii) yield DNA that is positive for human repetitive sequence Alu when probed by Southern blot analysis (Table 1B), and iii) contain cells that were dispersed and reisolated as human cell lines with human karyotype and similar isozymes to the TBE-series (TBE-1t) (Table 1B). These combined results indicate that delayed growth of TBE-1 cells in irradiated Balb/c mice results in the long latency period of development before TBE-tumors grow progressively. In addition, subcultured cell suspensions from primary TBE-1 and TBE-1SA tumors resulted in the isolation of human cell lines TBE-1t and TBE-1SAt with isozymes similar to those produced by TBE-1 cells (Table 1B). Although an 11 month latency period elapsed before any of the mice inoculated with early passage TBE-1 cells were first observed to carry progressively growing tumors, the isolation of the TBE-1t cell line from a tumor of this type indicates that this primary tumor resulted from the growth of TBE-1 cells in the test animals (Table 1B).

The operational determination that 0.7 cm diameter was to be considered the cutoff point for the positive identification of tumorigenicity in experiments in which TBE-series cell lines were inoculated is indicated by the results of growth observations presented in Fig. 1. The growth of TBE-1P30 (passage 30) and TBE-1SA cell lines after sc inoculation of irradiated athymic nude mice was plotted from weekly observations of the estimated tumor (or nodule) size vs time. The selection of anchorage independent cell line TBE-1SA results in a progeny cell line of TBE-1 with significantly more tumorigenic phenotype than TBE-1: i) the frequency of positive observation of tumors at the primary site of inoculation was observed to be 90% (0.9), and the latency period preceding progressive growth approximately 12 weeks (Fig. 1). Inoculation of TBE-1P30 (TBE-1 at passage 30) cells resulted in an increased frequency in the total number of animals that were positive for tumors when compared to TBE-1 experiments at early passage (Table 1A). However, the frequency (0.75) and latency period of TBE-1P30 tumors (20 to 24 weeks) indicate a significantly less tumorigenic phenotype at the sc site when compared to TBE-1SA (Fig. 1). These data indicate that selection of the TBE-1SA cells by growth under soft agar results in isolation of TBE-sublines that have increased ability to express tumorigenic properties in the nude mouse assay (Fig. 1, Table 1).

Upon necropsy of test animals the internal organs were examined to determine the presence of tumors at sites that were distal from the primary sc tumor (Table 1). The observation of metastatic tumors in animals that were inoculated with TBE-1, and TBE-1SA cells suggests that the TBE-series of cell lines has a measurable ability to metastasize from the sc site. When internal organs were observed to contain tumors, it was notable that a variety of

346

Table 2. Tumorigenicity in Balb/c nu/nu Mice
from NIH Contract Source

Cell Name	1 (unirradiated)	2 400Gy	3 LDH G6PD	4 Latency Period (weeks)
TBE-1SA	0/9			
		3/5	human B	11

A total of 18 Balb/c nude mice purchased from the NCI-Frederick
Cancer Research Facility were evaluated as test animals for
tumorigenicity assays. The irradiated animals were exposed to
400 Gy of gamma-radiation 24 hours prior to subcutaneous inoculation over the right shoulder with 2 x 10^7 TBE-1SA cells,
and the growth of tumors observed as described in Materials and
Methods for a period of 9 months. Tumor tissue was removed and
assayed for the presence of human isozymes to identify TBE-cells
in tumor tissue.

organs were affected in different animals including: the liver, spleen, and
kidney of animals inoculated with TBE-1 or TBE-1P30 (Table 1B). The tumor
tissues removed from these animals with metastatic tumors were positive for
histopathological identification as morphologically similar to primary TBE
tumors, human isozymes, and human repetitive sequences on Southern analysis
(Table 1B).

The xenograft transfer of TBE-tumor tissue between test animals successfully transferred progressively growing TBE-tumors between animals.
Xenograft transfer of TBE-tumor tissue results in a shorter latency period
than tumors from inoculations of TBE-cell suspensions (Table 1B). In each of
these cases tumor tissues were analyzed to determine: i) the presence of
human isozymes, and ii) positive hybridization on Southern analysis for human
repetitive sequences (Table 1B). However, it should be noted that although a
representative number of these animals were tested between transfer one and
transfer nine, and the latency period of transfer tumors was shorter (4-8
weeks) (Table 1B), necropsy observations of twenty animals bearing transfer
tumors that were 1.5 to 2.0 cm in diameter revealed no gross tumors in
internal organs (Table 1B).

The ability of tumorigenic variant lines of TBE-1 to grow as primary sc
tumors in the nude mouse assay was tested by inoculating TBE-1SA cells in
irradiated and unirradiated Balb/c nude mice obtained from the NIH extramural
contract animals facility at Frederick, MD, and maintained for an observation
period of 9 months at the Albert Einstein College of Medicine sterile animal
facilities during the determination of tumorigenic properties. The irradiated group of test animals received a 400 Gy dose of radiation from a Cs^37
source 24 hours prior to injection similar to the irradiation used in Table
1. The results in Table 2 indicate that TBE-cell lines that have determinable tumorigenicity when inoculated in animals bred and maintained at the
nude mouse facilities of Hazelton Labs, Rockville, MD, (Table 1), are also
tumorigenic when inoculated in a similar fashion in irradiated (400 Gy)
Balb/c nude mice obtained from the animal contract facility at Frederick, MD.
The observation of tumors in 3 of 5 irradiated mice inoculated with 2 x 10^7
TBE-1SA cells and observed for 9 months indicates that tumorigenicity of TBE-

cell lines may be studied in irradiated Balb/c nude mice obtained from the NIH contracted facility. In addition, the observation of 0 tumors in 9 test animals that were not irradiated indicates that additional immunosuppression by irradiation is important to the survival of TBE-1SA tumors (Table 2).

Ha-ras Hybridization Analysis of DNA Isolated from TBE-1 and Tumor Cell Lines Reisolated from Tumors in Athymic Nude Mice

The hybridization of DNA isolated from cell lines that were obtained by clonal isolation of human cells from tumors that were harvested during tumorigenicity experiments (Table 1). The tumor cell lines were characterized for TBE-human karyotype[26]; the expression of human isozymes LDH and G6PDH that are consistent with TBE-cell origin[25] (Table 1); the presence of human Alu repetitive DNA (Table 1); and growth in conditions without growth factor supplements to maintain NHBE cells but known to be adequate for growth of TBE

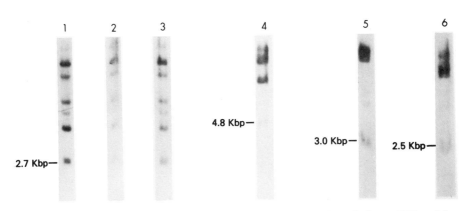

Fig. 2. Southern hybridization analysis of DNA isolated from TBE-cell lines hybridized at high stringency[25,27] with ^{32}P-labeled 700 base pair Ha-ras fragment probe. The 2.7 Kbp fragment released by digestion with HindIII from vHa-ras transfected NHBE cells (TBE-1; TBE-1T tumor cell line; and TBE-1SAT anchorage independent tumor cell line) in lanes 1 through 3 indicate that the transfected vHa-ras locus has been maintained through all three phases of malignant progression: i) transformation of NHBE to attain the capacity to continue growth in cell culture after vHa-ras transfection TBE-1 (lane 1); ii) a cell line recovered from long latency (13 month) tumor grown in athymic nude mouse resulting in early phase long latency period tumors TBE-1T (lane 2); and iii) selection of TBE-1 by plating under soft agar to derive a cell line that has attained the capacity to grow in anchorage independent conditions and subsequently grow more vigorously as a subcutaneous tumor in athymic nude mice with short latency (4-6 months) TBE-1SAT (lane 3). The 4.8 Kbp Bgl II fragment observed in TBE-1SAT (lane 4) indicates that one of the two BglII sites in the transfected vHa-ras segment transfected has been lost to mutagenesis or recombination. However, the observation of the predicted 3.0 Kbp HincII fragment (lane 5) and 2.5 HincII/BglII fragment (lane 6) from TBE-1SAT DNA that hybridize with the 700 bp Ha-ras fragment probe, indicate that the coding sequence for the structural gene of vHa-ras has been maintained intact during this series of biologically selective steps that challenge the capability of the cell to grow and express a malignant phenotype.

cell lines[25] (Table 1). DNA was isolated and digested with restriction enzymes that release fragments of the vHa-ras gene that identify the segment transfected into NHBE cells when TBE-1 was originally obtained by vHa-ras transformation of NHBE cells[25].

The digestion of DNA from TBE-1, TBE-1T, and TBE-1SAT with HindIII released a similar 2.7 Kbp fragment that hybridizes with ^{32}P-labeled 700 bp Ha-ras probe (Fig. 2, lanes 1-3 respectively). The 2.7 Kbp HindIII fragment has no common allele in the background of human cellular ras genes that will hybridize with the 700 bp Ha-ras fragment probe[25,27]. Thus, the 2.7 Kpb originates from the transfected vHa-ras segment released from the genome of each of three different derivatives of TBE-1: i) the original NHBE transformant TBE-1; ii) a TBE-cell line recovered from a tumor occurring with low frequency and after a long latency period of 13 months (Table 1) TBE-1T; and iii) a TBE-cell line selected first for anchorage independent growth under soft agar (SA) and recovered from a tumor with intermediate latency period of 6 months (Table 1) TBE-1SAT (Fig. 2). The digestion of DNA from TBE-1SAT with BglII released a 4.8 Kbp but not the expected 1.8 Kbp fragment, indicating the loss of one BglII site from the segment of DNA originally transfected[25] (Fig. 2, lane 5). However, the release of a 3.0 Kbp fragment when TBE-1SAT DNA is digested with HincII indicates that an intact coding sequence for the transfected vHa-ras gene has been maintained in this cell line. This is additionally indicated by the observation that the BglIIHincII double digestion of TBE-1SAT DNA released a slightly smaller 2.5 Kbp segment of DNA indicating that the BglII site near the coding sequence of the transfected vHa-ras gene remains intact (Fig. 2, lane 6). Thus, through three phases of malignant transformation of NHBE cells with the vHa-ras gene purported by the different biological properties of TBE-1, TBE-1T and TBE-1SAT the presence of a vHa-ras structure that includes an entire structural gene linked with its expression elements has been maintained without the requirement that a co-linked selective marker such as neo[R] or gpt be used to maintain the presence of the transfected vHa-ras gene in the TBE genome.

DISCUSSION

The transformation to cells with a determinant malignant phenotype after transfection of NHBE cells with the viral Harvey ras oncogene[25], indicates that study of the properties of TBE cell lines will provide an in vitro model for the carcinogenic function of vHa-ras in an important progenitor cell type for human lung cancer[26]. The observation that continued cell proliferation in cell culture leads to increased tumorigenicity is consistent with the reports from carcinogenesis models developed using human tumor cells[2,9,11,14]. In addition, the frequency of tumor occurrence increased, and the latency period shortened when the TBE-1SA cell line was selected from TBE-1 by growth in soft agar. This is consistent with the interpretation that the anchorage independent growth phenotype is associated with the properties required for vHa-ras transformed NHBE cells to develop the malignant phenotype. Also the rigors of selection experienced when human cells are inoculated in athymic nude mice to ascertain their properties as cells capable of growth as tumors in an animal, did not lead to loss of the transfected vHa-ras allele, or substantial mutation of the gene and loss of a functional structural element. This implies that the original transformant of NHBE leading to the derivation of a transformed epithelial cell with determinable capacity to become malignant after vHa-ras integration and expression was capable of retaining the gene without reestablishment of normal differentiation properties of epithelial cells that would result in slower growth and senescence to terminal phase.

The observation that TBE-1 and its derivatives produce type IV collagen, secrete laminin, and maintain laminin receptors on the cell surface indicates that vHa-ras transformed NHBE cells have attained several functions that are likely to be essential for malignancy[26]. The association of these biochemical properties with malignancy has been established for human tumor cell lines in several laboratories[4,7]. It is therefore notable that this is a consistent observation when the biochemical properties of TBE-1 and its malignant variants were analyzed. The general pattern of observations for the tumorigenic properties of TBE-derived cell lines is as follows: i) very early after transformation the cells are either non-tumorigenic or only very weakly tumorigenic requiring the inoculation of large numbers of mice and long latency period to attain positive tumor observations; ii) after continued growth in cell culture, selection for growth in soft agar or xenograft transfer of tumor tissue, tumorigenicity increases. The frequency of positive observations at the subcutaneous site and the shorter latency period of these tumors, provide strong indication that these properties may be positively selected from the original cell line. However, the ability to metastasize to distal sites of inoculation was inversely associated with the observation of tumorigenicity at the primary site of inoculation. Therefore the more rapidly a tumor developed at the site of inoculation, the less likely gross malignancy was observed in distal sites upon necropsy. This may simply have reflected a difference in the growth time, and microscopic malignancies may have gone undetected in animals that rapidly developed tumors at the primary site.

The results establish a basis for the development of an in vitro model for human lung cancer and the role of Ha-ras genes with malignant properties. The observation of biological and biochemical properties of TBE-derived cell lines that are consistent with previous observations of human tumor cell lines indicate that investigation of the tumorigenicity of Ha-ras transformed NHBE cells provides model cell lines that may be studied to reveal the action of other functions associated with malignancy in a progenitor cell type of human lung cancer. The observation of metastasis of TBE-primary tumors to distal sites, from the subcutaneous inoculation site of athymic nude mice, indicates that the metastatic properties of TBE derived tumor cell lines may also be studied.

REFERENCES

1. S. Paget, The distribution of secondary growth of cancer of the breast, Lancet 1:571-579 (1889).
2. I. J. Fidler, Selection of successive tumour lines for metastasis, Nature (New Biol) 242:148-149 (1973).
3. I. J. Fidler and M. L. Kripke, Metastasis results from pre-existing variant cells within a malignant tumor, Science 197:893-895 (1977).
4. I. J. Fidler, Tumor heterogeneity and the biology of cancer invasion and metastasis, Cancer Res. 38:2641-2660 (1978).
5. E. V. Sugarbaker, Patterns of metastasis in human malignancies, Cancer Biol. Rev. 2:235-278 (1981).
6. I. R. Hart and I. J. Fidler, The implications of tumor heterogeneity for studies on the biology and therapy of cancer metastasis, Biochim. Biophys. Acta 651:37-50 (1981).
7. L. A. Liotta, U. P. Thorgeirsson, and S. Garbisa, Role of collagenase in tumor cell invasion, Cancer Metastasis Rev. 1:277-288 (1982).
8. I. R. Hart, "Seed and soil" revisited: Mechanisms of site specific metastasis, Cancer Metastasis Rev. 1:5-16 (1982).
9. L. Ossowski and E. Reich, Changes in malignant phenotype of human carcinoma conditioned by growth environment, Cell 33:323-333 (1983).
10. L. A. Liotta, K. Tryggvason, and S. Garbisa, et al., Metastatic potential correlates with enzymic degradation of basement membrane collagen, Nature 284:67-68 (1980).

11. R. H. Kramer and G. L. Nicolson, Invasion of vascular endothelial cell monolayers and underlying matrix by metastatic human cancer cells, in: "International Cell Biology", H. G. Schweiger, ed., Springer-Verlag, New York and Heidelberg (1981).

12. L. A. Liotta, R. H. Goldfarb, and V. P. Terranova, Cleavage of laminin by thrombin and plasmin: α-thrombin selectively cleaves the chain of laminin, Thromb. Res. 21:663-673 (1981).

13. U. P. Thorgeirsson, L. A. Liotta, and T. Kalebic, et al., Effect of neutral protease inhibitors and a chemoattractant on tumor cell invasion in vitro, J. Nat. Cancer Inst. 69:1049-1054 (1982).

14. L. A. Liotta, C. N. Rao, and S. H. Barsky, Tumor invasion and the extracellular matrix, Lab. Invest. 49:636-649 (1983).

15. L. Eisenbach, S. Segal, and M. Feldman, Proteolytic enzymes in tumor metastasis: II. collagenase type IV activity in subcellular fractions of cloned tumor cell populations, J. Nat. Cancer Inst. 74:87-93 (1985).

16. E. S. Hujanen and V. P. Terranova, Migration of tumor cells to organ-derived chemoattractants, Cancer Res. 45:3517-3521 (1985).

17. K. M. Miner, T. Kawaguchi, and G. W. Uba, et al., Clonal drift of cell surface, melanogenic and experimental metastatic properties of in vivo-selected brain meninges-colonizing murine B16 melanoma, Cancer Res. 42:4631-4638 (1982).

18. G. L. Nicolson and S. E. Custead, Tumor metastasis is not due to adaptation of cells to new organ environment, Science 215:176-178 (1982).

19. J. C. Murray, L. A. Liotta, and S. Rennard, et al., Adhesion characteristics of murine metastatic and nonmetastatic tumor cells in vitro, Cancer Res. 40:347-351 (1980).

20. C. N. Rao, I. M. Margulies, and T. S. Tralka, et al., Isolation of subunit of laminin and its role in molecular structure and tumor cell attachment, J. Biol. Chem. 257:9740-9744 (1982).

21. J. Mollenhaur, J. A. Bee, and M. A. Kazarbe, et al., Role of anchorin C11, a 31,00-mol-wt membrane protein, in the interactions of chondrocytes with type II collagen, J. Cell. Biol. 98:1572-1578 (1984).

22. M. Kurkinen, A. Taylor, and J. Garrels, et al., Cell surface-associated proteins which bind native type IV collagen or gelatin, J. Biol. Chem. 259:5915-5922 (1984).

23. P. H. Hand, A. Thor, and J. Schlom, et al., Expression of laminin receptor in normal and carcinomatous human tissues as defined by a monoclonal antibody, Cancer Res. 45:2713-2719 (1985).

24. V. P. Terranova, E. S. Hujanen, and D. M. Loeb, et al., A reconstituted basement membrane measures cell invasiveness and selects for highly invasive tumor cells, Proc. Natl. Acad. Sci. USA 83:465-469 (1986).

25. G. H. Yoakum, J. F. Lechner, E. W. Gabrielson, B. E. Korba, L. Malan-Shibley, J. C. Willey, M. G. Valerio, A. M. Shamsuddin, B. F. Trump, and C. C. Harris, Transformation of human bronchial epithelial cells transfected by Harvey ras oncogene, Science 227:1174-1179 (1985).

26. G. H. Yoakum, L. Malan-Shibley, W. Benedict, S. Banks-Schleigel, U. P. Thorgeirsson, R. Roeder, M. Schiffman, L. Liotta, and C. C. Harris, Tumorigenic and biochemical properties of human bronchial epithelial cells transformed by Harvey ras oncogene, in preparation.

27. N. Rave, R. Crkvenjakov, and H. Boedthker, Identification of procollagen mRNAs transferred to diazobenzyl-oxymethyl paper from formaldehyde agarose gels, Nucleic Acids Res. 6:3559-3568 (1979).

DISCUSSION

Daniel Longnecker: I was interested in the fact that the cells are expressing monocyte epitopes. It reminded me that several pancreatic carcinoma cell lines have shown evidence of monocytic cell functions. Do you know of other examples where transformed cells exhibit monocyte characteristics? Do monocytes secrete type 4 collagenase?

George Yoakum: I don't know the answer to the last question with certainty. I think in the case of human small cell carcinoma, there are markers of monocytic type.

Joe W. Grisham: Dr. Yoakum, this is very beautiful work. One thing that has concerned me and I'm sure others about the transfection studies, is how do you distinguish the effect of the inserted gene itself and the damage that multiple insertions cause to the genome? Are there any controls to help you resolve them? How do you break those apart, or isn't it possible?

Yoakum: I think that you've hit upon a natural caveat in this approach of doing experiments. One has to accept that anything one does in vitro is in an artificial system. Secondly, I will say that if you are using a protoplast fusion system as opposed to a DNA precipitation method, where insertion of redundant copies at many sites occurs frequently, you are highly likely to wind up with a very few number of inserts. In this case (TBE-1) we know that the insert number is below three. So three possibly damaged sites out of 100,000 genes is not too bad as a worst case analysis. That's one argument that I would give you. But that is as good as Duesberg's statistics that run the other way, and it is probably about as well as one can do. Now the reason that that happens is that in protoplast fusion you are actually putting in circular molecules. You are not putting in a series of linear molecules ligated together on a backbone.

Grisham: Do you know where the insertions are, and are they consistent?

Yoakum: We have not mapped the chromosomal location. We have tried to follow on Southern blots how much this locus is changing, and there does appear to be some change, but we haven't seen a loss during in vitro or tumor tissue transplantation.

Lance A. Liotta: Have you observed any differences in the histologic subtype of the tumors produced from your transfections? Are they all undifferentiated carcinomas? Do you ever get any...

Yoakum: No, we've seen several different types. The undifferentiated carcinoma type is the most frequent type of TBE tumors observed. The other morphology is a fibrosarcomatous tumor that is very frequent upon sub cutaneous inoculation of TBE cells.

Liotta: Did you ever find a loss of the ability to express the activated p21 with retention of the metastatic and tumorigenic capability in your transfectants as you followed them along? In other words, hit and run phenomenon?

Yoakum: We know that the gene is present. We know that there is enough of the structural gene to be expressed. We know the RNA is made. But we have not done p32 labelling assays on every one of those cell lines. So the answer to your question is we don't know at this point. However, we have no data yet to demonstrate the "hit and run" mechanism involved in the development of TBE cell lines.

Liotta: In answer to the question about monocytes and whether they make type 4 collagenase, it was Dr. Longnecker that asked that, they do make it transiently when they are exiting into inflammatory and they differentiate into macrophages in the inflammatory exudate, but as circulating monocytes they don't make it.

Thomas Pretlow: It is very exciting work. My question is, aside from the

LDH data, are there other data to suggest, or are you certain that all those chromosomes in the tumor from the nude mice are human chromosomes?

Yoakum: Ward Peterson has done an excellent job of looking at a lot of karyotypes. We do find occasional fusions with mouse chromosomes included in the karyotypes of the cells lines that we derive from TBE tumors. Of course, we have put those back in the freezer. They are not terribly interesting, so the ones that I have shown you have human chromosomes only, at least at the karyotypic level. We haven't scored to see whether there are any mouse genes present by looking at repetitive sequences.

Joan R. Shapiro: Are there any karyotypic similarities such as albinos reported through the transformation of renal tubule cells? We found that all the transformants involved a chromosome 21 in some manner. Do you find any similarities for a given chromosome?

Yoakum: Although I have a huge raft of karyotypic data, I'll have to be honest with you and tell you that I have not done the huge amount of work of sifting through all those chromosomes to try to find a pattern. If I could get a computer that would take a picture of these things and come back with a graphic analysis for me, I'd be happy, but otherwise I think I'd go crazy. There are so many!

Shapiro: I'm very sympathetic to that.

G. Tim Bowden: You didn't tell us very much about the immortalization step. How effective is the adeno early region in terms of immortalization of the cells?

Yoakum: We've done some experiments with this cell type in this culture system for immortalization, and adenovirus E1a does not work very well. However, SV40 large T is an excellent gene for immortalizing this cell type and can be introduced from a promoter that does not express the small T and does not have an SV40 origin of replication. In the case of these two cell lines that I have derived, E1a is not involved. It is not present. There are enough differences between these two cell lines that I would say that the immortalization step has a) clearly nothing to do with ras per se -that's the first thing- it has something to do with the genetic background of those specific cells, because lot's of patients were transfected that did not derive a transformed cell. So the immortalization step is operationally critical to be able to continue to do the experiments on the ras, but in this case we've been able to get that without introducing a second gene. If you introduce E1a later on, you've shut down phenotypic expression of certain products that ensue from the transformation, such as the type IV collagenase gene, according to the work of Greg Goldberg in St. Louis.

Sanford Markowitz: I'm just confused about a technical aspect of this work. Were all the ras transfectants neo selected or are we only seeing the few of them that actually made it and grew out?

Yoakum: Neo was not used in this experiment. Neo was used in an experiment to derive the transfection efficiency, and that was done for a variety of reasons. Number one, the neomycin selection itself has a mutagenic potential for cells. It may counteract the oncogene effect. Therefore, we used transformation related function to develop two types of selection. One is continued growth, and one is resistance to serum which contains ß-TGF, and stimulates these cells to differentiate. Both of those work.

Moderator (George Michalopoulos): Do these cells make TGF-α at any point of their evolution?

353

Yoakum: Not that I know of. I think that's been looked at by John Lechner and I don't think they make TGF-α. They thought at one time they might make TGf-α, and they thought at one time they might make interleukin-2, and both of those turned out negative. NHBE cells make something that autogenously stimulates their own differentiation at high density growth conditions.

Michalopoulus: Would you care to speculate how a gene which makes a g-protein like a ras gene would lead to the karyotypic instability of these cells?

Yoakum: I think there are a couple of possibilities. The ones I favor are not the obvious ones. The most obvious one is that by affecting ATP metabolism mutant ras gene product may generate aberrant, mitotic spindles that lead to an altered karyotype. Another is that ras, as a differentiation regulatory gene, may be involved in some specific sort of recombination step that stages cells into a certain direction when it is overexpressed or expressed in mutated form. Thus, mutated or overexpressed ras may cause increased gaps and breaks. We're looking at inducible ras genes with regulable promoters such as metallothionine to see if we can stimulate biochemically higher degrees of one or the other of these two phenomena. I think that will give us a better clue.

Paul Duray: You may have mentioned this in your slides and I missed the data, but the source of the bronchial epithelium at autopsy -any characteristics about the donors? Were some of them smokers, non-smokers, that type of thing?

Yoakum: We have very limited information about the epidemiology in either of these two cases. In the first case, I know the person was black, I know the person was 27 years old, and I don't really know anything else other than he died of trauma. In the second case, I know the person was white and I know the person had no history of smoking but, as I said, we don't have the detailed epidemiology profile in either case.

IDENTIFICATION AND CHARACTERIZATION OF DIFFERENTIALLY

EXPRESSED GENES IN TUMOR METASTASIS: THE nm23 GENE

Patricia S. Steeg, Generoso Bevilacqua,
Mark E. Sobel, and Lance A. Liotta

Laboratory of Pathology
National Cancer Institute
National Institutes of Health
Bethesda, MD 20892

Tumor metastasis is a complex process involving tumor cell invasion, locomotion, intravasation and extravasation of the circulatory system, angiogenesis, colony formation, and avoidance of host immunological responses.
Two premises have guided our investigation into the genetic influences on
tumor invasion and metastasis. First, if the metastatic process is regulated, at least in part, by the activation and deactivation of specific
genes, then the multiplicity of cell functions in metastasis dictates that
many genes are involved. Second, the biochemical nature of molecules regulating and executing each of the tumor cell functions in metastasis is incom-
pletely understood. Because of the tedious purification process, it is
likely that many metastasis regulatory and effector compounds and the genes
encoding them are presently unknown. Based on these premises, we initiated
differential colony hybridization experiments to identify genes associated
with the tumor invasion and metastatic process. This technique identifies
genes either activated or deactivated between tumor cells of low and high
metastatic potential. It can therefore identify metastasis-related genes in
advance of conventional biochemical purification and DNA cloning. This paper
describes the identification and characterization of one such gene, nm23.

IDENTIFICATION OF THE nm23 GENE

We utilized seven cell lines derived from a single K-1735 murine melanoma
(Fidler et al., 1981; Fidler, 1984; Kalebic et al., 1988) to identify genes
differentially expressed in tumor metastasis. These cell lines exhibit significant differences in spontaneous and experimental metastatic potential
(Steeg et al., 1988). In vitro translations of total cellular RNA from each
K-1735 cell line are shown in Fig. 1. Most of the bands are uniformly expressed between the low metastatic potential clones 16 and 19 and the high
metastatic potential M2, M4, Tk, Tk-Eve and Tk-liver lines, confirming their
similarity. However, several bands were differentially expressed between the
low and high metastatic potential cell lines. These differentially expressed
bands indicate that specific mRNAs are differentially regulated in the K-1735
system.

A 40,000 component cDNA library was constructed by G/C tailing K-1735
cDNAs into the PstI site of pBR322. Duplicate filters of the library were

hybridized to ^{32}P-labeled mRNAs from a low metastic (clone 19) and a high metastatic (Tk-Eve) K-1735 cell line. Of 24 cDNA clones identified that were differentially expressed between these two K-1735 cell lines, the biology of one clone, pnm23, has been extremely interesting. Expression of the nm23 gene has been consistently down-regulated in tumor cells of high metastatic potential in four experimental models of tumor metastasis.

K-1735 MURINE MELANOMAS

The nm23 RNA levels of seven K-1735 melanoma cell lines were determined by Northern blot hybridization (Steeg et al., 1988). nm23 RNA levels in the low metastatic K-1735 clone 16 and clone 19 cell lines are 10-fold higher than in five related, high metastatic cell lines (Fig. 2). In situ hybridization experiments indicate that virtually all tumor cells express high nm23 RNA levels, as opposed to a subpopulation of cells (Steeg et al., 1988).

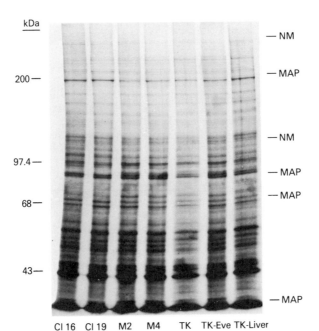

Fig. 1. In vitro translation of K-1735 melanoma RNAs. One microgram of total cellular RNA from each K-1735 cell line was translated in vitro using ^{35}S-methionine and a rabbit reticulocyte lysate, and the translation products electrophoresed on a 7% SDS-PAGE gel, a fluorograph of which is shown. Indicated by MAP are metastases associated proteins, bands expressed to a greater degree by the high metastatic potential M2, M4, TK, TK-Eve and TK-liver lines than the low metastatic clones 16 and 19 lines. Conversely, nm denotes bands expressed to a greater degree by the low metastatic K-1735 lines.

To confirm that the pnm23 cDNA clone was associated with the tumor metastatic process, nm23 RNA levels were determined in additional metastasic
experimental systems. nm23 RNA levels in three low metastatic, c-Ha-ras +
adenovirus 2 E1a cotransfected rat embryo fibroblast (REF) lines were 2- to
8-fold higher than in three control, high metastatic c-Ha-ras-transfected REF
lines (Steeg et al., 1989). Again, in situ hybridization experiments indicated that the high nm23 RNA levels observed in the low metastatic REF

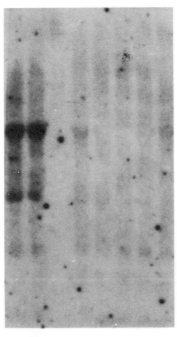

K1735 Cell Line:	Clone 16	Clone 19	M2	M4	TK	TK-Eve	TK-Liver
Median Experimental Pulmonary Metastases:	6	0	113	63	258	143	189
Range:	0-10	0-1	72-225	29-115	204-303	100-255	102-200

Fig. 2. Northern hybridization of pnm23 cDNA
insert to K-1735 murine melanoma cell
line RNAs. Total cellular RNA from
each K-1735 cell line was hybridized
to the PstI insert of pnm23 (Steeg et
al., 1988). To determine spontaneous
metastatic potential, 2 x 10^4 cells
of each K-1735 line were injected
subcutaneously along the backs of
C3H/NeN$^-$ mice. Seven weeks
postinjection, the animals were sacrificed, and gross pulmonary metastases were quantitated (p < 0.05 by
Mann-Whitney).

ras+E1a-cotransfected line were due to relatively uniform expression by each tumor cell.

NITROSOMETHYLUREA (NMU)-INDUCED RAT MAMMARY TUMORS

Nonmetastatic NMU-induced rat mammary tumors contained nm23 RNA levels an average of 1.7-fold higher than did metastatic primary NMU-induced tumors, and 3.2-fold higher than did pulmonary metastases of NMU-induced tumors (Steeg et al., 1988). The observation that nm23 RNA levels exhibited only a 2- to 3-fold difference between the various NMU tumors may be the result of contaminating normal stromal cells, lymphocytes, and endothelial cells, which all express relatively high nm23 RNA levels.

MOUSE MAMMARY TUMOR VIRUS (MMTV)-INDUCED MAMMARY TUMORS

Balb/c mice carrying the C3H strain of MMTV develop mammary tumors of relatively high metastatic potential, while Balb/c mice carrying the RIII strain of MMTV develop low metastatic mammary tumors (Basolo et al., 1987). Total cellular RNA was extracted from C3H and RIII strains of primary breast tumors, and their nm23 RNA levels determined by Northern blot hybridization.

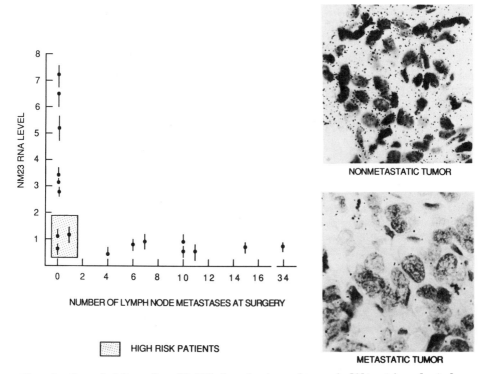

NONMETASTATIC TUMOR

METASTATIC TUMOR

NUMBER OF LYMPH NODE METASTASES AT SURGERY

HIGH RISK PATIENTS

Fig. 3. Correlation of nm23 RNA levels in primary infiltrating ductal breast carcinomas and the patient's number of involved (tumor bearing) lymph nodes at surgery (Bevilacqua et al., 1989). nm23 RNA levels were determined by in situ hybridization. Two representative hybridizations are shown at right to a tumor from a 0-involved lymph node patient (top) and a 4+ involved lymph node patient (bottom).

The low metastatic RIII tumor contained approximately three-fold higher nm23 RNA levels than did the C3H tumor.

HUMAN PRIMARY BREAST CARCINOMAS

nm23 RNA levels were determined in seventeen human primary infiltrating ductal breast carcinomas using in situ hybridization, and compared to the patient's number of involved (tumor-bearing) lymph nodes at surgery (Bevilacqua et al., in press) (Fig. 3). Lymph node involvement is the best indicator of metastatic potential. In a ten-year study, the number of women who developed metastases with 0, 1-3 or 4+ involved lymph nodes was 24%, 65%, and 86%, respectively (Fisher et al., 1975). As shown, all primary tumors from women with involved lymph nodes contained low nm23 RNA levels, consistent with high metastatic potential. Within the 0-involved lymph node group, two statistically significant ($\alpha=0.05$) groups emerged. Approximately 75% contained relatively high nm23 RNA levels, consistent with low metastatic potential. However, 25% of these tumors contained relatively low nm23 RNA levels. Those tumors with low nm23 RNA levels were also <10% positive to cytoplasmic estrogen receptor, negative for nuclear estrogen receptor and of poorly differentiated cytology, all characteristics suggestive of high metastatic potential. Thus, nm23 RNA levels are differentially expressed in human primary breast tumors, and low nm23 levels are consistent with a histopathological prediction of high metastatic potential.

LOSS OF GENE EXPRESSION IN METASTASIS

Our data indicate that in four rodent experimental metastasis model systems, and human breast carcinomas nm23 RNA levels declined in tumor cells of high metastatic potential compared with control non- or low metastatic tumor cells. Metastasis is therefore accompanied by a loss of expression of a specific gene. Whether changes in nm23 gene expression accompany or cause metastatic behavior will be determined by transfection experiments. The DNA sequence of a 700 bp pnm23-1 cDNA insert was novel compared with Genebank sequences (Steeg et al., 1988). The cDNA insert contains an open reading frame from its 5' end to double stop codons at position 498. Within this reading frame, three potential initiating methionines are found, of which one is surrounded by optimal translation initiation sequences and would encode a protein of approximately 16.7 kDa. This putative nm23 gene product contains no sequences indicative of a membrane or secreted protein, and may be an intracellular protein. Characterization of the nm23 protein, and its function in normal and metastasizing tumor cells will proceed by raising antisera to predicted nm23 peptides.

REFERENCES

F. Basolo, A. Toniolo, G. Fontanini, and F. Squartini, 1987, Lung colonization and metastasis of murine mammary tumors: relationship to various characteristics of pulmonary tumors, Invasion and Metastasis, 7:275-283.

G. Bevilacqua, M.E. Sobel, L.A. Liotta, and P.S. Steeg, 1989, Association of low nm23 RNA levels in primary infiltrating ductal breast carcinomas with lymph node involvement and other histopathological indicators of high metastatic potential, Cancer Res. 49:5185-5190.

I.J. Fidler, 1984, The Ernst W. Bertner Memorial Award Lecture: the evolution of biological heterogeneity in metastatic neoplasms, in: "Cancer Invasion and Metastasis: Biologic and Therepeutic Aspects," G. L. Nicolson, L. Milas, eds., Raven Press, New York.

I.J. Fidler, E. Gruys, M.A. Cifone, A. Barnes, and C. Bucana, 1981,
    Demonstration of multiple phenotypic diversity in a murine melanoma of
    recent origin, J. Natl. Cancer Inst., 67:947-953.
B. Fisher, N. Stack, and D. Katyich, 1975, Ten year follow-up results of pa-
    tients with carcinoma of the breast in a cooperative clinical trial
    evaluating surgical adjuvant therapy, Surg. Gynecol. Obstet.
    140:528-538.
T. Kalebic, J.R. Williams, J. E. Talmadge, C.-S. Kao-shan, B. Kravitz, K.
    Locklear, G.P. Siegal, L.A. Liotta, M.E. Sobel, and P.S. Steeg, 1988,
    A novel method for selection of invasive tumor cells: derivation and
    characterization of highly metastatic K1735 melanoma cell lines based
    on in vitro and in vivo invasive capacity, Clin. Exp. Metastasis,
    6:301-308.
R.M. Muschel, J.E. Williams, D. R. Lowy, and L.A. Liotta, 1985, Harvey ras
    induction of metastatic potential depends on oncogene activation and
    the type of recipient cell, Am. J. Pathol., 121:1-8.
R. Pozzati, R. Muschel, J.E. Williams, R. Padmanabhan, B. Howard, L.A.
    Liotta, and G. Khoury, 1986, Primary rat embryo fibroblasts
    transformed by one or two oncogenes show different metastatic
    potentials, Science, 232:223-227.
P.S. Steeg, G. Bevilacqua, L. Kopper, U.P. Thorgeirsson, J.E. Talmadge, L.A.
    Liotta, and M.E. Sobel, 1988, Evidence for a novel gene associated
    with low tumor metastatic potential, J. Natl. Cancer Inst.,
    80:200-205.
P.S. Steeg, G. Bevilacqua, R. Pozzatti, L.A. Liotta, and M.E. Sobel, 1989,
    Altered expression of nm23, a gene associated with low tumor meta-
    static potential, during adenovirus 2 E1a inhibition of experimental
    metastasis, Cancer Res., 48:6550-6554.
P.S. Steeg, G.B. Bevilacqua, A. Rosengard, M.E. Sobel, V. Cioce, and L.A.
    Liotta, Altered gene expression in tumor metastasis: The nm23 gene,
    in: Cancer Metastasis: Molecular and Cellular Biology, Host Immune
    Responses and Perspectives for Treatment, V. Schirrmacher, R.
    Schwartz-Albiez, eds., Springer-Verlag, Heidelberg, (In press).
U.P. Thorgeirsson, T. Turpeenniemi-Hujanen, J.E. Williams, E. H. Westin, C.A.
    Geilman, J.E. Talmadge, and L.A. Liotta, 1985, NIH3T3 cells transfect-
    ed with human tumor cDNA containing activated ras oncogenes express
    the metastatic phenotype in nude mice, Mol. Cell Biol., 5:259-262.

DISCUSSION

Sanford Markowitz: Have you looked at other lesions in the breast, some of
the benign lesions, and seen whether you have expression of this nm23 RNA
message?

Liotta: Yes, we do have nm23 expression in benign fibroadenomas of the
breast, and it is lost in the highly metastatic tumors. This was all done by
careful and laborious Northern and in situ hybridizations.

Unidentified Speaker: If you've looked at preliminarily, let's say, 20
breast carcinomas that are lymph node positive, how many would you find that
are negative or below a certain level with the nm23?

Liotta: In all the individual cases, virtually every one of them that had
positive lymph nodes had a reduced expression of this message. Now, what
causes the loss of this gene we don't know, and with Dr. Callahan's
collaboration we are hoping to identify the chromosomal location of this gene
to understand whether it is deleted or downregulated.

Yoakum: Would you care to speculate about what you think the function of
this potential suppressor gene that you've isolated is?

Liotta: It could relate as a DNA binding protein. It could be like EIA and up regulate and down regulate other genes by binding to appropriate DNA loci. We don't know for sure.

Unidentified Speaker: Going back to the preceding question, was the correlation you just made with the primary tumors or the metastatic foci?

Liotta: Most of the work was done with the primary tumors, and then we looked at the number of metastases associated with the primary tumor and that individual axillary dissection.

Unidentified Speaker: But the correlation was with the mRNA in the primary tumor site?

Liotta: That's correct. In other words, those tumors that Dr. Steeg studied that had more aggressive behavior in terms of lymph node metastases had a reduced expression. We did observe markedly reduced expression in the metastases themselves, but the data I presented was in the primary tumor. I want to emphasize that we can't really separate this prediction from the differentiated state of the tumor. We didn't get any correlation really with estrogen receptors, although a lot of the more aggressive tumors were negative for estrogen receptor. Most of the aggressive tumors were poorly differentiated as nuclear grade and histologic grade, so we don't know whether this is a differentiation correlate or not.

EDITOR'S NOTE

Progression to the metastatic phenotype may involve the loss of genes normally involved in development, morphogenesis, or differentiation. We have obtained the full length human cDNA NM23 (Rosengard et al, Nature, 342:177-180, 1989) for which RNA levels are reduced in high metastatic potential murine and human tumor cells. We have no identified the 17kDa protein product of this gene and find that the protein is virtually identical to the awd protein involved in Drosophila development. Mutation or allele loss associated with NM23 may lead to a disordered state favoring malignant progression. NM23 allele loss has been identified in a variety of human tumors. Loss of NM23 expression in breast cancer correlates with metastasis (Bevilaqua et al Cancer Research 49, 5185-5190, 1989) and is associated with a highly significant reduction in survival (Barnes et al manuscript submitted). Transfection of nm23 cDNA leading to augmented NM23 protein production abrogates metastasis by a non immunologic mechanism in rodent models (manuscript submitted). Recent studies indicate that at least one form of NM23 had NDP kinase activity (Liotta and Steeg Jour. Natl. Cancer Inst. 82, 1170-1172, 1990) may bind to G proteins, including ras p21, and may regulate receptor mediated activation through CDP-GTP exchange.

Enzyme-altered foci, 264
Epidermal carcinogenesis, 3, 5, 6, 8, 12
Epidermal cell proliferation, 20
Epidermal hyperplasia, 20
Epidermal growth factor, 222, 224, 279, 284
Epidermal growth factor receptor gene, 223
Epidermoid carcinomas, 286
Epigenesis, 301, 310
Epigenetic factors, 101
Epigenetic processes, 301
Epigenetic system, 302, 307
Epiphenomena, 62
Epithelial-mesenchymal interactions,   333
erbB-2 gene, 213
ES cells, differentiation in vivo, 90
ES cell line, 90, 92, 93
ES cells, origin, 85
Esterase D, 172, 177
Esterase D locus, 172
Estrogen receptor, 359, 361
Ethylnitrosourea (ENU), 8, 11, 23, 24, 26
Ethylphenylpropiolate (EPP), 21, 135-138
Explanted embryos, 101
Extracellular matrix, 51, 54

F9 EC cells, 86
Familial aggregation, 171
Familial bowel cancer, 224
Fanconi's anemia, 224
Feeder cells, 72
Feeder layer, 85
Fertilized embryos, 90
Fibroblasts, abnormal phenotype, 330
Fibroblasts, density-dependent cell migration, 330
Fibromatosis, aggressive, 105, 107
Fibromatosis, non-aggressive, 105, 107
Fibrosarcomas, 105, 106, 110
Filaggrin, 22, 23, 26
First "hit", 172
Flow cytometry, 265, 266, 299
Fluocinolone acetonide (FA), 35, 36
Foci, 281, 284
Foci
  asynchronous, 271, 273, 274
  clonal model, 272, 273
    heterogeneity, 274
    progression-linked, 272
  persistent, 272

Foci, (continued)
  remodelling, 273
    reversion-linked, 273
    darwinian postulation, 274
    preneoplastic, 264
    heterogenous, 267
Foci, autoradiographic labeling, 306
Foci, transformed, 302-305, 307
Foci-within-foci, 5
Focus formation, 303-306, 309
fos, 162, 163, 167
Fos proto-oncogene, 12
Free radical scavengers23
Free radicals 21, 23, 24

Gamma glutamyltransferase, 22
Gamma-glutamyl transpeptidase, 264, 279, 281
Gene amplification, 21
Gene and chromosomal translocations, 4
Gene deletions, 4
Gene amplification, 4, 208, 157-168
Genetic changes, 333, 334
Genetic events, 301, 302, 308, 309
Genetic instability, 221, 223, 224, 230, 241, 263, 268, 274, 279, 295
Genetic instability, parameters to measure, 230
Genetic probes, 216
Genomic instability, 71, 233-236
Genotypic alterations, 179
Germ cells, 83, 88, 90
Germ layers, 71, 72, 75
Germ line, 83, 89-91, 93, 98, 99
Germ line chimaeras, 89, 90
GGT activity, 284, 293
GGTP-Feulgen stain, 266
GLC, 279, 287, 289-291, 293
Glial tumors, 177
Glioblastomas, 177, 181
Glioblastomas multiforme, 254, 258
Glucose 6-phosphate dehydrogenase, 279, 284, 286, 293
Glutathione (GSH), 24, 25
Glutathione-S-transferase, 264
gpt, 340
Grafted embryo, 84
Growth factors, 4, 12, 73-75, 79, 145, 146, 184, 185, 189, 190, 195, 323, 324, 325
Growth factor receptors, 184, 189, 190
Growth factors, production, 74
Growth regulating factors, 71
Growth regulators, 72, 73
Growth regulatory genes, 155